HAZARDOUS MATERIALS CHEMISTRY

Second Edition

HAZARDOUS MATERIALS CHEMISTRY

Second Edition

Armando S. Bevelacqua

DELMAR

THOMSON LEARNING

Australia   Canada   Mexico   Singapore   Spain   United Kingdom   United States

THOMSON

DELMAR LEARNING

Hazardous Materials Chemistry, Second Edition

Armando S. Bevelacqua

Vice President, Technology and Trades SBU:
Alar Elken

Editorial Director:
Sandy Clark

Acquisitions Editor:
Alison Weintraub

Development Editor:
Jennifer A. Thompson

Marketing Director:
David Garza

Channel Manager:
Dennis Williams

Marketing Coordinator:
Mark Pierro

Production Director:
Mary Ellen Black

Production Editor:
Barbara L. Diaz

Editorial Assistant:
Maria Conto

Library of Congress Cataloging-in-Publication Data

Bevelacqua, Armando S., 1956–
    Hazardous materials chemistry / Armando S. Bevelacqua.—2nd ed.
        p.   cm.
    Includes bibliographical references and index.
    ISBN 1-4018-8089-4 (alk. paper)
1. Chemistry. 2. Hazardous substances.   I. Title.
    QD33.B48 2005
    604.7—dc22

2005008634

NOTICE TO THE READER

Publisher does not warrant or guarantee any of the products described herein or perform any independent analysis in connection with any of the product information contained herein. Publisher does not assume, and expressly disclaims, any obligation to obtain and include information other than that provided to it by the manufacturer.

The reader is expressly warned to consider and adopt all safety precautions that might be indicated by the activities herein and to avoid all potential hazards. By following the instructions contained herein, the reader willingly assumes all risks in connection with such instructions.

The publisher makes no representation or warranties of any kind, including but not limited to, the warranties of fitness for particular purpose or merchantability, nor are any such representations implied with respect to the material set forth herein, and the publisher takes no responsibility with respect to such material. The publisher shall not be liable for any special, consequential, or exemplary damages resulting, in whole or part, from the readers' use of, or reliance upon, this material.

Dedication

This book is dedicated to my parents, Salvatore and Lillian, for their constant reminder of educational values; to my children, Brianne and Nicholas, for which I provide the same; to my wife Michelle, for her criticism, review, patience, and support during the long hours I spent writing and reviewing material; and to the emergency response community. To this last I say, more so now than ever before, know your threats.

Contents

Foreword

Many years ago, when I was in high school, my educational advisor told me that if I wanted to pursue a career in chemistry I should major in engineering—in hindsight my advisor was right on all counts. He had several reasons for offering this advice—all of which I learned through my experiences as an employee with a chemical manufacturer and as an emergency responder. Learning and understanding the basic theories of chemistry equips you with the understanding of the principles of detection, and thus provides for effective response. Altogether, we learn best through the application of those theories, and it is the repetitive application of those theories that we call experience. Spills and releases do occur; understanding how these materials react allows responders the capability to search for a root cause and effectively manage the incident. However, without the basics of chemistry, these foundation strategies are not so obvious, and it is through your experience that you will discover the importance of what you will learn in *Hazardous Materials Chemistry*.

I have come a long way since those early days, and over time have met a number of emergency responders. Each responder has given me a different perspective on chemical incidents, and how one must use experience to mitigate them. Each understood the mechanics of the situation, the basic chemistry involved, and the reasoning toward mitigation. It is this logical process of deduction that *Hazardous Materials Chemistry* provides for the first responder. It is the understanding of the chemistry, application mechanics, and experience that leads us toward the strategy. In several cases, it requires more than the basic understanding of chemistry; in some cases the answers are embedded deep within the chemistry itself, requiring input from a chemist. You will never know what to expect as an emergency responder, so it is best to be prepared. Consider the following incidents, for example:

- A shipload of direct reduced iron entering port with a glowing hold was the result of the "hydrogen shift reaction," in which iron is the catalyst for water plus carbon monoxide, generating hydrogen, carbon dioxide, and a great deal of heat. The ship was the first of three such incidents to be saved.

- A long-term and annoying fire in a stump dump was tied to the same reaction as above and was controlled by starving it from water.

- A train derailment and fire in a tunnel under the center of Baltimore drew hundreds of responders. A team of industry personnel monitored the streets above for many hours. The railroad provided the list of car sequence and contents within an hour. Industry advisors determined that only one car could be the problem: it could not BLEVE, and other nearby cars were inert or contained slow-burning products. What was initially thought to be a 3-to-4-week project to douse the fires, remove the wreckage, restore the track bed, shut off a leaking water main, and resume rail traffic took less than 5 days.

I also discovered that as fire service trainees learn to deal with fires by putting them out, so should hazmat trainees learn by training with live chemicals. If potential technicians can neutralize a multigallon spill of oleum (fuming sulfuric acid) or thionyl chloride, they can learn to respect the chemical and not fear anything, even WMD. When properly equipped with education, sensing devices, tools, and common sense, we can make a well-informed decision. We call this "risk-based response." There is a growing trend in the hazmat community towards greater technology, in which case experience and knowledge can make the difference. It is also important to continue to acquire knowledge. New technology, such as infrared spectroscopy (see Chapter 7), can assist responders in accurate identification of dangerous goods.

Firefighters responding to hazmat incidents need to know basic chemistry principles. The keys are interpretation of the situation and knowing when to call in the appropriate resources. We are the eyes and the ears of the chemist when we are looking for mitigation strategies in the hazardous materials environment. Together we make a difference.

Gene Reynolds
HazMat 10
Baltimore County HazMat
Baltimore City HazMat

Preface

INTENDED AUDIENCE

The material presented in this book is not the pure chemistry found in the college classroom, but rather an understanding of some basic chemical concepts and how they relate to the hazardous materials scene. It is intended for firefighters law enforcement, and EMS who wish to specialize in hazardous materials, and can be used in the firehouse or academy settings, as well as presented in community college courses.

Most important is the nomenclature of compounds described throughout this text. Chemistry is difficult for many because when taken in a normal classroom setting, not only is theory is being learned, but also the language and concepts of application. In *Hazardous Materials Chemistry, Second Edition,* you will learn the naming process and the basic principles that are applied in hazardous materials emergency response. Division Chief Keith Williams of the Tampa Fire Department told me many years ago, "What hazmat chemistry does for the first responder is to provide them with a means of enemy identification." The goal, therefore, is to understand the theory of chemistry in order to understand how things work, and, more importantly, the naming tools for identification purposes.

At heart, of course, this is a real chemistry book. But it is not your typical chemistry book. Rather, it is a focused look at what is needed by the first responder, hazardous materials technician, specialist, and incident commander. Company officers, law enforcement personnel, and dispatchers will also benefit from the knowledge of these basic principles and naming sequences. The material herein is not intended to make a chemistry major out of anyone. For the most part, I have taken great liberties in order to explain certain concepts. As first responders, we are interested in the basic principles rather than the application of chemistry. What this textbook offers over and above the typical hazmat chemistry text is the direct application of naming, process, and utilization. Over the years of teaching this subject, there has been one question that students continually ask: "How is this knowledge going to help me in the real world?" My answer is simple, and is the main principle carried throughout this text: By understanding the nomenclature and properties of chemistry, anyone can evaluate health and safety issues, level of entrance protection, mitigation, and decontamination procedures. By understanding the basic principles of chemistry, critical decisions surrounding hazardous materials events can and will become less difficult.

Intent of This Book

In 2000, when the first edition of this book was being written, the hazardous materials field was moving forward faster than it ever had before. When the events of September 11, 2001, occurred, the hazmat community was catapulted into a wider service delivery with tremendous demand on chemical expertise. Knowledge of chemical principles has always been the crux of the hazmat responder. But it wasn't until the tragic events of 2001 and the anthrax attacks that soon followed that the true character of hazardous materials teams was defined.

During the past 28 years of my career, I have had the distinct opportunity to see a tremendous growth in the service delivery of hazardous materials. In the early days, delivery in the fire service was fragmented, education was limited, and decisions were largely based on the experience or gained knowledge handed down from seasoned veterans. As time went on and rescue delivery increased, separate units were created and trained to handle chemical incidents, then known as chemical units. Over the years, chemical units became known as hazardous materials teams. The chemical units have transformed into another level of hazardous materials response not only responsible for traditional chemical releases, but also for intentional chemical, radiological, and biological releases.

In the early days, education was generally basic in nature, and identifying needs was difficult. We did not truly understand exactly what chemical aspects we were trying to identify. Information was purely scientifically based, and translated to "tactical worksheets," which gave us the comfort that we knew what we should do. Over time some responders went back to school to gain a formal education in the sciences, and others relied on knowledge gained through trial and error. Some lived; others died or became severely injured.

Technology has developed to the point of using detection and monitoring techniques that were once thought of as only laboratory tools. These tools have made their way to the front line of hazardous materials response. This has added a new dimension to the knowledge base. A hazmat responder today must not only understand these new technologies, but also be able to converse in a variety of professional disciplines.

Even with this increase of technology, service delivery, and knowledge of intentional release, the basics of hazmat are found within seven potential outcomes:

1. Corrosive solid, liquid, or gas may be produced.
2. The material may have toxic qualities or disease potential.
3. The reaction may produce a flammable environment.
4. An exothermic reaction may cause pressurization or vaporization.
5. An endothermic reaction may cause condensation, increased viscosity, or solidification.
6. An explosive catastrophic event may occur.
7. A combination of two or more of the above-mentioned events may occur.

In addition to these problems, the hazard exists that a radiological isotope may be present. Adding the component of radiation, we must be aware of possible chemically antagonistic or synergistic reactions that a chemical family may have. These antagonistic or synergistic effects should not be identified solely as possible toxic events, but rather as effects that may change the entire outcome of the incident in terms of explosive potential (pressurization, vaporization), reactivity (intrinsic chemical behavior such as pyrophoric, or self-heating), toxicity, and corrosiveness.

It has been nearly 30 years since "hazmat" became a common phrase in the fire service, and in that time many great advances have been realized. Who would have thought 20 years ago that hazmat teams would carry with them technology to vibrate molecules in order to "view" them, be able to produce antigen/antibody reactions for biological detection, or have the ability to regenerate DNA from a single strand? Twenty years ago you would have thought I was talking about a new science fiction movie coming out next year, but this is actually what hazmat is starting to look like. Hazmat is now a mobile laboratory with a sophistication that would make any scientist envious.

The understanding of chemistry (and biology) is an absolute necessity more than ever before. Hazmat has become synonymous with chemistry, biology, physics, and mathematics. Emergency responders must embrace the understanding of chemistry. Decisions about a hazmat event are difficult and challenging. These decisions are based at times on limited facts about the incident and the overall understanding of the chemical involved. Emergency responders must take the initiative and educate themselves to a level that promotes safety and enables them to make critical decisions that must be made in times of need.

Chemistry can be, and is for many, a confusing and bewildering subject. For most individuals, the topic leads to great apprehension. It is because of these feelings and countless others that I became interested in education and have instructed hazmat teams in the basic nomenclature of chemistry.

As chemistry increasingly intrudes into our everyday lives, the understanding of this exacting science becomes more important and demanding. The emergency response agencies, including fire, EMS, and police, have great responsibility when a chemical emergency exists. Part of that responsibility is to understand chemical principles and the nomenclature in order to mitigate the incident safely and effectively. It is to this end that *Hazardous Materials Chemistry* was written.

HOW TO USE THIS BOOK

Hazardous Materials Chemistry is arranged in a logical sequence, with influencing subject matter presented at various points in the text. For example, after discussion of the atom and its structure, radiation and detection principles are introduced. Selected answers to the questions in the text are found at the end of each chapter to enhance comprehension as one progresses through the text.

- *Chapter 1* identifies the chemical and physical aspects of chemical compounds. The material presented is considered within the realm of a significant spill or release. Some statements refer to what is identified in the chemistry lab, and others consider how the property affects the emergency incident.

- *Chapter 2* introduces and reviews atomic structure as it applies to the chemistry of inorganic behavior. Inorganic chemistry is discussed from the naming point of view, identifying the hazards in each inorganic chemical class. Such topics as pH and radiation are addressed in detail. These discussions are directed to mitigation techniques and the hazards that can intrude on the remediation of the event.

- *Chapters 3, 4, and 5* identify the behavior and nomenclature of the organic chemicals. These are the most frequently encountered chemicals and represent the largest category of transported commodity. The material in these chapters facilitates categorizing chemicals into organic families and covers the hazards each can present. Additionally, a section on chemical warfare agents (CBRNE) is introduced, explaining the chemistry and toxicological aspects of these compounds.

- *Chapter 6* is a compilation of reference material and resources. Measurements and conversions are covered to enable the reader to move between different units of measurement and to learn about the application of theory as presented in the body of the text. Discussion centers on referencing application, environmental considerations, PPE, decontamination, and weaponized biological warfare agents. The resource section identifies books, databases, and Web sites that can be used to enhance the science officer's capability at the scene of a hazardous materials incident.

- *Chapter 7* is a new chapter in this edition. It looks at the science behind infrared technology. The spectra that are found in this book are real spectra that were generated by Smith Detection's HazMat ID™ and HazMat ID. Real interpretations are included for these spectra, such that the responder can look at real examples for comparison. This was done due to the wide application of this technology in the first response arena, and more specifically the use of the TravelIR and HazMat ID. This chapter goes into street chemistry along with the details of the technology and how to interpret infrared spectra. It ends with a step-by-step analysis of spectra. The philosophy is to understand the basic components of each spectrum such that when the spectra of an unknown is presented, all the possibilities are considered, rather than relying on a computer's statistical match.

Key Features of This Book

There are several features throughout the book that add to the strategic decision-making process:

- *Key Terms* identify important terminology that a responder should know in order to understand and communicate effectively with others on the scene of an incident. A *Glossary* is also included at the end of the book.

- *Notes* capture important facts that responders need to know in order to accurately identify hazardous chemicals.

- *Safety and Caution* boxes integrated throughout the text warn responders of hazardous chemicals and how to safely respond to dangerous goods.

- *Problem Sets* provide practice for properly identifying chemicals in hazardous materials response.

- *Acronyms* at the back of the book provides a handy reference for self-study and response.

NEW TO THIS EDITION

- Section on *Biological Materials:* A hazardous materials responder does not have a clear line when responding to a call, and must be prepared for any situation—whether chemical or biological.

- *Chapter 7:* Many technologies are being introduced to the hazmat arena. Detection devices have the most influence on the everyday hazmat responder—everything from a simple four-gas detector to a gas chromatography-mass spectroscopy (GC-MS) device. New technologies also include positive protein analysis, antibody/antigen reactions, ion mobility spectroscopy, and infrared technology. It was difficult to draw a line at some point and stop discussing new technologies and how they apply to chemistry. I chose to include infrared technology as a chapter due to the emerging widespread use of this technology in the fire service. Infrared technology's ease of use and understanding is much less than that of the GC-MS. Several other technologies are being introduced into emergency response, but as stated, this is street chemistry and I wanted to maintain that theme. Street chemistry has taken on a whole new complexion since the emergence of this new technology in emergency response, but basic chemistry factors still remain. If we can categorize the chemical into its basic state of matter and its general chemical family, our decisions on a hazmat incident are a given.

SUPPLEMENTAL MATERIAL

This second edition is part of an educational package that includes the textbook, an Instructor's Guide on CD-ROM, and a Field Operations Guide (FOG).

- The *Instructor's Guide on CD-ROM* is a comprehensive package that provides instructors with the necessary materials to prepare for classroom instruction:

  - The *PowerPoint slides* can be used in two ways. First, as a presentation format for the instructor. If the instructor is comfortable with the depth of knowledge that is required, only the PowerPoint slides can be used as placeholders during the discussion. The second method is as a self-paced presentation that the student can use. Along with the textbook, activities, and questions, the motivated student can use this as a guide through the chemistry, or as a distance-learning tool.

- ***Answers to Review Questions*** are provided for instructor reference when assigning the review questions at the end of each chapter to students.

- ***Case Studies*** can be distributed to students in class and used as a basis for class discussion or as exercises in developing critical-thinking skills.

- ***Quizzes, the Mid-Term, and the Final Exam*** provide instructors with the necessary material to evaluate student progress and comprehension of the material.

- The ***Student Manual*** is for the instructor to use for the student who may have a difficult time organizing the quantity of material that is presented during the classroom presentation. It will help those students that need some organizational direction, and at the same time focus on the subject matter in the same order in which the text is presented. It is a printed outline of the text, so the student can focus on the material rather than the organization of notes.

(Order No.: 1-4018-8091-6)

- The ***Field Operations Guide (FOG)*** is an addition that came out of several class presentations. It is organized in a manner that translates the subject matter learned in the classroom into the field. It presents all the information that is found in the textbook in an abbreviated fashion. It provides the student with a guide to follow outside the learning environment and in the real world. It contains all the referencing components presented in the text.

(Order No.: 1-4018-8090-8)

Acknowledgments

Some 30 years have passed since the inception of hazardous materials response. In the early days, a general understanding of commonly used chemicals was the extent of our chemistry education. Commanders wanted us to memorize these common chemicals so that we could quickly attack a problem if one should arise. The frequency of hazmat incidents has increased over the years, and we also learned that a quick attack is not always the best approach. With this increase, and to some extent because of limited education, firefighter injuries and deaths began increasing in the hazmat field. Classes such as chemistry started to come to the forefront, along with air monitoring, mitigation strategies, personal protective equipment, decontamination, and so on. With this influx of education, one issue has remained: street-level application of the science. It is this one concept that has remained in my mind during the long hours of research and writing. It is a concept that in some cases has been forgotten, a concept that I have tried to maintain throughout the first edition and now the second.

However, this concept of street-level application, sprinkled with scientific fact and theory along with experience, has not come just from me, the author. It truly is a collective effort of the extremely smart and experienced individuals who make up the hazardous materials response community. It is this group, the hazmat responders, who have contributed to this body of work over the years.

As you can imagine, producing a book such as this takes a lot of time to write as well as countless hours of review. Many of the individuals who contributed to this book are people behind the scenes, including the multitude of students that I have had over the years. Many of the individuals are nameless to me, but without their critical commentary or suggestions, this book could not have achieved such high caliber.

I would like to thank the countless responders that I have taught who have posed questions either based on experience or out of curiosity. These individuals and their questions are the true reason for this book and the revision of the second edition. It is their interest and inquiry that has led to the format and direction that this project has taken. I want to especially thank the crews that I have worked with or have had under my command, and in particular the Orlando Hazmat Team. It is due to these individuals that the technical applications of street knowledge and years of experience culminated in the direction and tactical application of the chemical principles contained herein.

In my tours throughout the country educating hazardous materials team members, I have had the opportunity to meet and talk with responders from around the country and abroad. Many of these individuals took the time to review my work, comment on my lectures, and share their experiences with me. They are remarkable individuals who share the enthusiasm for emergency work that I have had since my entrance into the emergency arena. These professionals have in their own way helped me with this project; these individuals include: Dieter Heinz, Tim Regan, Greg Noll, Charlie Onesko, Bill Hand, Mike Hildebrand, Cid Ceresa, Chris Hawley, Mike Piland, Tommy Erickson, Mike Callan, Jan Kuzma, Malcolm Trigg, Tommy Moffett, Glen Joseph, Jeff Borkowski, Doug Wolfe, and John Wilson.

Special appreciation goes to the following for their review and criticism during the long revision period; their hard work is much appreciated:

Al Verioti
Director of Training/Safety Officer
Waterbury Fire Department
Waterbury, CT

Robert Marchisello
Battalion Chief
Philadelphia Fire Department
Philadelphia, PA

Joe Gorman
HazMat Technician
Fairfax County HazMat
Fairfax, VA

Chris Waters
Director
Charleston County Emergency Management
Charleston, SC

Carter Davis
HazMat Captain
Honolulu, HI

Gene Reynolds
Responsible Care Coordinator
FMC
Baltimore County/City, MD

William Spencer-Strong
Captain, HazMat Administrator
Baltimore County Fire Department
Baltimore, MD

Danny Peterson
Professor
Arizona State University
Mesa, AZ

Glen Rudner
Hazardous Materials Officer
Dept. of Emergency Management
Humfries, VA

Mark L. Norman, Ph.D.
Application Scientist
Smith Detection, Emergency
   Response Division
Danbury, CT

Frank Docimo
Docimo and Associates, LLC
Stamford, CT

Cris Aguirre
HazMat Lieutenant
Miami Metro Dade, FL

Kristina Kreutzer, Ph.D.
Research Chemist
DuPont Experimental Station
Wilmington, Del

Thank you to Mark Norman and Missy Robinson, both of Smith Detection, Emergency Response Division, for their help with the graphics in this publication. Mark, thanks for the endless e-mails on technical points of IR theory application and spectra, and thanks to you, Missy, for your assistance with very specific photographs and graphics.

I am proud to say that the essence of this book is dedicated to the street practitioners, and to those at Thomson Delmar Learning, who were instrumental in the development of this project. Thank yous go out to all the responders who have used this work in their hazmat education, and to all those who will. A special thanks to Jennifer A. Thompson from Thomson Delmar Learning, who would patiently answer my calls about protocol, my constant questions on procedure (which I can never remember), and generally acted as my sounding board for format and design.

The acknowledgments would not be complete without the recognition of close friends such as Rick Stilp, who although he did not have the time to contribute to this work, saw and understood its need in the response community. Thank you to Lieutenants Roger Herota and Bob Coschignano and District Chief B.K. Will, who said that the chemistry is beyond them, but who have nevertheless sat in the hazmat command seat and have had to make very difficult decisions. Thanks to my family, who have been a constant source of encouragement throughout this project and others like it. To my son Nicholas, who would ask, "Why does it have to work this way?" I would explain the concept and add, "Read and learn." To my daughter Brianne, who would shrug and say, "That's cool." And to my wife Michelle, who had the pleasure (or curse) of reading this manuscript and listening to my many ideas numerous times.

ASB

About the Author

Armando S. Bevelacqua has been involved with emergency first response since 1975. He currently serves as District Chief—Terrorism Task Force and Special Operations with the City of Orlando Fire Department, Orlando, Florida. His assigned duties make him responsible for hazardous materials response and technical rescue. He has developed many hazardous materials and EMS educational programs for the Orlando Fire Department, along with programs instituted in surrounding departments and across the nation. He has been actively involved with hazardous materials and technical rescue training, quality assurance programs, documentation review, community emergency response teams, and EMS activities.

Armando also teaches at local colleges, instructing students in fire science, hazardous materials response, hazmat medicine, and EMS-related topics. He writes freelance articles and educational textbooks, with published topics including report writing for EMS providers, emergency medical response to hazardous materials incidents, and he has authored a hazardous materials field guide and terrorism handbook for operational responders geared for the first responder. He has presented nationally on several controversial issues in the disciplines of EMS, hazardous materials, and management. Armando lectures to fire departments throughout North America and is an adjunct instructor for the National Fire Academy, the International Association of Firefighters, and several federal agencies.

He is involved with many emergency-planning entities, including the Local Emergency Planning Committee (LEPC). He is an active member on several federal committees dealing with first response to WMD events and hazardous materials response. He has worked with Rocky Mountain Poison Control on the development of standardized medical protocols for WMD events. Additionally, he is an active member of the State of Florida Operations Working Group for Equipment and Training in the Bio-Chem Environment and the EPA's Air Monitoring and Detection Initiative, which is geared towards limitations of detection systems in the first response arena. Armando holds a bachelors degree, and is constantly addressing the educational needs in the first response community.

FESHE Correlation Grid

Crosswalk between *Hazardous Materials Chemistry, Second Edition* and the National Fire Academy's *Chemistry for Emergency Response*

Hazardous Materials Chemistry 2ED			**Chemistry for Emergency Response** (NFA)	
Chapter	**Subject**	**Page**	**Unit**	**Page**
1	Structure of Atoms	2	2	IG 2-21
1	Periodic Table	4	2	IG 2-11
1	Chemical Bonding	11	2	IG 2-27
1	Properties of Matter	16	7	IG 7-1
1	Toxicology	28	10	IG 10-1
2	Ionic Bonding	46	2	IG 2-47
2	Symbols and Structure	47	2	IG 2-47
2	Binary Compounds	51	3	IG 3-17
2	Metal Oxides	54	3	IG 3-23
2	Inorganic Peroxides	55	3	IG 3-41
2	Covalent Bonding	56	5	IG 5-5
2	Ionic/Covalent Properties	57	2, 5	IG 2-47, IG 5-5
2	Oxysalts	58	3	IG 3-49
2	Acid/Base Reactions	61	7	IG 7-41
2	Metal Hydroxides	63	3	IG 3-32
2	Inorganic Acids	64	4	IG 4-8
2	Binary Acids	65	4	IG 4-10
2	Hydrates	66		
2	Binary Nonsalts	68	4	IG 4-5
2	NonMetal Oxides	69	4	IG 4-7
2	Other Inorganic Compounds	71	3, 4	IG 3-61, 4-15
2	Combustion	73		
2	Radiation	78	8	IG 8-3
3	Principles of Bonding	90	2	IG 2-27
3	Organic Structure and Nomenclature	92	5	IG 5-7
3	Alkanes	94	5	IG 5-17
3	Alkenes	99	5	IG 5-23
3	Alkynes	103	5	IG 5-28
3	General IUPAC Naming	103	5	IG 5-12
3	Aromatics	108	5	IG 5-30

(continues)

Crosswalk between *Hazardous Materials Chemistry, Second Edition* and the National Fire Academy's *Chemistry for Emergency Response (Continued)*

Hazardous Materials Chemistry 2ED			Chemistry for Emergency Response (NFA)	
Chapter	**Subject**	**Page**	**Unit**	**Page**
3	Kinetic Molecular Theory	110	7	IG 7-3
3	Branching and Its Effects	115	7	IG 7-3
4	Organic Functional Groups	136	6	IG 6-7
4	Alkyl Halides	140	6	IG 6-21
4	Amines	143	6	IG 6-26
4	Nitrogen Groups	147	6	IG 6-38
4	Ethers and Epoxides	151	6	IG 6-64
4	Organic Peroxides	155	6	IG 6-68
4	Polymers	156		
4	Branching and Its Effects	157	7	IG 7-3
5	Alcohols, Phenols, Sulfur Analogs	170	6	IG 6-81
5	Carbonyl Groups	176	6	IG 6-99
5	Aldehydes	176	6	IG 6-85
5	Ketones	178	6	IG 6-83
5	Esters	179	6	IG 6-90
5	Organic Acids	181	6	IG 6-88
5	Acyl Halides	182		
5	Amides	183		
5	Cyanide Salts	183	4	IG 4-12
5	Isocyanates	184		
5	Branching and Its Effects	184	7	IG 7-3
5	Chemical Warfare Agents (CBRNE)	184		No separate section
6	Decision-Making Process	206		
6	Measurements and Conversions	209		
6	Scene Considerations	212		
6	Environmental Influences	215		
6	Chemical Influences	220		
6	Basic Chemical Applications	235		
6	Toxicological Considerations	226		
6	General Monitoring and Detection	237	9	IG 9-11
6	PPE	238		
6	Decontamination	240		
6	Chemical Warfare Agents (CBRNE)	243		
6	Biological	247		
7	Infrared Spectroscopy	263		

Fire and Emergency Services Higher Education (FESHE) Course Correlation Grid

Name	Chemistry for Emergency Response (NFA 0233) Chemistry of Hazardous Materials (NFA 0234)	*Hazardous Materials Chemistry, 2nd Edition Chapter Reference*
Course Description	This course is designed to prepare the responder to function safely at the scene of a hazardous materials incident by understanding the potential hazards. This is accomplished by gaining recognition of chemical nomenclature and basic principles of chemistry in order to assess risks to responders and the public. The course seeks to convey to first responders or prevention officers a sound understanding of the basic chemistry of hazardous materials to permit them to correctly assess the threat posed by hazardous materials incidents that may occur accidentally or through intentional means. Problem-solving sessions and interactive discussion cover topics such as salts and inorganic nonsalts, hydrocarbons, hydrocarbon derivatives, and hydrocarbon radicals. Application of chemistry to thermodynamics, volatility, and combustion provides real-world examples. An understanding of basic chemistry is helpful to receive maximum benefit from the course.	
Prerequisite	None	
Outcomes	*Value how an understanding of chemistry is important to the hazard and risk assessment process.*	1
	Describe parts of the atom, atomic number, atomic structures, and related properties.	1
	Practice drawing atomic structures to reinforce the concept of orbitals.	1
	Identify symbols, names, and atomic numbers of elements with reference to the periodic table.	1
	Define isotope and identify what constitutes an isotope.	1
	Demonstrate knowledge of ionic bonding by naming salts correctly.	2
	Given formulas or names, determine balanced salt formulas.	2
	Identify the seven types of salts (binary, oxide, peroxide, hydroxide, oxy, cyanide, and ammonium) and their hazards.	2
	Given a description of a substance involved in an incident, identify the type of salt and the potential hazards involved.	2
	Identify elements and compounds that tend to be water-reactive because of chemical reactivity or physical state.	1, 2
	Identify the hazards of oxidizers as individual chemicals when combined with other chemicals, and be able to determine worst-case scenarios.	1, 2
	Demonstrate knowledge of covalent bonding.	2, 3
	Given formulas or structures, determine the identity of inorganic compounds.	2
	Given a description of a substance involved in an incident, identify type of inorganic nonsalt and the potential hazards involved.	2
	Describe the impacts of pH and the concept of acid–base reactions.	1, 2, 6
	Correctly identify and analyze a hydrocarbon by determining its structure, isomers, bonds, and shape.	3
	From formulas, determine whether a particular hydrocarbon compound has a saturated, unsaturated, or aromatic bond.	3
	Explain the basic concepts of the International Union of Pure and Applied Chemistry (IUPAC) naming nomenclature.	3, 4, 5
	From the carbon hydrogen ratio in the formula, determine whether a particular compound has a straight, branched, or resonant arrangement.	3
	Identify specific hydrocarbons given a name, formula, or structure.	3
	Correctly identify the potential for, and impacts of, polymerization, decomposition, and slow oxidation reactions.	3, 4, 5
	Correctly identify the characteristics of alkene, alkyne, alkane, and aromatic hydrocarbons.	3
	Given a diagram, name the structure that represents a specific derivative.	4, 5
	From formulas, identify hydrocarbon derivatives and deduce the chemical characteristics that determine their hazardous properties.	4, 5

(continues)

Fire and Emergency Services Higher Education (FESHE) Course Correlation Grid (Continued)

Name	Chemistry for Emergency Response (NFA 0233) Chemistry of Hazardous Materials (NFA 0234)	*Hazardous Materials Chemistry, 2nd Edition* Chapter Reference
	Identify specific hydrocarbon derivatives given the name, formula, or structure.	4, 5
	Given the name, formula, or structure, identify the functional groups of the hydrocarbon derivatives.	4, 5
	Identify the hazards commonly associated with each of the functional groups.	4, 5
	Identify the key physical and chemical properties of hazardous materials.	1
	Identify how molecular weight, size, and polarity affect these physical properties.	1, 3, 4, 5, 6
	Define isotope and identify what constitutes an isotope.	2
	Identify the characteristics that contribute to whether or not an isotope will be radioactive and undergo decay.	2
	Name the three most common forms of ionizing radiation.	2
	List three measures for protection from radiation (time, distance, and shielding).	2
	Identify the various principles of detection of radio isotopes.	2
	Describe the thermodynamic properties and the associated hazards of compressed and liquefied compressed gases.	1, 3, 6
	Describe the thermodynamic properties and the associated hazards of cryogenic materials.	1, 3, 6
	Discuss the calculation of airborne concentrations based on maximum concentration formula.	1, 3, 6
	Summarize the general principles of toxicology.	1, 6
	Determine toxic events.	1, 6
	Identify factors influencing toxicity.	1, 6
	Describe the basic characteristics of explosive materials.	4
	Identify the general formula for explosives.	4
	Describe the three components of an explosion.	4
	Recognize common explosive compounds.	4
	Identify specific chemical reactions involved in a lab process based on the types of equipment and their setup.	5
	Given a list of organic and inorganic chemicals, identify their potential use in illicit activities and potential hazardous conditions their presence may create.	5
	Given an illicit activity, identify the potential chemicals present and hazardous conditions.	5

Chapter 1

Chemical Structure and Physical Effects

Learning Objectives

Upon completion of this chapter, you should be able to:

- Analyze the atomic structure and its components, listing relationships to the periodic table.
- Compare the relationship between the neutral atom and isotopes.
- Compare the groups, periods, regions, and families on the periodic table.
- Explain the relationship between the group's number at the top of the table and the chemical structure.
- Categorize the trends on the periodic table.
- Explain the formation of anions and cations.
- Explain the theory of resonance.
- Describe rate of reaction and the relationships to activation energy.
- Identify the physical properties of matter.
- Identify the chemical properties of matter.
- Describe the dose-response relationship.
- Define the route of entry and the modes.
- Differentiate the terms of toxicology.

THE STRUCTURE OF ATOMS

protons

positively charged fundamental particles that are part of the nucleus of an atom and carry a positive charge

neutrons

fundamental particles without electrical charge that are part of the nucleus of an atom

electrons

negatively charged fundamental particles orbiting the nucleus of an atom

nucleus

the center of an atom, containing the protons and neutrons

atomic weight

the mass of an atom, which equals the sum of protons and neutrons

In 400 B.C. a Greek philosopher by the name of Democritus was the first to describe a system of matter consisting of basic building blocks, which we now know today as the elements. However, it was not until the Middle Ages that people attempted to produce gold from common metals by means of chemical experiments. This primitive science was called alchemy and it was the forerunner of chemistry. Alchemists, although driven by the desire to manipulate matter, were the first to acquire a wealth of knowledge through observation.

Another 600 years would pass before the investigation of the properties of the elements would show that they are built up from atoms, which consist of even smaller particles called **protons**, **neutrons**, and **electrons** (see Figure 1–1). Atoms have been found to be largely empty space, with the center of the atom (the **nucleus**) containing most of the mass of the atom. This nucleus contains the protons and the neutrons, with the electrons orbiting around it at various distances.

The important properties of the atomic particles are their mass and their electrical charge. Electrons carry a negative charge to balance the positive charge of the proton in the nucleus. The positively charged particles are called protons, and the uncharged particles are called neutrons. Protons and neutrons both have a mass of 1 on the **atomic weight** scale, while electrons are considered to have essentially zero (1/1840 is the size as compared to the proton). The electron is the lightest of all the atomic particles. All elements have electrons. The sharing or transfer of electrons is the basis of chemical reactions and the bonding of elements. Electrons revolve so fast that if we could see such a particle, it would look like a cloud around the center nucleus.

■ Note

During radioactive decay, the particle that is identified as the alpha particle consists of two protons and two neutrons (helium nucleus). The beta particle is identified as the electron, and gamma is the energy liberated from decay.

■ Note

In some chemistry books, the atomic number is identified as the Z number and A indicates the mass number. This designation comes up in connection with radioactive isotopes.

---

isotope

one form of an element having the same number of protons but a different number of neutrons

Atoms of different elements vary in the number of their protons, neutrons, and electrons. The factor that determines the type of atom is the number of protons in the nucleus. The number of neutrons may differ depending on the **isotope** of the atom. All atomic nuclei (electronically neutral atoms) contain an equal number of protons in relation to the number of electrons that are in orbit.

Figure 1–1 *A graphic representation of the lithium atom. Here we have three protons, several neutrons, and three electrons representing an atom.*

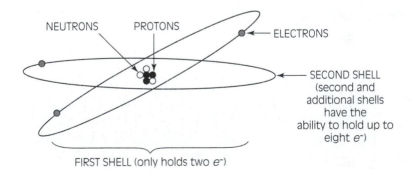

NEUTRONS   PROTONS

ELECTRONS

SECOND SHELL (second and additional shells have the ability to hold up to eight e⁻)

FIRST SHELL (only holds two e⁻)

atomic number
the number of protons in the nucleus of an atom

mass number
the total number of neutrons and protons in the nucleus of an atom

The number of protons dictates the number of electrons within a neutral atom. If the atom has fourteen protons, then there will be fourteen electrons in orbit around the atom. The number of protons determines the **atomic number**. The atomic number is used to catalog the order of the elements on the periodic table (see inside front cover).

All atoms except those of the most abundant form of hydrogen contain some neutrons in the nucleus. The neutrons do not affect the chemical properties of the atom, but rather its mass. Because the electrons are so light, the mass of an atom results from the number of neutrons and protons (**mass number**). The number of neutrons and protons is equivalent to one atomic mass unit (AMU).

Number of Protons + Number of Neutrons = Mass Number

Atomic Number = Number of Protons = Number of Electrons

Many elements contain two or more slightly different types of atoms called isotopes. Isotopes are one or more forms of an element that have the same number of protons (and electrons) but differ by the number of neutrons. They may be defined as atoms with the same atomic number but with different numbers of neutrons, and therefore with different atomic mass numbers. Nevertheless, isotopes of the same element have identical chemical properties. For example, hydrogen normally does not have a neutron in the nucleus but can have one or two neutrons in its respective isotopes. Therefore, hydrogen is said to have three isotopes (see Figure 1–2). Most elements that exist in nature have one predominant isotope; however, some of the elements contain several coexisting isotopes.

■ Note

There are ninety-two naturally occurring elements; the rest are man-made. After uranium (Z = 92), the majority of the rest are found only within the laboratory.

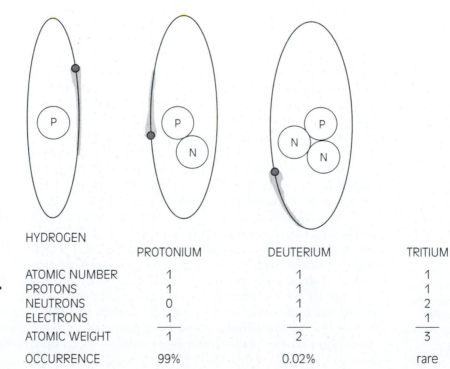

HYDROGEN

	PROTONIUM	DEUTERIUM	TRITIUM
ATOMIC NUMBER	1	1	1
PROTONS	1	1	1
NEUTRONS	0	1	2
ELECTRONS	1	1	1
ATOMIC WEIGHT	1	2	3
OCCURRENCE	99%	0.02%	rare

Figure 1–2 *The atomic weight of hydrogen is 1.0080, which accounts for its three isotopes.*

When we measure the atomic weight, we are dealing with vast numbers of atoms, and the value found can only be an average. If the element exists in the form of two or more isotopes, then the atomic weight will be an average of the masses of each isotope and the percentage of occurrence.

■ Note
We are discussing visual models that help us understand chemical behavior.

■ Note
Although hydrogen is in the same column as the alkali metals, it is considered a nonmetal, thus the line underneath it within the periodic table.

THE PERIODIC TABLE

In 1869 Dimitri Mendeleev discovered that elements progress in a logical order. It was due to this observation that Mendeleev predicted the elements that were at the time not yet discovered. In his original chart, Mendeleev based the known elements in groups and periods based on what was known in that day. He left gaps assuming that other elements not yet discovered would complete the table. His system of classification gave us the periodic table.

Families and Groups

The **periodic table** provides a general framework of the elements to show relationships between the chemical groups. This table arranges the chemicals according to atomic number. Looking at the chart, we see chemicals arranged in vertical columns called **groups**. Within each group, elements that have similar chemical properties are found. Looking down a group we see that the elements are arranged such that the top of the group represents a gaseous element (with a few exceptions such as carbon and beryllium, which are solids), and as we move down they become solids (exceptions are mercury and radon). Looking at each group as it appears on the periodic table also gives insight into the general properties of the elements. There are sixteen divisions represented by eight groups, each designated by a Roman numeral. The newer tables identify the groups from left to right as 1–18.

Each horizontal row is called a **period**, and the atoms are arranged from left to right in order of size, from larger to smaller. The first six periods end with a noble gas. The rare earth metals (or inner transition metals) are placed in rows 6 and 7. For easy reading these elements are pulled to the lower portion of the table.

Within each box the chemicals (atoms that represent elements) are represented by symbols. It is these symbols that we use when we write formulas or reactions to represent the elements (Figure 1–3).

We can further identify areas that represent certain qualities of chemicals by placing them into regions (Figure 1–4). The first two columns or groups and seven elements from groups IIIA and IVA (or 13–14, or IIIB–IVB) are termed the **representative metals**. The groups in the lower portion of the main chart are called the **transition metals** and are identified by the groups IIIB through IIB (or 3–12, or IIIA–IIB). Hydrogen and the elements on the right side of the chart are called the **nonmetals**. Between the nonmetals and the right-side representative metals are the **metalloids**. These have properties of both metals and nonmetals depending on the reaction the individual element undergoes. Underneath the

---

periodic table
a general framework of the elements that shows relationships between the chemical groups

groups
vertical columns of the periodic table containing families of elements of similar properties

periods
the seven horizontal rows of the periodic table; the first six periods end with a noble gas; the rare earth metals (or inner transition metals) are placed within rows 6 and 7; for easy reading these elements are pulled to the lower portion of the table

representative metals
a region of the periodic table that includes the first two columns or groups and seven elements from groups IIIB–IVB (or 13–14, or IIIA–IVA)

transition metals
a region of the periodic table that includes the groups in the lower portion of the main chart, identified by the groups IIIA–IIB (or 3–12, or IIIB–IIB)

nonmetals
a region of the periodic table that includes hydrogen and the elements on the right side of the chart

metalloids
a region of the periodic table that includes the elements between the nonmetals and the right-side representative metals

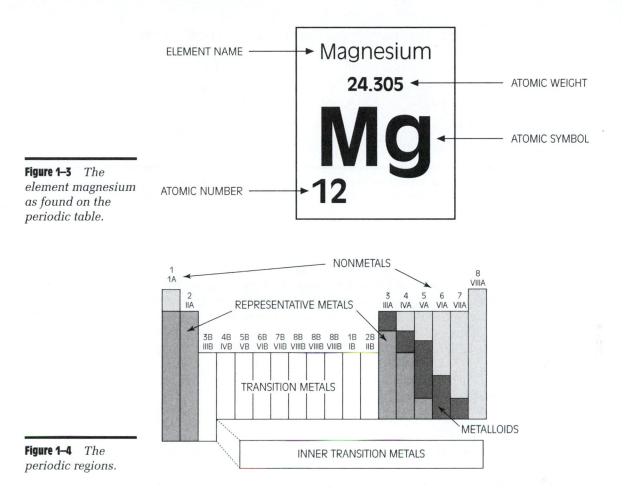

Figure 1–3 *The element magnesium as found on the periodic table.*

Figure 1–4 *The periodic regions.*

inner transition metals
a region of the periodic table underneath the transition metals that includes two main groups: lanthanides and actinides

transition metals and a part of this region are the **inner transition metals** or rare earth metals. This area has two main groups called the *lanthanides* and *actinides*. Lanthanides and actinides occur naturally, but the later portion of the actinides are man-made through neutron bombardment and are extremely rare.

Electronic Relationships

Atoms are normally electrically neutral; that is, there are as many electrons present (negatively charged particles, or anions) as there are protons (positively charged particles, or cations). When atoms are combined to form compounds, they may not have the same number of electrons as protons. Because the number of protons does not vary, the element does not change. However, if there are fewer electrons than protons, the atom will be positively charged, thus forming a positive ion. With more electrons than protons, the atom is negatively charged and forms a negative ion.

shell
the region in space where electrons are located around the nucleus of an atom; the principal energy level of an electron

The smaller the diameter of the atom, the higher the tendency to attach electrons, whereas the larger atoms have a tendency to lose electrons. It is this positive core that holds the electron cloud in place (opposites attract). Thus, if an atom is small, the attraction is close to the central positive core and the electron cloud is held close, gaining electrons when they are available. However, if the attraction is farther away, then the outer **shell** is loosely held, giving rise to the opportunity of losing an electron (see Figure 1–5).

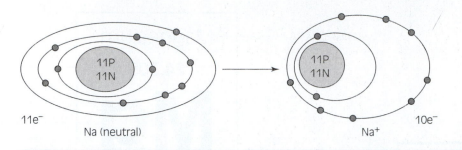

Figure 1–5   *The neutral sodium atom becoming a positive ion. Notice that the size has decreased once the electron has been given up and an ion has been formed. Remember, the outer shell is farthest away, making it difficult to hold on to that one electron.*

ionic bond
the electrostatic attraction of oppositely charged particles; atoms or groups of atoms can form ions or complex ions

valence electrons
the electrons in the outermost shell of an atom

covalent bond
the force holding together atoms that share electrons

octet rule
the observation that simple atoms achieve an electronic stability by attaining eight electrons in the outer shell through ionic or covalent bonding, each trying to attain the noble gas state of eight electrons in the outer shell

ionization
the process by which an electron is removed from the outermost shell of an atom

energy
the capacity to do work

ionization energy
the amount of energy required to remove an electron from the outermost shell of an atom

Normally, an atom of high electronegativity (an atom that has a tendency to attract electrons and therefore has a stronger negative quality than positive) and low ionization (ionization is the ability to lose or gain an electron) will react to form an **ionic bond**. This reaction occurs when metals, the elements from the left side of the periodic table, bond with nonmetals, the elements from the right side of the table. These bonded materials are commonly called *salts*.

> **● Caution**
> The reactions that occur between groups IA or IIA and group VIIB may be explosively violent.

The electrons in the outermost shell are called the **valence electrons** and are what we classify as the bonding electrons. It is this outer shell that shares electrons (**covalent bond**) or transfers electrons (ionic), enabling the atom to form a molecule and engage in chemical reactions. We are also going to show that the outermost shell would like to acquire eight electrons in order to make the bond stable (the exception is hydrogen, where two electrons are all that is required). These eight electrons represent the octet of the outer shell, called the **octet rule**, which governs stability within a bond.

Problem Set 1.1   For the first fifteen elements on the periodic table, calculate the number of shells and electrons within each shell.

Problem Set 1.2   For the first three elements in groups 1, 2, 3, 14, 15, 16, 17, and 18, calculate the electrons in the outer shell and look up the respective electronegativity (Hint: you should have already found the outermost electron configuration in Problem Set 1.1).

Ionization is the process by which an electron is removed from the outermost shell by placing **energy** into the system. The energy required is called the **ionization energy**. As shells increase in number, the outermost electron shell is farther away from the center nucleus, and thus easier to displace. Conversely, when we have an atom with fewer shells, greater energy is required to remove the electron. The higher the mass or atomic number, the more electrons there are, and thus the increase in the number of shells.

Looking at the periodic table we can see that the ionization energy requirements follow a regular pattern. Ionization energy increases from left to right across

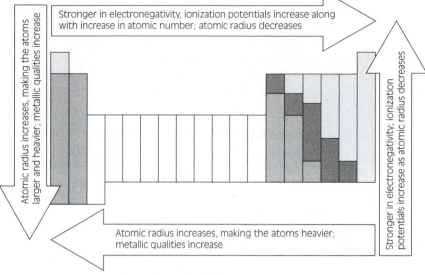

Figure 1–6
Generalized relationships across the periodic table.

a period and increases from the bottom to the top of a group (see Figure 1–6). It is this ionization energy that is used to determine the electronegativity of an element and is the principle behind monitors such as the photoionization detector (PID) and flame-ionization detector (FID). The halogens group (group VIIB), oxygen group (group VIB), and the nonmetals are all electronegative.

■ Note
When using a photoionization detector, it will be necessary to identify the proper ionization potential. Many substances have first, second, and third ionization potentials. Follow manufacturer's recommendations for the ionization potentials.

■ Note
Radiation is a form of ionization. There are four types of radioactivity: alpha, beta, gamma, and neutron. Alpha and beta particles are physical particles; a gamma ray is a high-energy photon, or light package, released when an atom "falls" from a high level of energy to a lower level. (See Chapter 2 for additional discussion of radiation.)

■ Note
Americium is used in ionizing smoke detectors.

---

ionization potential (IP)
the minimum energy required to release an electron or photon from a molecule; measured in electron volts (eV)

Electronegativity is a measure of how strong an atom holds onto its electrons or attracts electrons from other atoms. It is the atom's ability to attract and hold electrons. This attraction is measured by determining the energy used to pull an electron away from its shell (ionization energy). This is called the **ionization potential (IP)** and is expressed in electron volts (eV). It is also a way to describe a bond that is not primarily covalent or ionic, but rather has characteristics of each and is known as a polar covalent bond. This ionization potential is what is used by the PID and the FID. In these detector types, a light wave (PID) or a flame (FID) is used as the energy source for the removal of the electron. When we use one of these instruments, we look up the IP for that particular molecule. Once the monitor generates a number, we can use the manufacturer's relative reaction ratios and calculate the amount of the gas present.

The military's chemical agent monitor (CAM) utilizes the radioactive isotope of nickel as the energy source. This energy is placed on two walls of the corridor;

each side of the corridor thus has a positive and negative charge, respectively. Once the chemical agent enters the corridor, the radiological isotope generates energy, thus creating an ion. On one wall of the corridor there is a hole or gate through which the ion passes, and at the far end is a wall. When the ion passes through the gate and strikes the wall, a time of flight is registered. It is this time versus mass that indicates the presence of the molecules in the affected atmosphere. This is a good example of how chemical theory (ionization) is applied to the real world.

> ● **Caution**
>
> The alkali metals are very reactive and are not found freely in nature. These metals have only one electron in the outer shell, giving them a strong tendency to react and form an ionic bond.

General Characteristics

alkali metals (group IA, 1, or 1A)

the elements that are in the first column or group to the left on the periodic table: lithium, sodium, potassium, rubidium, cesium, and francium

Alkali Metals. Group IA, or group 1 or 1A, elements listed in the first column or group on the left on the periodic table are **alkali metals:** lithium, sodium, potassium, rubidium, cesium, and francium. They are highly reactive in the pure form. The elements in this group give up an electron easily because there is only one electron in the outer shell. Therefore, it is rare that any of this group is found in its pure form. In nature, all are salts (compounds).

> ■ **Note**
>
> The aluminum group is sometimes referred to as the "other metals." Although ductile and malleable like the transition elements, their valence electrons are from the outer shell. For use in this text the valence is +3, but these elements may have valences of +3, +4, −4, or −3. All are solids and are highly dense.

The reactions of these elements are so violent that only sodium, lithium, and potassium are used commercially. The chemicals in this group are highly water-reactive and prone to ignition on exposure to moist air. They produce strong caustic solutions that can result in severe burns to the eyes, skin, and respiratory system. In addition, they will produce hydrogen gas on contact with water.

alkaline earth metals (group IIA or 2)

the elements that are in the second column or group from the left on the periodic table: beryllium, magnesium, calcium, strontium, barium, and radium

Alkaline Earth Metals. Group IIA, or group 2, elements listed in the second column or group from the left on the periodic table are the **alkaline earth metals** beryllium, magnesium, calcium, strontium, barium, and radium. These elements appear to be gray with a metallic brilliance. Some of the common commercial uses of these elements relate to their radioactive qualities. Magnesium is used in metal alloys and beryllium is used in the hazmat industry for nonsparking tools. As with the first group, this second group is also very reactive. These elements have two electrons in the outermost shell and freely release them for bonding. They react with acids, releasing hydrogen gas that is extremely flammable. In addition, they are water-reactive, produce solutions that may be caustic, and have the potential to ignite in air if heated.

> ■ **Note**
>
> The Romans used lead in the water aqueduct system and to sweeten wine. It is thought that this intake of high quantities of lead actually caused the fall of the Roman Empire.
>
> ■ **Note**
>
> The nonmetals do not conduct electricity well. They exist in two or three states of matter at room temperature. Valence numbers are +4 or −4, but −3 and −2 valences are also possible.

> **■ Note**
>
> The metalloids (see Figure 1–4) are the elements that are between the metals and the nonmetals. They have properties of both the metals and the nonmetals.
>
> **■ Note**
>
> Hemoglobin within the blood contains an iron atom that readily accepts oxygen for transportation through the arterial cardiovascular system. The bonding of oxygen to the hemoglobin molecule forms oxyhemoglobin. This bond is weak and enables the red blood cell to carry oxygen to the cells. Once at the cell, the bond is broken. There are several other configurations that are stronger than the oxygen-to-hemoglobin bond, thus not allowing oxygen transfer; for example, sulfoxhemoglobin, carboxyhemoblobin, and methemoglobin.

Transition Metals. The next ten groups (groups IIIA–IIB, 3–12, or IIIB–IIB) are called the transition metals, which are stable elements. The stability of some of these elements is so great that these elements are sometimes referred to as the noble metals. The transition metals have a distinct ability to have several valence numbers because they can combine with other elements utilizing not only the outermost shell electrons but inner shell electrons as well. This is the reason that you will find several valence numbers for most of these elements. These are metals that conduct electricity and heat.

> **■ Note**
>
> In the early 1800s the effect of mercury was noted among those individuals who made hats (hatters). Mercury was used in the final stages of the hat production. The hatter was directly exposed to the mercury fumes, which caused a variety of nervous system dysfunctions and hyperactivity. Thus the phrase "mad as a hatter," which was highlighted in the book *Alice's Adventures in Wonderland*.

All of the elements in this area of the periodic table are true metals, some of which are referred to as the heavy metals. These elements pose health hazards that are frequently monitored for medical consequences.

aluminum family (group IIIB, 13, or IIIA)
a periodic-table group of metals and one nonmetal: boron, aluminum, gallium, indium, and thallium

Aluminum Family. Group IIIB, or group 13 or IIIA, on the periodic table of elements are the **aluminum family** of metals and one nonmetal: boron, aluminum, gallium, indium, and thallium.

Their physical state is solid for the most part (except gallium, which is a solid below 86°F). The general hazards of this group are poisoning and other varied effects depending on the element combination. Boron is used in flares and some of the pyrotechnic rockets; it gives fireworks their distinctive green color. Although not a potent poison, it may result in chronic toxic effects if ingested. Indium is primarily used for making low-melting alloys and is toxic through inhalation. Thallium is a toxic substance; the compound thallium sulfate was mostly used in rodenticides during the 1970s. Because of its highly toxic qualities, household availability of thallium is prohibited.

germanium family (group IVB, 14, or IVA)
a group of elements on the periodic table represented by carbon, silicon, germanium, tin, and lead

Germanium Family. Group IVB, or group 14 or IVA, of the periodic table represents the **germanium family** of elements: carbon, silicon, germanium, tin, and lead. These elements range from gray-black to silvery white and are all solids that have high densities. Germanium initially did not have many industrial applications; however, with the introduction of computers and the Internet, this element has found many useful industrial applications. It also has been found to be advantageous in fighting specific bacteria, which makes it highly useful in the medical field.

carboxyhemoglobin

the compound formed when carbon monoxide reacts with hemoglobin, depriving the hemoglobin of its ability to carry oxygen and thus causing chemical asphyxiation

nitrogen family (group VB, 15, or VA)

a group of elements on the periodic table consisting of nitrogen, phosphorus, arsenic, antimony, and bismuth

pyrophoric

a liquid or solid that will ignite within 5 min without an external ignition source after coming in contact with air

sulfur family (group VIB, 16, or VIA)

also known as the chalcogen family; a group of five moderately reactive elements on the periodic table: oxygen, sulfur, selenium, tellurium, and polonium

halogens (group VIIB, 17, or VIIA)

a group of elements on the periodic table consisting of fluorine, chlorine, bromine, iodine, and astatine

noble gases (group VIII or 18)

a group of elements on the periodic table made up of helium, neon, argon, krypton, xenon, and radon; sometimes referred to as inert gases

Lead is a toxic heavy metal that, much like mercury, can cause a variety of central nervous system disorders. The toxic effects of lead are long term, which is one reason why lead is no longer used in plumbing or as a tint in white paint.

Carbon and oxygen sometimes combine to produce carbon monoxide (CO). Once it is attached to the hemoglobin molecule, a very stable bond is created. The new compound is called **carboxyhemoglobin**. This molecule is so stable that oxygen can no longer attach to the red blood cell, causing chemical asphyxiation. General hazardous characteristics of this group are heavy-metal poisoning and other varied effects depending on the element combination.

Nitrogen Family. Group VB, or group 15 or VA, on the periodic table makes up the **nitrogen family**: nitrogen, phosphorus, arsenic, antimony, and bismuth. In this group nitrogen is the only element that is a gas; the rest are solids. The general hazards of this group are heavy-metal poisoning and other varied effects depending on the element combination. For example, phosphorus is **pyrophoric** in certain forms.

Sulfur Family. Group VIB, or group 16 or VIA, on the periodic table is the **sulfur family**, sometimes called the chalcogen family, and consists of oxygen, sulfur, selenium, tellurium, and polonium. The first two concern us most because they are the most prevalent. These elements are moderately reactive. The general hazards of this group are heavy-metal poisoning and other varied effects depending on the element combination.

Halogens. Group VIIB, or group 17 or VIIA, on the periodic table are the **halogens**, consisting of fluorine, chlorine, bromine, iodine, and astatine. This group of elements is one of the most widely used groups of chemicals. These chemicals are all toxic and produce a variety of serious health effects. This group is utilized in the production of the halons. They are corrosive, irritating, and strong oxidizers. Great care should be exercised during a hazardous materials operation involving these elements.

> ● **Caution**
> The halogens, or the salt-formers, have seven electrons in their outermost shell, giving them a valence of −1. Halogens are used in combination with methane or ethane resulting in a nonflammable, colorless gas that can be used as an extinguishing agent. In this configuration they represent an asphyxiation hazard.

Noble Gases. Group VIII, or group 18, on the periodic table is known as the **noble gases** and made up of helium, neon, argon, krypton, xenon, and radon. All the elements in this group are stable without the introduction of energy and are sometimes referred to as inert gases. The reason for their stability is that the outer shell contains eight electrons. Commercial use is focused on this extreme stability. The general hazard of this group is that these elements are displacement asphyxiants.

> ● **Caution**
> The noble gases, except helium, which has only two electrons in its outer shell, have eight electrons in the outer shell, making them very stable. However, they also represent an asphyxiating potential.

CHEMICAL BONDING

atomic radius
the distance halfway across an individual atom

ionic radius
the distance halfway across an individual ion

ion
an atom or group of atoms that have lost or gained one or more electrons, giving the entire structure an overall valence

cations
positively charged ions

anions
negatively charged ions

compound
a substance composed of two or more elements in a fixed proportion that are bonded chemically

molecule
composed of two or more atoms, the smallest unit of a substance that retains all the properties of the substance

The **atomic radius** and **ionic radius** are the distances halfway across an individual atom or **ion**, respectively. Because we have two types of atoms (or ions) that engage a chemical reaction, it is helpful to understand the relationship of these principles. As mentioned earlier, the size of an atom has an effect on the atom's tendency to maintain or give up the outermost electron. The distance from the nucleus is the primary factor. When the distance from the center is such that the attractive tendencies are weak at the outer edges, then the disposition is to lose an electron. Conversely, when the atomic radius is small, then the forces maintaining the outer shell are strong. The tendency is to maintain the electrons and gain more until the octet rule has been satisfied.

When an atom such as sodium has one electron in the outer shell, removing the electron produces a positive ion. If, for example, the atom has seven electrons in the outer shell, then the tendency is to gain an electron, and the result is a negative charge. Chlorine is a good example of this tendency. In each case an ion has been created.

It is the outermost shell or shells that are involved with the sharing or transfer of an electron. In some cases, an element has its outermost electron stripped from its orbit. Elements that have one or two fewer electrons are called **cations**, and they possess a greater positive charge. On the other hand, some elements gain one or two electrons, making them negatively charged. These are called **anions**. Collectively this group of atoms is called ions.

In the first example in Figure 1–7, sodium (Na) loses an electron and becomes a charged particle that pairs up with the chlorine ion (Cl). In the second example, the chlorine ion has seven electrons and needs to gain an electron to be satisfied. These two elements combine as table salt (sodium chloride or NaCl). The sodium has a positive charge, with eleven protons and ten electrons, and the chlorine has become negatively charged, with seventeen protons and eighteen electrons.

When looking at the periodic table, we can easily see how many electrons are in the outer shell by looking at the top of the table itself (by using the classical Roman numeral system for groups IA–IIA and IIIB–VIIB) (see Figure 1–8). The basic structural units of the compound are the elements. A **compound** is a substance that has two or more elements combined in a fixed proportion. For compounds to be formed, elements must attach themselves together and form a **molecule**. NaCl, or table salt, is an example of a compound. The attachment is an attraction that contains relatively moderate to strong forces holding the molecule together. This force is called the *chemical bond,* or the attachment between the elements to form molecules.

In terms of valence numbers, each group has the same numbers and types of electrons in the outer shell. This fact provides us with bonding information and the similarities of each chemical within a group (see Figure 1–8). Under normal conditions mercury and bromine are liquids. Fluorine, chlorine, nitrogen, oxygen, hydrogen, and the noble gases are gases at standard room temperature. The remaining elements are solids.

Figure 1–7   *Two molecules of table salt.*

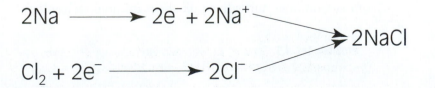

$$2Na \longrightarrow 2e^- + 2Na^+$$
$$\longrightarrow 2NaCl$$
$$Cl_2 + 2e^- \longrightarrow 2Cl^-$$

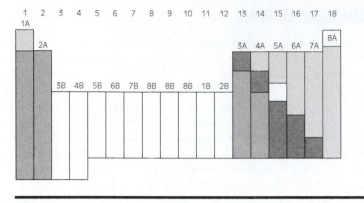

Figure 1–8   *The "A" groups can be identified as the valence numbers in columns 1A and 2A. These elements are going to give up one and two electrons, respectively. In group 3, three electrons are given up; however, the rest of the group can act differently given certain environments. Group 4A(14) (C, Si) shares four electrons, and the rest of the group can act differently given certain environments. Group 5A gains three electrons, Group 6A gains two electrons, and group 7A needs only one electron. Group 8A elements are noble and have their octets filled.*

Formulas

We utilize formulas and the symbols that represent the elements when describing a chemical as a shorthand for the visual idea of bonding and the compounds to be considered.

Each molecule has a structural plan that nature follows. This system has its own geometry that provides a variety of molecules. Matter as we know it must be electrically neutral. Thus, the negatives must equal the positives. When ions come together to create a compound, a transfer of electrons takes place. Once the electrical neutrality is satisfied, a compound is generated. Ionic compounds are three-dimensional lattice structures, extremely strong, and continuous. Thus, most are solids at normal room temperature. In order to break this bond, a tremendous amount of energy must be placed into the compound.

Not all compounds are held together by ionic bonds. A covalent bond is the chemical bond created by the sharing of electrons in order to form the molecule. The sharing requires that two or more elements get close enough together to allow the outer shells to share a pair of electrons. Additionally, there must be available space around the molecule in order for the reaction (combination) to take place. In this case, the bond creates a stable molecular configuration. The more stability a compound possesses, the more energy is required to break the compound apart.

Lewis dot structure
a visual representation of elements, atoms, and ions

To understand how electrons move, we employ several methods by which we can visualize the outermost shell. The **Lewis dot structure** (see Figure 1–9) is the most convenient, but it is laborious. In an attempt to simplify this approach, we will use the stick drawing (H-O-H for water, H_2O, as an example). Although each molecule has a specific set of geometric rules, these rules are not going to be considered for our use. However, one must realize that it is this geometry that has an effect on the reaction of a molecule, as we will soon see in the discussion of reaction rates and when talk about infrared technology.

Problem Set 1.3   Draw Lewis dot structures for the first three elements in groups 1, 2, 13, 14, 15, 16, 17, and 18.

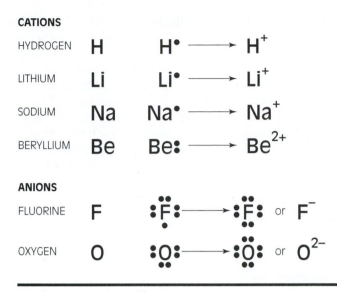

CATIONS

HYDROGEN

LITHIUM

SODIUM

BERYLLIUM

ANIONS

FLUORINE

OXYGEN

Figure 1–9   *Here for simplicity's sake we utilized Lewis dot structures to identify the valence electrons. We can see that in those elements that give up electrons (cations) the overall charge became positive and correlates with the group number they are in. The anions need electrons. You can subtract their group number from the octet of eight to determine how many electrons they need.*

Lewis dot structures are a format used to represent the theoretical observations made through experimentation. In this depiction, all the valence electrons are counted and grouped into pairs. Each pair of electrons represents a bond between the atoms to form a molecule. Each pair in association with the atoms that are being bonded should have a complete set of electrons representing the octet of the outer shell. The completion of the octets of the atoms may be accomplished by adding electrons from either another atom or the atoms that are within the given structure. In some cases double or triple bonds may exist.

So far what we have described are single bonds. For some elements a double bond can exist due to the electron configuration or structure of the complex ion. In Chapter 3 we investigate single, double, and triple bonds. For now, realize that the configuration of the bonds is dependent on several factors that are beyond the scope of this book; however, the bonds are covalent and occur between nonmetals as a general rule.

Resonance

When Lewis dot structures do not provide a rational structure that enables us to visualize what chemical geometry looks like, then we must consider resonance bonding. For example, in the sulfur dioxide molecule (see Figure 1–10), the only way to maintain the octet rule with the given structure is to have one single and

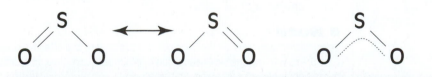

Figure 1–10   *A resonance structure (right) is a combination of structures (center and left), not really a single bond and not really a double bond.*

resonance
a description of a molecule consisting of a combination of more than one structural representation

one double bond. In experiments, however, the bonds between the sulfur and the oxygen are the same length. Since a double bond and a single bond have two different bond lengths, how can this be? The answer is **resonance**. The bonds between the sulfur and both oxygen atoms are neither a single nor double bond. Each bonding electron contributes to the bonds within the entire structure. This phenomenon can be represented by two structures. The first structure has a double bond on the left side, and the second structure has a double bond on the right. The double-headed arrow between each structure indicates that the structure is actually something between what is depicted on paper. The nitrate anion follows this configuration of representation, as does the benzene ring.

Problem Set 1.4   Draw the resonance structures for: NO_3^-, NO_2, and NO.

Rate of Reaction

When we speak about the rate of a reaction, we are talking about the speed at which the reaction occurs. If a reaction has a slow rate, the molecules combine at a rate that advances the end products over a long period of time. If a reaction occurs rapidly, the rate at which the molecules are combining is fast—and it may be a dangerous reaction.

When we speak of reaction rate, not only are we talking about energy (energy can be another chemical, heat, or light, or a combination thereof) being placed into the system (the reaction), we are also discussing the effectiveness of collision. Looking at a normal reaction, it can be seen that two molecules colliding can either strike each other head-on, from either side, or from behind. If one molecule strikes the other molecule from behind, the energy may be displaced and could push the second molecule farther away. A side strike causes a situation that displaces the molecules at right angles to each other. However, if the moving molecules strike each other in an effective fashion, an effective collision results.

Why are effective collisions important? Bonds must be broken in order for new bonds to be created. The breaking of bonds takes enormous amounts of energy. This energy is gained from an effective collision. However, for a reaction to take place, several factors are important, including the concentration of a substance. If there is more of one particular substance than another, the chances increase that an effective collision will take place. Likewise, if we neither increase our substance nor put enough into our reaction vessel, then the frequency of collision would be low and the possibility of an effective collision would be small.

■ Note
The appropriate geometry of the molecules promotes reaction. We can force this by increasing the temperature, which, in turn, increases collisions. By virtue of increasing collisions, the right geometry can be achieved, and thus, reaction.

■ Note
Colorimetric tubes work on the principles of reaction rate and effective collisions. The reagent in the tube reacts with the chemical in question to produce a stain. Through graduations on the tube, parts per million or percentage can be established.

polymerization
the chemical reaction (extremely violent if uncontrolled) that occurs when smaller compounds are linked together to form larger compounds

Temperature also affects the frequency of collision. When we raise the temperature of a substance, the molecules increase their relative speed. By increasing their speed, the frequency of an effective collision increases also.

If we were to add pressure to our system, then we would see that the volume in which the molecules have to move around is decreased. Once the room for the molecules to move around in is limited, the chance of an effective collision also is increased. Obviously, if all three forces are applied to a system, the chance of a collision that would produce a combination would be increased.

The appropriate geometry and orientation of the molecules promotes reaction. Those factors can be affected by increased concentration of the substance, temperature, or pressure. Although geometry and orientation are not directly observable at a hazmat incident, they are factors in the decision-making process. Does the chemical involved have the ability to achieve the correct orientation, activation energy, and effectiveness of collision to produce a reaction? For example, vinyl chloride within a container will undergo **polymerization** in air, so contact with oxidizers must be avoided.

!Safety
Metal containers involving the transfer of 5 gal or more of vinyl chloride must be grounded and fitted with self-closing valves, flame arrest, and vacuum atmosphere to avoid polymerization.

---

standard temperature and pressure (STP)
a temperature of 32°F (0°C or 273 K) and a pressure of 1 atm (760 mm Hg or 760 torr) for measured volumes of gas

Most of what we see in the documented reference literature states pressures and temperatures at which reactions take place. These conditions are standard, and so are called **standard temperature and pressure (STP)**. Any other set of conditions, as it relates to temperature and pressure, will be identified in the reference literature.

When describing flammable or combustible conditions, not only will the reference literature identify temperature and pressure, but also the container in which the materials were tested. Two types of containers are described: open cup (O.C.), and closed cup (C.C.). With the open cup, the vapors are allowed to evolve above the solution and into the testing chamber. With the closed cup, the vapors are contained within the container holding the material.

■ Note
To convert degrees Fahrenheit to degrees Celsius, the following formula is used: °F = 1.8°C + 32

---

activation energy
the amount of energy required for reactants to combine, progress over the hill, and generate the final product

catalyst
a substance that increases the rate of reaction without being consumed in reaction

inhibitor
a substance that controls or delays a reaction

Reactions are like hills in that you must be able to cross a certain point in order to reach the other side (see Figure 1–11). That "certain point" is called the **activation energy**, the minimum amount of energy needed for the collision to be effective. If the hill is low, then the energy (or additional energy) required is small and the reaction will progress with relative ease. If the hill is large, then the reaction will need a lot of energy. This reaction will need to be forced through with an additional supply of energy (additional temperature, pressure, or substance concentration that increases frequency of collision). A **catalyst** is a material that lowers the activation energy (decreases the hill and becomes part of the product) so that the reaction can proceed faster. An **inhibitor** is a chemical that increases the activation energy (increases the height of the hill), thus slowing down the main reaction.

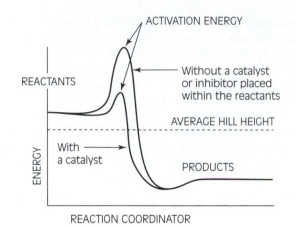

Figure 1–11 *You can see from the graph that the average hill height is high enough such that we would have to increase the temperature, concentration of reactants, or pressure in order for this chemical process to move toward completion or the formation of products. Looking at the reaction on the top we can see that to make it over the hill we are going to need help in the form of a catalyst. If we do not want the reaction to occur, we place an inhibitor in it. Likewise the bottom reaction shows us how reactions need energy, but in this case the catalyst has lowered the activation energy.*

PROPERTIES OF MATTER

We share our world with a variety of substances. These substances occupy space, have mass, and are referred to as *matter*. Scientists study these substances, their interrelationships, and their reactions. From this investigation scientists are able to create new substances.

Matter has phases that are commonly called states of matter: solid, liquid, and gas (see Figure 1–12). A solid can take up space and has certain physical properties, the most distinct of which is retaining its shape with or without a container. This shape is maintained through a defined system of attraction that permits a rigid structure. Inherently, a solid may or may not conduct electricity, reflect light, or be hard or brittle, for example. However, the melting and boiling points can be measured. In a solid the molecules are very close in proximity to each other. Although the atoms are vibrating very fast, the actual movement cannot be seen.

A liquid conforms to the shape of the container it occupies. Although it has attractive forces, these forces are not as strong as in the solid substance. This limited force allows the substance to have shape only in the container and to flow to

SOLID   LIQUID   GAS

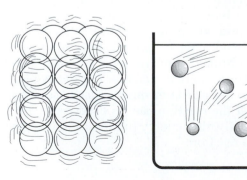

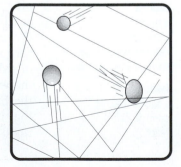

Figure 1–12 *All matter has three states or forms.*

the lowest points. In the liquid state, molecules are free to move, slipping by one another. This action is observed when a liquid flows out of its container.

viscosity
a measure of flow

Viscosity is a measure of flow. It describes the thickness of liquids or how well they flow. A low-viscosity liquid will flow like water. The lower the viscosity, the higher the tendency for the liquid to spread. Conversely, the higher the viscosity, the slower the spread (for example, cold molasses).

When referring to hydrocarbons, primarily combustible and flammable liquids, viscosity can relate to the production of static electricity. Some hydrocarbons have polarity; in other words, they have a negative and positive end, very much like a magnet. As these tiny magnets move over each other they create a small charge of static electricity. Once this electricity finds a ground, a spark can be created and if the conditions are right the vapors can ignite.

● Caution

The movement of liquids and gases can produce static electricity, a potential ignition source.

■ Note

Some chemicals are characterized by persistence; that is, the physical state is maintained, thus allowing the chemical to remain in the environment for long periods of time.

---

gas
a chemical that exists as a molecular identity in air

A **gas** also takes up space and conforms to the container in which it is carried. However, the attractive forces are much less than those within the liquid substance, thus allowing the gas to move about more freely. In the gaseous state, molecules are the farthest apart, colliding with other molecules and the container itself. In each state of matter, an increase in temperature increases the molecular movement; likewise, decreasing the temperature decreases the molecular movement.

Condensation is the conversion of a gas (or vapor) into the liquid state. Energy is released into the surrounding environment as the gas moves into a liquid state. This phenomenon can best be illustrated when steam (heated water vapor) condenses on the skin; the energy that is released is fairly high, causing a severe burn.

vapor
the diffused state of matter that is released from a liquid substance, which when combined with air forms a potential ignitable mixture

---

■ Note

The Five Step Field ID Method™ and HazCat Kit™ utilizes principles of chemical and physical properties to identify families of chemicals. From this deduction further analysis can be performed to give logical identification possibilities.

---

dusts
fine particles of matter, such as coal dust, sawdust, and grain dust

As the state of matter changes, its physical appearance may change, but its chemical composition will remain the same. For example, if we were to put an ice cube on a hot surface, we would see the solid ice melt into a liquid, and when enough thermal energy is applied, the liquid ice (water) changes into a gas (technically a **vapor**).

mists
liquids that have been atomized, for example, spray paint, high-pressure oil leaks, and aerosolization of nerve agents; a mist and aerosol can be thought of as the same

---

■ Note

Gases are the natural state of compounds such as carbon monoxide, propane, and nitrogen, as examples. Vapors are the "gases" that are given off from flammable or volatile liquids. Gasoline and acetone are examples. **Dusts** are fine particles of matter such as coal dust, saw dust, and grain dust. **Mists** are liquids that have been atomized; for example, spray paint, high-pressure oil leaks, and aerosolization of nerve agents. A mist and aerosol can be thought of as the same.

We discuss the particular properties of state change later. For now remember that these characteristics are observations and calculations made through experimentation. These properties are both chemical and physical.

Physical properties describe the outward physical changes or attributes a substance may have. They are the properties that can be measured or observed without changing the true identity of the matter. They are characteristic or intrinsic to the material. **Chemical properties** describe the types of reactions that a particular substance may undergo. The true identity of the matter is changed. Thus, the only way the chemical properties of a particular substance can be identified is through chemical experimentation.

physical properties
intrinsic characteristics of a substance that can be observed and measured, such as appearance, melting point, density, and so on

chemical properties
intrinsic characteristics of a substance described by its tendency to undergo chemical change, such as heat of combustion, reactivity, and so on

heterogeneous
a material or solution in which there are visual differentiating parts

homogeneous
a material or solution having no differentiating parts

!Safety

The gauge on a pressure tank is calibrated to read zero when the tank is empty, and is designated as psig (lb/in^2 gauge). Absolute pressure is the gauge pressure plus atmospheric pressure, and is designated as psia (lb/in^2 absolute).

Matter can additionally be characterized as heterogeneous or homogenous (see Figure 1–13). **Heterogeneous** describes matter existing as a mixture of pure substances having different chemical and physical properties. These properties will remain separate from each other. The mixture is not the same at all points. Picture a jar with steel balls and glass marbles. At some points you would observe a bunch of steel balls or glass marbles. A **homogeneous** mixture is a substance having the same chemical and physical properties throughout. It is sometimes referred to as a *solution*.

■ Note

Salt crystals dissolved in water is an example of a solution. The salt ions evenly distribute through the entire solution, and the physical and chemical properties are the same at all points within the solution, making it homogeneous.

Physical Properties

The physical properties of an element or compound are useful in describing the material involved. They are the properties that have qualitative numbers, their own fingerprint within nature. Each substance has its own set of chemical and physical properties. A sulfur dioxide molecule, for example, will always possess an individual set of chemical and physical properties. So if a sulfur dioxide molecule melts, boils, looks, and freezes like a sulfur dioxide molecule, it must be sulfur dioxide.

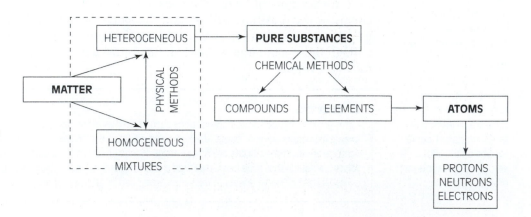

Figure 1–13 *The relationships of matter and definitions presented.*

Appearance, melting point, boiling point, density, specific gravity, vapor density, solubility, vapor pressure, viscosity, freezing point, and pH are all examples of physical properties of matter. These properties are characteristic of a particular element or molecule. They are qualities measured without changing the chemical makeup of the compound or molecule.

appearance
the form, size, and color of a material

Appearance is the form, size, and color of a material. The form may be described as a solid, liquid, or gas, and the particulate size can be powder, dust, or fumes. The form of the hazardous material will dictate the management strategies toward incident stabilization. For example, if the material is a liquid, leak and spill control may be the tactics of choice. When a gas is involved, limiting its release or changing its physical form may be the tactical procedure. If the material is in the solid form, confinement may be the action chosen in order to minimize exposures. Remember that solids in particulate form and gases are easier to decontaminate than liquid materials or slurries (a mixture of a liquid and finely divided particles).

Because some substances have characteristic states of matter and color, color can give clues regarding the substance involved in an incident. Consider that smoke is a primary indicator of a reaction. Thick, black smoke is characteristic of benzene and petroleum products; white smoke of hydrochloric acid and phosphorus-type products; gray-white-brown smoke of nitrocellulose and nitric acid; greenish-yellow smoke of chlorine; and gray-brown smoke of iodine.

melting point
the temperature at which a material changes from a solid to a liquid

The **melting point** is the temperature at which a material changes from a solid to a liquid. If a material has a low melting point, we would expect it to become a liquid at average temperatures and pressures. This material could possibly even become a gas if the boiling point is also significantly low (high and low are references to the ambient temperature). That presents inherent problems of containment, confinement, and decontamination. Liquids flow freely in the absence of a container at the scene of a hazmat incident.

> **■ Note**
> Surface area exposed to the atmosphere has a relationship to how much vapor is released into the environment. The greater the surface area, the higher degree of vapor production.
>
> **■ Note**
> Melting point and freezing point can be thought of as the same thing: the point at which phase change occurs.

boiling point
the temperature at which a liquid's vapor pressure equals the atmospheric pressure

The **boiling point** is the temperature at which the atmospheric pressure and the vapor pressure of a liquid are equal. At this point the liquid will transform from the liquid state into the gaseous state.

> **❗Safety**
> Hazardous materials that are in a liquid state with a very low boiling point must be kept under pressure or they will boil and change form.

Hazardous substances in a liquid state possessing a low boiling point usually have a relatively higher vapor pressure (see Figure 1–14). This type of material presents potential fire, reactivity, and health hazards. Conversely, the high–boiling-point liquids have relatively low vapor pressures. These liquids would need an active energy source (such as fire) to convert them from the liquid state to the vapor state. Substances with high boiling points are generally

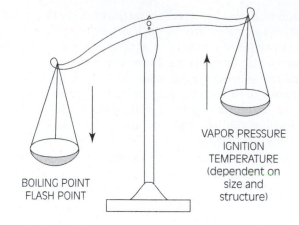

Figure 1–14 *The boiling point (BP)–vapor pressure (VP) and flash point (FP)–ignition temperature (IT) relationships. They are inverse functions.*

density

the mass of a substance per a given volume

safer than those with low boiling points. However, there are a few exceptions to this rule; for example, sarin has a very high boiling point but does release toxic vapors even at room temperature.

Density is a relationship and can be applied to all three states of matter. It is the mass of a substance per a given volume. For solids and liquids it is expressed as grams per cubic centimeter; for gases, as grams per liter.

$$\text{Density} = \frac{\text{Mass}}{\text{Volume}}$$

The density of a substance is also a function of temperature in that raising the temperature decreases the density. Density, therefore, is a function within its specific phase. This will occur only at a certain temperature. See critical temperature/pressure and phase diagrams in Chapter 3.

For example, 1 gal (volume) of water weighs 8.35 lb (mass). A container holding 3 gal of water weighs roughly 25 lb. We can figure the density of water by entering that information into the formula:

$$\text{Density} = \frac{25 \text{ lb}}{3 \text{ gal}}$$

$$= 8.333 \text{ lb/gal}$$

Therefore, if we know that a chemical has a volume of 4 gal and a density of 19.8 lb/gal, then through algebraic manipulation we arrive at 79.2 lb of the substance—roughly 80 lb of a hazardous substance. The tools and appropriate level of personnel can now be anticipated for the mitigation process.

Both specific gravity and vapor density are applications of this principle for liquids and vapors, respectively. They both look at the mass per unit volume. From these calculations, predictions on the movement of the liquid or vapor can be made. For example, specific gravity is used to predict whether a substance will sink or float on water. Vapor density, like specific gravity, will predict whether a vapor will settle or rise in air.

■ Note

Suppose a railcar of phosphoric acid overturned. The consist tells us that the railcar is 198,900 lb. Utilizing the specific gravity of 1.69 times 8.35 H_2O/1 g H_2O and dividing it into the weight of the acid, we figure 14,095 gal of acid:

$$198,900 \text{ lb}/1.69 \times 8.35 = 14,095 \text{ gal}$$

specific gravity

the weight of a solid or liquid as compared to an equal volume of water

Specific gravity is the weight of a solid or liquid compared to an equal volume of water. Water has a value of 1, a dimensionless quality, in relation to the compared material. In other words, there is no unit of measurement. If the specific gravity of the tested material is less than 1, the material will float in water; if greater than 1, the material will sink in water. When dealing with liquid materials in a hazmat situation, we must know if the product will float or sink in order to choose the proper spill control techniques. If, for example, the material floats on water, then the material may also produce a vapor release, leading to yet another problem. However, if the material sinks, then containment of the gaseous product is confined; by using water or foam as a barrier, the potential of fire is reduced. In either case we must remember the environmental impact of such a decision.

> ### ■ Note
> The use of underflow and overflow damming depends on the specific gravity of the substance being dealt with. Overflow dams are used when substances are heavy and insoluble within the water environment. When the contaminant is floating on the water, underflow damming is used.
>
> ### ■ Note
> Prediction of a vapor cloud depends on several factors: density of the gas; wind, wind direction, ambient temperature, and humidity; cloud cover or the lack thereof; height, amount, and duration of the discharge; and the temperature of the vapor.

vapor density (VD)

the weight of a vapor or gas as compared to an equal volume of air

liquefied petroleum gas (LPG)

a compressed and liquefied mixture of alkanes and alkenes: propane, propylene, butane, isobutane, and butylenes

Vapor density (VD) is a comparison like specific gravity, but of a gas or vapor to the ambient air. Vapor density is the weight of a vapor or gas as compared to an equal volume of air. Air is given a density of 1. If the relationship indicates a number greater than 1, then the vapor will drop or settle. If the comparison yields a number less than 1, then the vapor will rise, creating a vapor cloud. This vapor may or may not dissipate depending on the density of the gas, wind, humidity, and other related factors.

> ### ■ Note
> In the *NIOSH Pocket Guide to Industrial Hazards*, vapor density is identified as RgasD, or relative gas density, which is referenced to air where air = 1. Molecular weight above or below 29 will yield the same result.

solubility

the ability of a material to blend uniformly with another to form a solution

solvent

a substance that has the capability of dissolving another substance to form a solution

solute

the substance added to a solvent

miscibility

the ability of materials to dissolve into a uniform mixture

Vapor density and vapor pressure are two of the properties considered when dealing with plume dispersion models (weather conditions and topography are also a part of plume dispersion modeling). Under certain conditions the water vapor in the air must also be considered. This is especially true when dealing with the lighter-than-air gases in high-humidity environments; for example, methane has shown qualities very much like those of **liquefied petroleum gas (LPG)**. If we consider that vapor density is the effect of the partial pressure, molecular weight, and mass present, and combine the effect of atmospheric conditions, we can see that inversion layers in the environment will shift the physical properties. The inversion layer within the environment traps the normally lighter-than-air molecule, thus giving it outward properties of heavier-than-air gases.

Solubility is the ability of a material to blend uniformly with another material to form a solution. The material present in the greater amount is called the **solvent**, and the material present in the lesser amount, usually the additive, is called the **solute**. Certain materials are miscible in any proportion, and others are not. **Miscibility** is the ability of materials to dissolve into a uniform mixture. Usually

polarity

the quality of possessing two opposing tendencies, as in the existence of a negative and a positive end on a molecule

concentration

the relative amount of the lesser component of a mixture or solution

we think in terms of a water-miscible substance; when we say a substance is miscible in water, we mean it is infinitely dissolvable in water. The factors affecting solubility are the **polarity** and **concentration** of the materials in solution, and whether the solute is a liquid or a solid.

> **■ Note**
>
> Polar compounds in the liquid state create hydrogen bonding and thus tend to stick to one another. This "stickiness" reduces their tendency to vaporize.
>
> **■ Note**
>
> Polarity has a lot to do with the solubility of a solute in a solvent.

Polar substances have a positive end and a negative end (see Figure 1–15), while nonpolar substances do not have an electronegative or electropositive end. Like will dissolve in like: polar in polar, and nonpolar in nonpolar. If the solute is a solid, the polar/nonpolar dissolvability character still holds true; however, the solvent can only dissolve a limited portion of the solid. It is not infinite, as it is with liquids. Each solvent will have a saturation point. Above this point the added solute will not dissolve.

> **■ Note**
>
> The chemical VX can produce enough vapors within a room to create a toxic environment in a matter of minutes. Sarin, however, will produce a toxic environment in a few seconds. The difference lies in the volatility of the chemical rather than its vapor pressure. A highly volatile substance will vaporize at relatively low temperatures.

Whenever we talk about a solution (or solubility), we are discussing a homogeneous mixture: All the parts of the end mixture are composed of the same material. If we have an acid diluted with water, the runoff will be a mixture of the acid and the water. This condition is partly due to the polarity of each compound.

> **■ Note**
>
> Boiling point, size of the molecule, and polarity all affect how well a compound will vaporize or volatilize. In general, heavyweight compounds produce small amounts of vapor, and lightweight compounds produce significant vapor.

However, if we wanted to dilute carbon tetrachloride, water would not be our choice because water and carbon tetrachloride do not mix, partly due to the polarity difference. Carbon tetrachloride is miscible in alcohol, such that the alcohol mixture would work well for decontamination, provided that heat generation does not occur. The carbon tetrachloride decontamination solution containing alcohol would, of course, have to be properly disposed of.

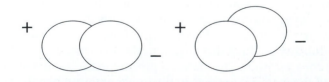

Figure 1–15 *Molecules can have positive and negative ends, depending on the individual atom's electronegativity. Each molecule acts like a tiny magnet. We will see how this impacts infrared spectra in Chapter 7.*

● **Caution**

Care should be exercised when using alcohol as a decontamination solution because it can decrease body temperature with as little as 10% of body surface area exposed. It can also increase absorption of a chemical.

neutralization

the process of bringing an acid or alkali back to a pH of 7

The concept of dilution also has its basis in the solubility of the decontamination solution. Although water is the primary solution for decontamination, the amount of water needed in some cases may be more than would be expected. For example, when dealing with corrosive materials, dilution as a method of **neutralization** would require an extensive amount of water. One gallon of a fluid that has a pH of 1 would require a total volume of 100,000 gallons of water to move the pH to around 6 (see the sections on pH in Chapter 2, and later in this chapter).

Some substances are polar, and when added to water (also polar), produce a total polar solution. Such is the case when a gallon of alcohol is on fire and an attempt to extinguish it using water is made. The result would be that the total volume of the alcohol and the water that was used would be on fire. This sort of situation is, in fact, a dilution problem. It takes a lot of water to dilute the alcohol to a point where vaporization no longer occurs, roughly ten times more.

Solubility also comes into play when a toxic substance is placed into solution for dispersion of the material. Not only does its solubility give the material greater distribution power, it also increases the health hazard in terms of skin and mucosa absorption.

● **Caution**

If a substance has a low water-solubility point, it may be soluble in lipids or proteins, adding to the health hazard.

vapor pressure (VP)

the pressure exerted by a vapor; in particular, the pressure a gas exerts against the sides of an enclosed container

Vapor pressure was mentioned briefly in the discussion of boiling point. To elaborate, **vapor pressure (VP)** is the pressure exerted by a vapor; in particular, the pressure a gas exerts against the sides of an enclosed container (see Figure 1–16). It is the pressure of the material as it boils off within a container (similar to evaporation). Each atom within the liquid material is bouncing about until it reaches enough velocity to escape the liquid. Once this molecule has escaped the liquid form, it has been changed into a molecule traveling through the air space, producing a gaseous state. It is this movement of atoms in the gaseous state that is measured as vapor pressure. All liquids have a measurable vapor pressure.

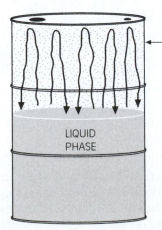

HEAD SPACE OF
THE DRUM SATURATED
WITH VAPOR RELEASED
FROM THE SURFACE OF
THE LIQUID

LIQUID
PHASE

Figure 1–16   *Vapor pressure.*

sublimation

a physical change in which a substance passes from a solid into the gaseous state without having a liquid state

outage

an amount expressed in percentage by total volume of the container, by which a container of a liquid falls short of being completely filled

flash point

the minimum temperature at which the vapor of a liquid or solid ignites when in contact with an ignition source, but there are not enough vapors to sustain combustion

freezing point

the temperature at which a liquid becomes a solid

Vapor pressure is measured in millimeters of mercury (mm Hg). It is the weight of the atmosphere in a defined space, where stated air pressure is 760 mm Hg or 14.7 lb/in^2, and it is temperature dependent. Water, for example, has a VP of 21–25 mm Hg. Naphthalene (and its derivatives), camphor, dry ice, and iodine crystals are solids that have a vapor pressure due to a characteristic called **sublimation**. In this process the material goes directly to the vapor state from the solid phase without appearing to have gone through the liquid state. (Under certain conditions, vapors can move into the solid state without passing through the intermediate liquid phase.)

> **● Caution**
> High vapor pressures have the potential to lead to significant respiratory injuries.

If we are dealing with a closed container, then the high–vapor-pressure materials have a potential to breach the container under high temperature conditions (the explanation for **outage**, which occurs with the expansion of liquids). If this same container has flame impingement, the potential further increases. When we look at vapor pressure, the higher the vapor pressure, the lower the boiling point.

Flash point is a component of vapor pressure. Remember that the flash point of a material is that minimum temperature under which a liquid will give off vapors to form an ignitable mixture in air, provided that the flammable limits are within favorable perimeters and that an ignition source is present.

Freezing point and melting point can be thought of as synonyms. Which term applies depends on the context of the chemical reaction. If, for example, the chemical is moving from a solid to the liquid phase, then we use the term melting point. If the product is going from the liquid phase into a solid phase, then we use the term **freezing point**. The amount of heat required to move the chemical from the solid state to the liquid state, or the amount of heat that must be removed to move the liquid to a solid depends on the chemical itself. The specific substance and the state of matter in which it is being contained will have an impact on these two principles.

> **■ Note**
> If you have an acid without a water component, the reading from pH paper is meaningless.

pH

the scale used to measure how acidic or basic a substance is based on hydrogen ion concentration

acid

a compound that forms hydronium ions in water

base

a compound that forms hydroxide ions in water

The **pH** (positive hydronium ion) scale is used to measure how acidic or basic a substance is. It is a relative term. There are several theories that support the principles of acid–base reactions. In principle, an acid–base reaction is an ionization reaction. For strong acids (pH 0–4) and strong bases (pH 10–14), the **acid** will give up or break apart to create a H$^+$ ion (hydronium) within a solution, and the **base** will give up or break apart to create a OH$^-$ (hydroxyl) within a solution. It is the relative strength of a substance in water that we are discussing.

> **● Caution**
> It is possible to have an acid stronger than a pH of 0 or a base stronger than a pH of 14. These superacids and superbases are extremely dangerous.

When we talk about weak acids (pH 3–6) and weak bases (pH 8–10), we are discussing acids and bases that do not break apart or give up the H$^+$ or OH$^-$ readily. In fact, part of the original compound remains the same within the solution. So, the higher the tendency to give up a hydronium or hydroxyl ion, the stronger the acid or the base, respectively. The lower the tendency to give up a hydronium or hydroxyl ion, the weaker the acid or the base, respectively.

The pH scale starts at 0 and ends at 14. Seven is considered to be neutral, whereas 0 to 6.9 is acidic and 7.1 to 14 is considered as basic. It is a measurement of the hydronium ion concentration within the solution and is established from a measurement called mole per liter. Because of the measurement parameters, each number on the pH scale represents a tenfold difference. For example, when we determine the pH of distilled water, we find that the H^+ concentration within the solution is 10^{-7} mole/liter, or 0.0000001. If we count the number of decimal places back to 1, we would count seven decimal places. This means that the pH of distilled water is 7. At the scene of an acid spill we determine that the moles/liter of the acid is 0.1 or 10^{-1} mole/liter. Counting the decimal places for 0.1, we see we move the decimal one place in order to make it a whole number, thus the pH of the chemical in question is 1, or severely acidic. We would perform the same operation if we had 0.00000000001 or 10^{-11} mole/liter, or a pH of 11—severely basic (sometimes called alkali). (See page 231.)

■ Note

Residue in tanks permits a 1% content of material. The following is a simple relationship of how much could be within a tank based on size versus 99% "empty": 50,000 = 500 gallons; 30,000 = 300 gallons; 5,000 = 50 gallons; 3,000 = 30 gallons; 1,000 = 10 gallons.

Chemical Properties

As previously stated, chemical properties are the intrinsic characteristics of a substance described by its tendency to undergo chemical change. These characteristics include properties such as heat of combustion, reactivity, and flash point. All chemicals try to reach a state of equilibrium. It is this desired condition that creates the need to undergo a chemical reaction. The reaction takes place because of the specific architecture of the elements or compounds within the surrounding environment.

ignition temperature
the minimum temperature at which a material will ignite and sustain combustion without a continuing outside source of ignition

Ignition temperature is the minimum temperature at which a material will ignite and sustain combustion without a continuing outside source of ignition. The heat of combustion relates to a reaction in which the products are completely oxidized or in which complete combustion takes place. It is the heat liberated during the reaction. This reaction is with the oxygen (any material that can act as an oxidizer will provide this side of the fire triangle or fire tetrahedron) within the atmosphere or surrounding material. Heat and light are the primary by-products of this process.

As stated before, when a liquid produces enough vapor to flash when an ignition source is available but not sustain combustion, this condition is called its flash point. Flash point and ignition temperatures can be in close proximity on the temperature scale. (This concept is most important when dealing with organic compounds.)

flammable ranges or limits
represent the minimum and maximum concentration of a mixture in air favorable for ignition

---

● Caution

When the flash point is lower than the ambient temperature, vapors are produced. A lower boiling point equals a low flash point and higher possibility of a dangerous atmosphere.

Flammable ranges or limits represent the minimum and maximum concentration of a chemical in the air that is favorable for ignition to occur. When a gas is released into the environment, the wind, ambient temperature, topography, and availability of an ignition source play roles in the chemical's ability to ignite. If we could measure the gas as it is traveling through the air, we would discover three distinct areas of concentration. First, the vapor would be "rich," within this atmosphere or at a high concentration, around the container that originally held the material. Even if an ignition source were available, the gas would not burn in this area. As we continued our measurements moving away from the container, we would find an area that has the perfect percentage of air and chemical vapor concentration to permit ignition. Farthest from the source, we would measure the concentration of the chemical and find it to be "lean"; there would not be enough vapor to ignite in relation to oxygen. It would not burn even if there were an ignition source available. However, one must also remember that around the flammable limits may lay the toxic ranges.

As shown in Figure 1–17, the range of flammability is not linear; temperature versus the degree of vapors mixing within air defines the range of flammability. Additionally, 10% of the **lower explosive limit (LEL)** is below the temperature and vaporization intersection. As temperature increases, so does vaporization, and thus concentration within air, creating the flammable range.

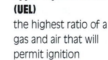

lower explosive limit (LEL)

the lowest ratio of a gas and air that will permit ignition

■ Note
The wider the flammable limits, the more dangerous the chemical.

■ Note
MEC stands for minimum explosive concentration.

■ Note
In general, LELs for flammable liquids are 1–3%; for flammable gases, 3–6%.

These concentration bounds or limits are called the flammable or explosive limits and are different for each chemical compound. The difference depends on the vapors that are produced, and thus the range of flammability. If, for example, the **upper explosive limit (UEL)** range were, say, 98%, the explosive potential would be very real due to the wide flammable range that this particular chemical may have. Above this 98% point, the concentration of the chemical vapors within the atmosphere may be too high or rich to burn. However, one would have to transcend through the flammable range in order to reach this rich environment.

upper explosive limit (UEL)

the highest ratio of a gas and air that will permit ignition

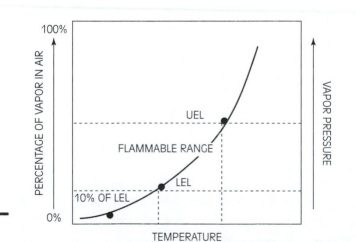

Figure 1–17
Flammable range.

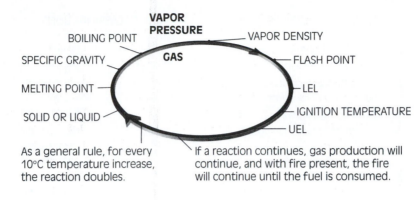

Figure 1–18 *The fire cycle as a replacement for the fire triangle.*

Consider ethylene oxide as an example, with an explosive range of 1–99%, or acetylene at 2–100%. The UEL in these cases cannot be exceeded. They have potentially extremely hazardous environments before the upper explosive limit is reached, and reaching the UEL becomes hazardous. All compounds with wide flammable ranges pose this potential problem.

The fire cycle (see Figure 1–18) as a replacement for the fire triangle explains the continuous motion and characteristics of fires involving hydrocarbons and solid materials. If, for example, a material is a solid, heat impinging on the material will cause it to move from a solid to a liquid and then into the gas phase. Once gas, which has a vapor pressure and density, is produced, a mixture must be attained for it to then move into a flash point state through its LEL and UEL. At some point, if not cooled, it will continue to repeat the cycle, becoming more intense with every repetition until the fuel or oxygen is consumed, or heat is reduced.

Heat Transfer. Some basic principles that apply in connection with heat and fire are conduction, convection, and radiation (see Figure 1–19). **Conduction** is the mechanism by which heat is transferred between materials in contact with one another. Conduction is affected by the material's total thickness, cross-section, and ability to transfer heat. It can be thought of as a balance between the thermal energy that is being given to the material and the amount of heat that can be absorbed. Increasing surface area around which heat is being liberated can increase conduction. **Convection** is the mechanism of heat transference by the

conduction

the mechanism by which heat is transferred between materials that are in contact with one another

convection

the mechanism of heat transference by the movement of a gas or liquid

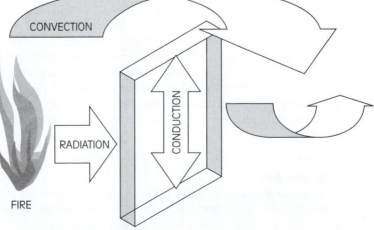

Figure 1–19 *The heat from a fire can travel in three ways, individually or in combination: by convection, conduction, or radiation.*

FLAMMABLE LIQUIDS CODE:

Flammable compounds	Class IA: Flash point less than 73°F, boiling point less than 100°F
	Class IB: Flash point less than 73°F, boiling point greater than 100°F
	Class IC: Flash point greater than 73°F, boiling point less than 100°F
Combustible compounds	Class II: Flash point greater than 100°F, boiling point less than 140°F
	Class IIIA: Flash point greater than 140°F, boiling point less than 200°F
	Class IIIB: Flash point greater than 200°F

DOT CLASSIFICATIONS:

Divisions	Definition
2.1	A gas at 68°F or less or with a boiling point of 68°F or less, which is ignitable in a mixture of 13% or less or has a flammable range of 12% or greater
2.2	Nonflammable, nonpoisonous, compressed gas, including liquefied gas, pressurized cryogenic, asphyxiating
2.3	Poisonous gas at room temperature with a boiling point of 68°F or less
2.4	Corrosive gas (Canada only)
3.1	Flash point less than 0°F
3.2	Flash point between 0°F and 72°F
3.3	Flash point between 73°F and 141°F
Combustible	Flash point between 141°F and 200°F

Figure 1–20   *Comparison of National Fire Protection Association (NFPA) 30 and DOT.*

radiation

the mechanism by which heat is transferred between two materials that are not in contact through the outward movement of heat from the source via electromagnetic waves

movement of a gas or liquid. It is technically the heat that transfers through a gas or liquid medium at the point of contact with the container. **Radiation** is the outward movement of heat from a source via electromagnetic waves. It is the mechanism by which heat is transferred between two objects that are not touching but are in relatively close proximity to one another. See Figure 1–20 for classification identification.

TOXICOLOGY

Toxicology, for our purposes, must be thought of as the study of a biochemical reaction: A chemical compound that is foreign to the human body is reacting with the biometabolism and causing an effect. That effect may be slight or it may be catastrophic. Most of the time when we interact with chemicals the dose is very small, allowing the body to compensate for the exposure. However, if the insult is large, that is, the exposure is of a greater magnitude than what the body can metabolize, an effect from the exposure is observed. This effect is called the **dose-response** (see Figure 1–21).

dose-response

the range of effects that can be observed within a given population that has been exposed to a chemical

There are thousands of chemicals in our daily environment, and millions within our environment as a whole. Some are naturally occurring and some are man-made (synthetic). Because of this vast quantity of chemicals, it is sometimes hard to link the cause and effect to an exposure. Did the exposure (the cause) lead to the medical problem (the effect)? The fundamental reason this relationship is so hard to establish is our difficulty in understanding the human (and animal) physiological functioning on the biochemical (cellular and tissue) level.

■ **Note**

Phillipus Paracelsus was the first to recognize the dose-response relationship in the sixteenth century as to the quantity of a poison that is harmful. He is considered the father of toxicology.

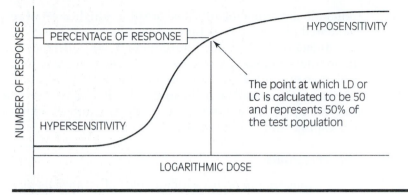

Figure 1–21   *From the hazmat perspective, we are not sure where on the curve an individual may have a response. Factors such as age, previous exposure, race, sex, and immunity all play a role in the dose-response.*

Route of Entry

There are four basic considerations when it comes to exposure:

1. The amount of the chemical, which includes the concentration (How much are we exposed to?)

2. Rate of absorption, which has to do with the shape and polarity of the chemical (absorption and distribution)

3. Rate of detoxification, which depends on the organism's metabolism

4. Rate of excretion, which is conditional to the end result of the metabolic pathway (phase I or II) (see Chapter 6—page 226)

Exposure also deals with the route by which the chemical may enter the body, or *routes of entry*. These routes can share components; for example, inhalation will include a certain degree of absorption quality. We can generalize exposure routes into four basic modes:

1. Absorption

2. Ingestion

3. Inhalation

4. Injection

Absorption commonly refers to dermal contact, but it can occur in any organ. Inhalation injuries include absorption across the alveolar membrane through or into the bloodstream. Thus the agent is transported throughout the body, affecting numerous organs and tissues as it passes. Again, absorption is normally thought of as the transmission of a chemical through the skin or mucous membranes; however, it may occur within the respiratory tree. For most workers, the absorption route of entry involves the highest percentage of injury.

Ingestion is the intake of a solid or liquid into the gastrointestinal tract. We commonly refer to this route as *oral ingestion*. Here, as elsewhere in the body, a certain degree of absorption does take place because ingestion is actually an absorption relationship within the gastrointestinal tract. At a hazardous materials incident, the process is usually seen during the rehabilitation and decontamination phases of the incident; for example, eating, drinking, or smoking prior to effective decontamination facilitates ingestion.

Inhalation is the intake and absorption of a gas (fume, aerosol, particle, dust, and so forth) into the lungs, where it is distributed to the rest of the body. Inhalation is the most common exposure route. Because the lungs can move a

absorption

the incorporation of one material into another, or the passage of a material into and through the tissues; the ability of a material to draw within it a substance that becomes a part of the original material

ingestion

the incorporation of a material into the gastrointestinal tract

inhalation

intake and absorption into the body via the respiratory system tremendous volume of air as we breathe, a high level of respiratory protection is mandated in hazmat situations.

Injection is the forced introduction of a substance into the body. Although the incidence of such an exposure at a chemical accident is rare, the possibility is always present. Usually we see this type of exposure during the entry phase. Perhaps a team member is not aware of a sharp object (or the open valve in a high-pressure cylinder), and it penetrates the suit and the skin, introducing the chemical via injection into the body.

injection
the forced introduction of a substance into the underlying body tissue

Types of Exposure

Acute exposure refers to a sudden onset of symptoms or an exclusive short-term episode. It relates to the hazardous materials exposure as a single event that causes an injury. This single exposure is usually of short duration. It is sometimes classified as nonpredictable. This single dose either occurs within a 24 hr period, or as a constant exposure for 24 hr or less. In certain cases, it could be multiple exposures within the 24 hr time frame. Acute also refers to the rapid onset of symptoms after such exposure.

Subchronic and subacute are terms that have been used interchangeably in the literature to describe the same type of exposure. However, in order to be correct, the appropriate term is **subchronic exposure**. This type of exposure is a recurring acute exposure that, in total, happens during approximately 10% of the organism's life span.

Chronic exposure is a long-term exposure, usually recurring during 80% of the total life span of the organism. Chronic effects are much harder to establish than the acute effects because of the many factors involved. Our understanding of toxicological responses and knowledge of biochemistry is limited. Chronic exposure may affect not only the exposed organism, but its offspring as well. In addition, the cumulative effect of chronic exposure may lead to hyper- or hyposensitivity.

acute exposure
a short-term exposure or the rapid onset of symptoms after an exposure or several exposures occurring within 24 hr

subchronic exposure
an exposure that recurs during approximately 10% of the organism's life span

chronic exposure
a long-term exposure, usually recurring during 80% of the total life span of the exposed organism

Exposure Limits

In the United States, the recommended threshold limit values (TLVs) are established by the American Conference of Governmental Industrial Hygienists (ACGIH). The recommended exposure limits (RELs) are established by the National Institute for Occupational Safety and Health (NIOSH). Workplace environment exposure limits (WEELs) are established by the American Industrial Hygiene Association (AIHA), with permissible exposure limits (PELs; also seen as published or personal exposure limits) established by the Department of Labor Occupational Safety and Health Administration (OSHA).

In Europe and across the former Soviet block nations, different standards exist. For example, the German Research Society (GRS) has a maximum allowable concentration value (MAC or MAK) analogous to the ACGIH standard.

The numbers that each United States organization utilizes are roughly the same. In some instances the testing procedures may vary; however, for the most part, the TLV standards of the ACGIH, a trademark for the ACGIH, are customarily used as the industry standard. The committee that introduced these levels has stated that the TLVs are *not* intended for the following uses:

- To describe the level of hazard as it relates to the toxicity for a chemical or group of chemicals
- For the evaluation of air pollutants
- For the decision of an extended work period within an environment or when estimating the toxic capacity of a chemical

- To describe the cause and effect of a chemical substance
- If the working conditions are greatly different than those within the United States work environment

ACGIH's only intended purpose was to establish safe working conditions in the United States for those individuals who may become exposed on a daily basis. They were not intended to be used by emergency response personnel at the scene of a hazardous materials accident. However, they are the only numerical indicators of exposure limits that we have for chemicals. Remember that these standards were developed for the long-term exposure found in industrial processes and relate to the worker who becomes involved with these chemicals. These numbers should not be viewed as absolutes, but rather as indicators of an estimated level of harm. *All* toxicological data used should be scrutinized during the decision-making process. By utilizing these values in association with all other toxicological data, and through our understanding of chemistry, we can plan our safe level and manage the incident with one objective: to reduce the possibility of exposure to response personnel.

■ Note

Although many of the chemical toxic values are given in mg/m^3, some of the literature will give conversion factors for mg/m^3 to ppm. For a ballpark estimate we can use the following formula: $mg/m^3 = (ppm \times MW)/24.5$, where ppm is parts per million and MW is the molecular weight.

■ Note

Parts per million (ppm) can be visualized as 1 oz of cream in 10,000 gal of coffee, or 1% is equal to 10,000 ppm. A ppb, or part per billion, is a drop of cream in a 22,000 gal cup of coffee, or 1% is equal to 10,000,000 ppb.

We can utilize these numbers with associated information because most of them were established for safe working conditions without the need of personal protective equipment (PPE). At an incident, some level of PPE is nearly always worn, thereby increasing the margin of safety. However, allowances must be made for all of the values, considering the testing parameters under which they were derived.

The threshold limit value is the level of exposure that starts to produce an effect. Below this level, there will not be any discernible effects if a worker is exposed to a chemical repeatedly. Most of these values refer to airborne concentrations and are based on exposure during a standard workweek. The ACGIH's TLV-TWA is based on an 8 hr day, 5 days a week to give a 40 hr workweek. OSHA also considers this time frame as the exposure standard. In both cases they refer to this level as the TLV-TWA, in which TWA is the time-weighted average: the average concentration of a chemical that most workers will be exposed to during the normal 40 hr workweek, at 8 hr a day. NIOSH denotes a TLV-TWA based on a 10 hr work day, 4 days a week, to give the 40 hr workweek.

Based as it is on the traditional 8 hr day and 40 hr workweek, the TLV-TWA cannot be applied to conditions of nontraditional schedules, such as one 8 hr day a week plus two 16 hr days on weekends to arrive at 40 hr. Its values were designed to take into consideration exposure to a chemical for the length of a day, provided that the maximum TLV-TWA was not reached, and time off to allow the body to eliminate any stored chemicals. So, in 1976, the ACGIH committee that designed the TLV came up with short-term exposure limits (STELs), a weighted 15 min event. It is sometimes referred to as the emergency exposure limit (EEL).

■ **Note**

Because of the chemical agents used during World War I in Europe, soldiers were forced to use chemical protective equipment. During that same time period in the United States, no attempts were made to institute the use of personal protective equipment in the industrial setting.

■ **Note**

The relationship between industrial manufacturing and health concerns was recognized by Dr. Alice Hamilton. In 1919, after being accepted to the faculty of Harvard Medical School, she began a campaign to identify and educate the public on chemical exposures.

■ **Note**

Medical Surveillance is a program by which the health of responders is reviewed within an occupational health setting. Only those responders exposed to a chemical for more than 30 days in a year and above the OSHA standard, or members of a hazardous materials response group, are required to be a part of a Medical Surveillance program.

■ **Note**

The ability to detect certain smells such as the bitter almond smell of cyanide is a sex-linked recessive trait. Only 60–80% of the general population can smell the aroma of cyanide, at a ratio of three men to every one woman.

● **Caution**

A 1% decrease of oxygen in air equals a 50,000 ppm displacement. Remember that oxygen represents 1/5 of the atmosphere. A significant displacement!

The STEL refers to an exposure that occurs for only 15 min and is not repeated more than four times a day. Each 15 min exposure event is interrupted by a 60 min nonexposure environment. Thus, STEL applies to an individual exposed to a chemical for 15 min with a 1 hr break between exposures, not to exceed four times within an 8 hr day. More recently, the ACGIH recommended the use of excursion limits (ELs), a weighted average with the exposure time not to exceed five times the published 8 hr TWA. This exposure can only occur for 30 min within any one workday that is 8 hr in duration. Therefore, ELs are more realistic than the STEL.

The ACGIH has defined other limit values: TLV-s and TLV-c, the threshold limit values for absorption through the skin (s), and for ceiling levels (c). At ceiling levels, exposure ensures death if not properly protected. Ceiling levels should *never* be exceeded without evaluating the PPE required. This level is similar to NIOSH's immediately dangerous to life and health (IDLH) value.

OHSA, which is a branch of the Department of Labor, has, for the most part, adopted the ACGIH TLVs as their own PELs. OHSA and NIOSH define IDLH as the maximum airborne concentration that an individual could escape from within 30 min without sustaining *any* adverse effects. The IDLH (and TLV-c) give us a level at which the only form of protection is that of full encapsulation with self-contained breathing apparatus. In some cases, the skin absorption factor must also be evaluated. It is recommended that IDLH atmospheres not be entered unless the individual is properly trained and the appropriate protective garments are donned. Individuals caught unconscious within an IDLH environment will more than likely have died by the time the hazardous materials team can make entry.

Working with IDLH, TLV, or **LClo** values is problematic for several reasons. For one thing, it only applies to the healthy, young, active male working population. It does not account for the possibility of hypersensitive or hyposensitive

LClo

lethal concentration lo is the lowest concentration that resulted in organism death as recorded

individuals. Females can be sensitive to a variety of organic compounds that IDLH does not take into consideration. For another, even though its values are based on the effects that may occur if there happened to be a 30 min exposure, this criterion *does not* mean that one can stay in the atmosphere for 30 min without ill effect.

■ Note

American Conference of Governmental Industrial Hygienists exposure limits terminology:

> TLV (threshold limit value): The level of exposure to airborne contaminants that starts to produce an observable effect.

> TWA (time-weighted average): A refinement of TLV to take into account an 8 hr day, 40 hr workweek with repeated exposure without any adverse effects.

> STEL (short-term exposure limit): A TLV restricted to a 15 min event in which the worker is exposed to the chemical continuously. The event must not have any of the following effects: any irritation; chronic tissue damage; or the impairment of a self-rescue.

> EL (excursion limit): An average exposure not to exceed the published 8 hr TWA, and not to occur for more than 30 min on any workday.

> TLV-s: Identifies a material that is absorbed through the skin.

> TLV-c: A ceiling level for exposure.

■ Note

Many discussions on toxicology use Haber's Law, which gives a dose over time with equivalents. For example, 10 min at 8 mg/m^3 = 2 min at 40 mg/m^3 = 1 min at 80 mg/m^3. All are equal to 80 mg-min/m^3.

Finally, NIOSH does not have values for all IDLH toxic chemicals. Any IDLH value should be used only as a guideline and not as an absolute level.

The Environmental Protection Agency (EPA) utilizes the values as presented from ACGIH, NIOSH, and OSHA. It is mostly concerned with the impact that a chemical may have on the environment and the organisms that make up that environment. The EPA has developed Emergency Response Planning Guidelines (ERPGs), as follows:

- ERPG-1: The maximum concentration in air below which it is believed nearly all individuals could be exposed for up to 1 hr without experiencing other than mild, transient adverse health effects or perceiving a clearly defined objectionable odor.

- ERPG-2: The maximum concentration in air below which it is believed nearly all individuals could be exposed for up to 1 hr without experiencing

---

■ Note

Occupational Safety and Health Administration exposure limits terminology:

> PEL (permissible, published, or personal exposure level): Has the same meaning as TLV-TWA.

> IDLH (immediately dangerous to life and health): The maximum airborne contamination that an individual could escape from within 30 min without any side effects.

> CL (ceiling level): Similar to TLV-c.

■ Note

NIOSH's REL (recommended exposure limit) are similar to PEL and TLV-TWA where RELs are often more conservative than the TLV.

or developing irreversible or other serious health effects or symptoms that could impair their abilities to take protective action.

- ERPG-3: The maximum concentration in air below which it is believed nearly all individuals could be exposed for up to 1 hr without experiencing or developing life-threatening health effects.

In addition to the ERPGs, the EPA has also developed TEELs (temporary emergency exposure limits). They are:

- TEEL-0: The threshold concentration below which most people will experience no appreciable risk of health effects.

- TEEL-1: The maximum concentration in air below which it is believed nearly all individuals could be exposed without experiencing other than mild, transient adverse health effects or perceiving a clearly defined objectionable odor.

- TEEL-2: The maximum concentration in air below which it is believed nearly all individuals could be exposed without experiencing or developing irreversible or other serious health effects or symptoms that could impair their abilities to take protective action.

- TEEL-3: The maximum concentration in air below which it is believed nearly all individuals could be exposed without experiencing or developing life-threatening health effects.

Because short-term and long-term exposure limits sometimes vary, the EPA has proposed a new threshold exposure limit called Acute Exposure Guideline Level (AEGL). This represents threshold exposure limits for the general public and is applicable to emergency response exposures for periods ranging from 10 min to 8 hr (AEGL-1, AEGL-2, and AEGL-3) and will be distinguished by varying degrees of severity of toxic effects (see Table 1–1). It is believed that the recommended exposure levels are applicable to the general population, including infants and children and other individuals who may be sensitive and susceptible. With increasing airborne concentrations above each AEGL level, there is a progressive increase in the likelihood of occurrence and the severity of effects described for each corresponding AEGL level. Although the AEGL values represent threshold levels for the general public, including sensitive populations, it is recognized that certain

Table 1–1 *Exposure limits summary.*

Organization	Exposure Abbreviation	Definition
ACGIH	TLV	8 hr time-weighted average
ACGIH	STEL	15 min time-weighted average
OSHA	PEL	8 hr time-weighted average
NIOSH	REL	10 hr time-weighted average
NIOSH	IDLH	Dangerous to immediate threat to life; escape is possible without permanent injury
EPA	ERPG	Planning guidance
EPA	TEEL	Planning guidance
EPA	AEGL	Guidance for emergency responders with exposure durations

individuals subject to unique metabolic responses could experience the effects described at concentrations below the corresponding AEGLs level. The AEGLs are defined as follows:

- AEGL-1 is the airborne concentration (in ppm or mg/m^3) of a substance above which it is predicted that the general population, including susceptible individuals, could experience notable discomfort, irritation, or certain asymptomatic, nonsensory effects. However, the effects are not disabling, and are transient and reversible on cessation of exposure. Airborne concentrations below AEGL-1 represent exposure levels that could produce mild and progressively increasing odor, taste, and sensory irritation, or certain nonsymptomatic, nonsensory effects.

- AEGL-2 is the airborne concentration (in ppm or mg/m^3) of a substance above which it is predicted that the general population, including susceptible individuals, could experience irreversible or other serious, long-lasting, adverse health effects, or an impaired ability to escape.

- AEGL-3 is the airborne concentration (in ppm or mg/m^3) of a substance above which it is predicted that the general population, including susceptible individuals, could experience life-threatening health effects or death.

The last group of exposure-limit toxicological terms that we are going to discuss are related to the dose-response of drugs as they relate to chemicals. If the chemical in question is airborne and is primarily an inhalation hazard, you would want to know its lethal concentration (LC). If the chemical poses a threat other than an inhalation injury, you would want to know its lethal dose (LD). Both quantities describe the effect on a percentage of the population. A 50% death rate after the introduction of a chemical inhalation hazard is denoted as lethal concentration fifty (LC_{50}). The lethal dose fifty (LD_{50}) is the calculation of an expected 50% death toll after an exposure to a chemical that may or may not include an inhalation injury. Similarly, an LC_{25} or LD_{25} is a 25% death toll within the tested population.

■ Note

Dose-response exposure terminology:

LClo: The lowest concentration of airborne contaminates that can cause injury.

LDlo: The lowest dose (solid/liquid) that can cause an injury.

LC_{50}: A level of concentration at which 50% of the test population died from the introduction of this airborne contaminate.

LD_{50}: A dosage level at which 50% of the tested population died from the introduction of this chemical as a solid or liquid.

LCT_{50}: A statistically derived LC_{50} (LDT_{50} is a statistically derived lethal dose—Haber's Law).

MAC: The maximum allowable concentration (European, also MAK).

RD_{50}: A 50 percent calculated concentration of respiratory depression (RD = respiratory depression) in response to an irritant, over a 10–15 min time frame.

The number that is given with the associated LC or LD is relative. In general, the smaller the number, the more toxic the chemical. Likewise the greater the number the less toxic it is. The reference literature that enumerates the LD and LC for humans actually use statistical extrapolations. They are derived from the mean lethal dose or the average lethal dose (ALD) or concentration. From there they are calculated to the human experience. At times this value is referred to as the toxic dose low (TDL), or the toxic concentration low (TCL). As previously noted, the concentration denotes the inhalation injury and the dose denotes absorption of a liquid or solid either through the skin or by ingestion.

Not everyone will react to a toxin in the same way. For this reason, 50% of the population is the standard for LDs and LCs. Also, the problem of extrapolation from an animal population to the human is an assumption. For example, if two chemicals have an LD_{50} of 1000 ppm and 10,000 ppm, respectively, the first chemical is more toxic. However, if a third chemical is observed with an LD_{50} of 4000 ppm but it has a higher percentage of death at the beginning of the curve, then hyposensitive individuals will become affected early on. Overall, even though the LDs may have escalating numbers, it is the beginning of the curve that truly designates the acute toxicity.

One must remember that for the most part these are values that are observed in the laboratory under controllable conditions; these tests are not done on human beings. However, there are a few cases in which the human experience has been documented. In these cases, the reference literature usually denotes this by placing the chemical-exposure animal in parentheses.

Summary

Chemistry is the science of matter, energy, and their reactions. Matter is anything that occupies space. It may be a solid, liquid, or gas. It also may be a pure substance or a mixture. Pure substances may be compounds or elements. Mixtures are physical blends of elements, compounds, or both. The elements are materials that cannot be broken down into simpler form by normal means and still retain the properties of that particular element.

Atoms consist of protons, found in the nucleus of the atom with an atomic weight of 1 and electrical charge of +1. The atom will possess the same number of protons within a given element. Neutrons with no electrical charge are also found in the nucleus of the atom. The atomic weight of the neutrons is 1 and the number will vary within the nucleus. Electrons are in the shell surrounding the nucleus and essentially have no weight; however, they have an electrical charge of −1. Chemical behavior is ultimately determined by the number of electrons in the outer shell.

The atomic number indicates the number of protons in the nucleus and is equal to the number of electrons. The elements on the periodic table are listed by increasing atomic number. The configuration of the chart results in repetition of certain common properties between the elements within groups. The atomic weight varies; however, the number of protons within the nucleus remains the same. The weight variations are due to the number of neutrons and the average of the isotopes that occur.

The periodic table is the basic tool for chemical identification and the potential hazards that may arise. Each box within the table represents an element. Each element is represented by a symbol. The position within the table indicates the physical and chemical properties of an element. As we move down the table, the elements become larger. Moving across the table from left to right, the elements become smaller and more electronegative.

Nature does not like anything to be in an unstable state; there is a continual desire for the elements to achieve a state of stability. In the case of atoms, that means the outer shell of an element must have the appropriate number of electrons. There are two ways that an atom can achieve this state of stability. One is by forming molecules whereby the whole structure will satisfy the rule of stability. The other is by the element or group of elements (molecule) forming an ion.

Ions may also be formed by a group that has collectively gained or lost electrons. The group will also have a charge equal to the number of electrons gained or lost. Negatively charged ions are called anions and positively charged ions are called cations. This concept is true whether or not the ion was formed by a single atom or a group. Once ions are bonded, the resulting compound must achieve an electrically neutral environment.

Ionic bonding occurs between a metal and a nonmetal. The large metals hold electrons very loosely; the relatively smaller (more electronegative) nonmetals pull very hard on the electron. Metals tend to give up electrons to nonmetals.

Chemical and physical properties allow us to understand the basic chemistry that a substance possesses. Knowledge of these properties enables us to use monitoring equipment, make intelligent decisions, and establish strategic priorities. By using our knowledge of chemical and physical properties, the hazards and mitigation techniques have a basis in the sciences.

Toxicology is a complicated science, much of which is still truly unknown. However, safe practices can be established and medical monitoring adhered to by applying a basic understanding. Although toxicological values are based on established scientific theory, their application to the hazardous materials incident is sometimes difficult to qualify and quantify. With new detection technologies and a deeper understanding of human physiology, we can quantify hazardous environments if we know and understand the biological hazards that are involved.

Before working on the review questions, memorize the following chemical symbols. Having them committed to memory will help with your research of hazardous materials at the incident scene.

Hydrogen	H	Helium	He	Lithium	Li	Beryllium	Be
Boron	B	Carbon	C	Nitrogen	N	Oxygen	O
Fluorine	F	Neon	Ne	Sodium	Na	Magnesium	Mg
Aluminum	Al	Silicon	Si	Phosphorus	P	Sulfur	S
Chlorine	Cl	Argon	Ar	Potassium	K	Calcium	Ca
Scandium	Sc	Titanium	Ti	Vanadium	V	Chromium	Cr
Manganese	Mn	Iron	Fe	Nickel	Ni	Copper	Cu
Zinc	Zn	Arsenic	As	Bromine	Br	Krypton	Kr
Silver	Ag	Cadmium	Cd	Tin	Sn	Antimony	Sb
Iodine	I	Xenon	Xe	Cesium	Cs	Barium	Ba
Platinum	Pt	Gold	Au	Mercury	Hg	Lead	Pb

Review Questions

1. All matter consists of three basic states. What are these three states?

    A. volume, mass, and liquid

    B. solid, liquid, and gas

    C. atoms, elements, and molecules

    D. liquid, elements, and gas

2. The basic unit identified on the periodic table is called:

    A. proton

    B. neutron

    C. element

    D. compound

3. Compounds, molecules, and elements have properties that are characteristic of that particular material. What are these properties known as?

    A. chemical properties

    B. physical properties

    C. reaction properties

    D. both a and b

4. The outward form of a hazardous material is a description of what property?

    A. vapor pressure

    B. melting point

    C. boiling point

    D. appearance

5. The temperature at which a material changes from a solid to a liquid is a definition of:

    A. vapor pressure

    B. melting point

    C. boiling point

    D. appearance

6. The temperature at which the atmospheric pressure and the vapor pressure of a liquid are equal is a definition of:

    A. vapor pressure

    B. melting point

    C. boiling point

    D. appearance

7. The particle that revolves about the center portion of an atom is called:

   A. electrino

   B. positron

   C. negatron

   D. electron

8. The center of an atom has two particles that make up the nucleus. What are they?

   A. protons

   B. electrons

   C. neutrons

   D. A and C

9. All atomic nuclei contain an equal number of protons in relation to the number of electrons that are in orbit.

   A. true

   B. false

10. The atomic number is used to group the elements on the periodic table.

    A. true

    B. false

11. Atoms of the same element may have a different number of neutrons within the nucleus. These classes of the same atom are called:

    A. isomers

    B. isotopes

    C. ionic

    D. ions

12. How many electrons are in the outer shell in group IA?

    A. 1

    B. 2

    C. 3

    D. 4

13. How many electrons can group IA accept or lose?

    A. 1

    B. 2

    C. 3

    D. 4

14. How many electrons are in the outer shell in group IIA?

    A. 1

    B. 2

C. 3

D. 4

15. How many electrons can group IIA accept or lose?

    A. 1

    B. 2

    C. 3

    D. 4

16. How many electrons are in the outer shell in group IIIA?

    A. 1

    B. 2

    C. 3

    D. 4

17. How many electrons can group IIIB accept or lose?

    A. 3

    B. 4

    C. 5

    D. 6

18. How many electrons are in the outer shell in group IVA?

    A. 1

    B. 2

    C. 3

    D. 4

19. How many electrons can group IVA accept or lose?

    A. 4

    B. 5

    C. 6

    D. 7

20. How many electrons are in the outer shell in group VA?

    A. 4

    B. 5

    C. 6

    D. 7

21. How many electrons can group VA accept or lose?

    A. 3

    B. 4

    C. 5

    D. 6

22. How many electrons are in the outer shell in group VIB?
    A. 4
    B. 5
    C. 6
    D. 7

23. How many electrons can group VIB accept or lose?
    A. 2
    B. 3
    C. 4
    D. 5

24. How many electrons are in the outer shell in group VIIB?
    A. 4
    B. 5
    C. 6
    D. 7

25. How many electrons can group VIIB accept or lose?
    A. 1
    B. 2
    C. 3
    D. 4

26. What chemicals do the following abbreviations represent?
    A. Li
    B. Na
    C. K
    D. Rb
    E. Cs
    F. Fr

27. What is the name for the first group on the periodic table?
    A. alkali metals
    B. alkaline earth metals
    C. transition metals
    D. nitrogen family

28. What is the name for the second group on the periodic table?
    A. alkali metals
    B. alkaline earth metals
    C. transition metals
    D. nitrogen family

29. What chemicals do the following abbreviations represent?
    A. Be
    B. Mg
    C. Ca
    D. Sr
    E. Ba
    F. Ra

30. What chemicals do the following abbreviations represent?
    A. Hg
    B. Cd
    C. Cr
    D. Tl
    E. Zr
    F. Zn

31. What chemicals do the following abbreviations represent?
    A. B
    B. Al
    C. Ga
    D. In
    E. Ti

32. What chemicals do the following abbreviations represent?
    A. C
    B. Si
    C. Ge
    D. Sn
    E. Pb

33. What chemicals do the following abbreviations represent?
    A. N
    B. P
    C. As
    D. Sb
    E. Bi

34. What chemicals do the following abbreviations represent?
    A. O
    B. S
    C. Se
    D. Te
    E. Po

35. What is the name given for group 7A?
    A. alkali metals
    B. alkaline earth metals
    C. transition metals
    D. halogens

36. What chemicals do the following abbreviations represent?
    A. F
    B. Cl
    C. Br
    D. I
    E. At

37. Which of the following is a true statement?
    A. Alkali earth metals do not react with water.
    B. Nitrogen oxides are nontoxic.
    C. All halogens are harmful.
    D. All the above.

38. How many electrons can the halogen-group elements accept or lose?
    A. 7
    B. 5
    C. 3
    D. 1

39. Which is the most electronegative element of the following choices?
    A. O
    B. Br
    C. F
    D. N

40. What is the name given for group VIII?
    A. alkali metals
    B. alkaline earth metals
    C. noble gases
    D. halogens

41. What chemicals do the following abbreviations represent?
    A. He
    B. Ne
    C. Ar
    D. Kr
    E. Xe
    F. Rn

42. Density is defined best by which statement?
    A. how hard a substance is
    B. how much of a substance there is
    C. the relationship between the mass and volume of the substance
    D. the mass per unit volume of the substance, which for solids and liquids is expressed as g/cm^3

43. What is a comparison of weight between water and the material being tested called?
    A. specific density
    B. density
    C. specific gravity
    D. vapor density

44. What is a molecule that has escaped a liquid traveling through the air space, giving it a gaseous state, called?
    A. vapor density
    B. vapor pressure
    C. vapor temperature
    D. vapor change

45. What is the minimum temperature under which a liquid will give off vapors to form an ignitable mixture in air without sustained combustion called?
    A. vapor pressure
    B. fire point
    C. ignition temperature
    D. flash point

46. If we could measure a gas released into the environment as it is traveling through the air, we would see three distinct areas of concentration. What are these areas called?
    A. flammable ranges
    B. explosive limits
    C. flammable limits
    D. all the above

47. Match the following abbreviations of specific toxicological values with their definitions and the associated agency:
    A. PEL
    B. MAK
    C. TLV
    D. EL
    E. WEEL

1. exclusion limit
2. workplace environment exposure limit
3. maximum allowable concentration
4. threshold limit value
5. permissible exposure limit
   a. AIHA
   b. ACGIH
   c. OSHA
   d. GRS
   e. NIOSH

48. Match the following definitions to the appropriate toxicological term:
   A. Based on an 8 hr day, 5 days a week, to give a 40 hr workweek.
   B. The letter "s" identifies that the material is absorbed through the skin.
   C. Based on a 10 hr workday, 4 days a week, to give a 40 hr workweek.
   D. Only occurs for 15 min and is not repeated for more than four times a day. Each 15 min exposure event is interrupted by a 60 min nonexposure environment.
   E. IDLH (immediately dangerous to life and health)
   F. The letter "c" denotes ceiling levels.
   1. NIOSH TLV-TWA
   2. STEL
   3. TLV-TWA
   4. EEL
   5. TLV-c
   6. TLV-s

49. Match the following ACGIH values:
   A. TLV-TWA
   B. TLV-STEL
   C. TLV-s
   D. TLV-c
   E. TLV-EL
   1. An average exposure not to exceed the published 8 hr TWA. This will not occur for more than 30 min on any workday.
   2. Identifies a material that is absorbed through the skin.
   3. Fifteen-min excursion in which the worker is exposed to the chemical continuously.
   4. A ceiling level.
   5. An 8 hr day, 40 hr workweek with repeated exposure without any adverse effects.

50. Match the following OHSA and NIOSH values:
   A. IDLH
   B. PK
   C. PEL
   D. REL
   E. CL
   1. Same as TLV-TWA
   2. Similar to TLV-c
   3. Same as PEL and TLV-TWA
   4. The maximum airborne contamination that an individual could escape from within 30 min without any side effects
   5. Different depending on the testing agency; the apogee of the daily allowable limit

51. Match the following:
   A. LClo
   B. LDlo
   C. LC_{50}
   D. LD_{50}
   E. LCT_{50}
   F. MAC
   G. RD_{50}
   1. The lowest concentration of airborne contaminates that can cause injury.
   2. Fifty percent of the test population died from the introduction of this airborne contaminate.
   3. A statistically derived LC_{50} (LDT_{50} statistically derived lethal dose).
   4. A 50% calculated concentration of respiratory depression in response to an irritant over a 10–15 min time frame.
   5. The lowest dose (solid or liquid) that can cause an injury.
   6. The maximum allowable concentration (European).
   7. Fifty percent of the tested population died from the introduction of this chemical, which may be a solid or liquid.

Problem Set Answers

Problem Set 1.1

Element	Shells	Shell Electrons			Total Electrons
Hydrogen	1	1			1
Helium	1	2			2
Lithium	2	2	1		3
Beryllium	2	2	2		4
Boron	2	2	3		5
Carbon	2	2	4		6
Nitrogen	2	2	5		7
Oxygen	2	2	6		8
Fluorine	2	2	7		9
Neon	2	2	8		10
Sodium	3	2	8	1	11
Magnesium	3	2	8	2	12
Aluminum	3	2	8	3	13
Silicon	3	2	8	4	14
Phosphorus	3	2	8	5	15

Problem Set 1.2

Element	Group	Outer Electrons	Electronegativity
Hydrogen	IA or IA or 1	1	2.20
Lithium	IA or IA or 1	1	0.98
Sodium	IA or IA or 1	1	0.93
Beryllium	IIA or IIA or 2	2	1.57
Magnesium	IIA or IIA or 2	2	1.31
Calcium	IIA or IIA or 2	2	1.00
Boron	IIIB or IIIA or 13	3	2.04
Aluminum	IIIB or IIIA or 13	3	1.61
Gallium	IIIB or IIIA or 13	5	1.81
Carbon	IVB or IVA or 14	4	2.55
Silicon	IVB or IVA or 14	4	1.90
Germanium	IVB or IVA or 14	6	2.01
Nitrogen	VB or VA or 15	5	3.04
Phosphorus	VB or VA or 15	5	2.19
Arsenic	VB or VA or 15	7	2.18
Oxygen	VIB or VIA or 16	6	3.44
Sulfur	VIB or VIA or 16	6	2.58
Selenium	VIB or VIA or 16	8	2.55
Fluorine	VIIB or VIIA or 17	7	3.98
Chlorine	VIIB or VIIA or 17	7	3.16
Bromine	VIIB or VIIA or 17	1	2.96
Helium	VIII or 0 or 18	8	—
Neon	VIII or 0 or 18	8	—
Argon	VIII or 0 or 18	8	—

Problem Set 1.3

Element	Group	Lewis Dot Structure
Hydrogen	IA or IA or 1	H
Lithium	IA or IA or 1	Li
Sodium	IA or IA or 1	Na
Beryllium	IIB or IIA or 2	Be
Magnesium	IIB or IIA or 2	Mg
Calcium	IIB or IIA or 2	Ca
Boron	IIIA or IIIB or 13	B
Aluminum	IIIA or IIIB or 13	Al
Gallium	IIIA or IIIB or 13	Ga
Carbon	IVB or IVA or 14	C
Silicon	IVB or IVA or 14	Si
Germanium	IVB or IVA or 14	Ge
Nitrogen	VB or VA or 15	N
Phosphorus	VB or VA or 15	P
Arsenic	VB or VA or 15	As
Oxygen	VIB or VIA or 16	O
Sulfur	VIB or VIA or 16	S
Selenium	VIB or VIA or 16	Se
Fluorine	VIIB or VIIA or 17	F
Chlorine	VIIB or VIIA or 17	Cl
Bromine	VIIB or VIIA or 17	Br
Helium	VIII or 0 or 18	He
Neon	VIII or 0 or 18	Ne
Argon	VIII or 0 or 18	Ar

All dot structures are based on the rough rule of using the group's valence.

Problem Set 1.4 NO_3

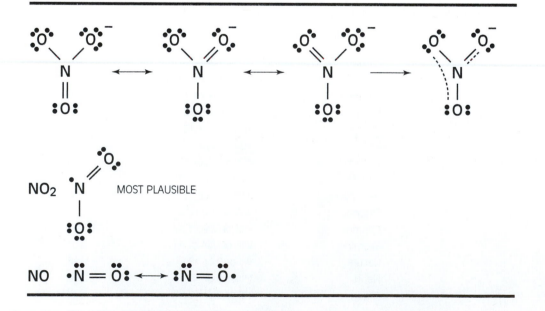

Chapter 2

Inorganic Compounds

Learning Objectives

Upon completion of this chapter, you should be able to:

- Identify ionic bonding and its relationship with electronegativity.
- Explain what constitutes a salt and a nonsalt.
- Construct Lewis dot structures and describe their application.
- Explain the octet rule.
- Reproduce the cross-valence technique.
- Designate the naming configuration and general hazards of various inorganic compounds.
- Categorize the Latin names that are used in chemistry.
- Formulate the relationship of hydronium and hydroxide ions.
- Contrast the scales utilized in acid–base reactions.
- Explain the hydrates versus hydroxides.
- Compare the fire cycle and the process of combustion.
- Identify the uses of extinguishment.
- List the process of combustion.
- List the three hazards of radiation.
- Identify the forms of measurement for radiation.

When two or more atoms come together to form a bond, several phenomena come into play simultaneously. We explored some of these variables in Chapter 1, when we spoke about rate of reaction, spatial relationships, and resonance bonding, and touched on acid–base reactions and the polarity of molecules. Although many different laws of nature cause reactions to take place, our discussion leaves out some of the technical understanding. We want to focus on how compounds are named and their associated hazards. Understanding how chemicals are named allows us to quickly identify the potential hazards a compound may possess.

■ Note

Temperature, pressure, volume, concentrations, and strength must all be considered in the decision-making and risk-assessment processes. Remember that most chemicals are reactive at some specific temperature and pressure. When given, look at the reaction temperature and pressure. Consider all the physical and chemical properties carefully.

！Safety

To understand the basic hazards that can be produced at a hazmat scene, a first responder needs to be able to identify chemical families. This ability depends on recognizing individual compounds.

The overall study of chemistry is divided into two separate but closely related branches: inorganic and organic chemistry. Because the naming of compounds follows certain rules, the first characteristic we need to identify is whether a compound is inorganic or organic. Inorganic chemistry is the study of substances that do not contain **hydrocarbons** (the major components are not carbon based), whereas organic chemistry is the study of carbon compounds. Both types of compounds have reactions that are distinct and individual to their specific chemistry. In particular, it is the bonds and their configuration that are important. For our purposes it is necessary to be able to identify the differences in the structures so that we recognize the parent families from which compounds are derived. Understanding the nomenclature enables us to recognize the ionic and covalent bonds that are representative of certain family classes. An understanding of the periodic table and the placement of the elements will thus prove to be beneficial.

hydrocarbons
compounds whose molecules are composed predominately of carbon and hydrogen

---

■ Note

The order of naming a salt is first the metal, then the nonmetal.

IONIC BONDING

Ionic bonding is the type of bonding that occurs when a metal (which loses electrons) is bonded with a nonmetal (which gains electrons). The resulting compound is called a salt. In this type of bond, one or more electrons will leave the outer shell of the metal atom and transfer into the ring of the nonmetal atom's shell. Because all compounds are striving to achieve a state of stability, the number of atoms that will make this transfer depends on the number of electrons that need to be gained or lost (see Figure 2–1).

An atom with low ionization energy has high **electronegativity** and usually produces an ionic bond. By definition, an atom in this configuration will be one

electronegativity
a measure of how strongly an atom holds on to its electrons or attracts electrons from other atoms

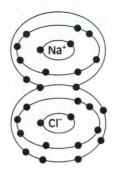

Or, using the outside orbital in our electron dot configuration:

Na$^+$ $\overset{\bullet\bullet}{\underset{\bullet\bullet}{\underset{\bullet}{\overset{\bullet}{Cl}}}}$$^-$

Or, as in the dash method:

Na–Cl

Figure 2–1 *Example of ionic bonding: correlating the drawing configurations.*

with a fairly small diameter, which makes it able to hold on to its outer electrons. This electronegative atom will look for an atom that is electropositive. When they meet, an ionic bond (compound) is formed. Ionic bonding usually takes place when an atom from the left side of the periodic chart combines with an atom from the right side.

So let us look at ordinary table salt as an example:

$$Na^+ + Cl^- \rightarrow NaCl$$

The sodium ion has a positive charge because it possesses eleven protons and ten electrons. It is a cation. The chloride ion possesses seventeen protons and eighteen electrons, making it an anion. Each element is trying to acquire a stable electronic state. By transferring electrons, each attains stability. We can say that they achieve a noble gas state, remembering that the noble gases all have eight electrons in their outermost shell and will not react. All elements are trying to attain that stable state with eight electrons in their outermost shell (the octet rule).

● **Caution**

Metals and nonmetals combine to form salts. Metals lose electrons and nonmetals gain electrons. In naming, the metal is named first, then the nonmetal. Be aware that salts:

☠ Conduct electricity   ☠ Dissolve in water   ☠ Do not burn under normal
☠ Can be toxic   ☠ Are mostly solids   conditions

■ **Note**

If you haven't already done so, memorize the names of the elements and their symbols as listed at the beginning of the "Review Questions" section of Chapter 1.

■ **Note**

The dash method for identifying chemical molecules is preferred and we will use this method primarily later, once an understanding of bonding has been established.

The salts that result from the ionic bonding of a metal and a nonmetal are compounds that are formed when the hydrogen of an acid is replaced by a metal or comparable substance. Because these compounds can exist in water solutions, their primary configuration in a solution is that of ions. However, they also exist in solid form. When this occurs the presence of water can cause a reaction to take place. When found in nature, in general they are usually solids, are water-soluble, and have a high melting point. Fortunately, they do not burn under normal conditions.

Nonsalts (hydrocarbons) are compounds that are made up of only nonmetals. They can be organic or inorganic and have any state of matter at normal temperatures and pressures. For the most part they are covalently bonded structures. Nonsalt compounds are addressed in Chapters 3, 4, and 5.

SYMBOLS AND LEWIS DOT STRUCTURES

To understand the reactions, their components (see Figure 2–2) have to be represented by symbols and formulas. Each element has a symbol, and a formula has two sides, like an equation. On the left side of the formula we have the reactants, and on the right, the products of the reaction. The arrow in the middle is used to express "yield," and is similar to an equals sign in mathematics. The law of

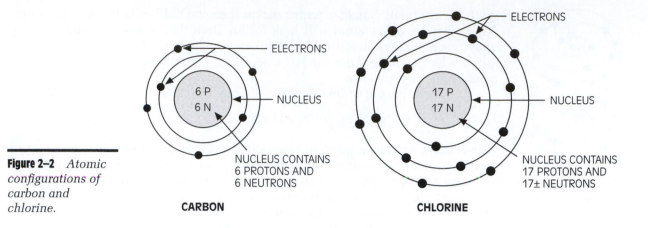

Figure 2–2 *Atomic configurations of carbon and chlorine.*

ELECTRONS

6 P
6 N

NUCLEUS

NUCLEUS CONTAINS
6 PROTONS AND
6 NEUTRONS

CARBON

ELECTRONS

17 P
17 N

NUCLEUS

NUCLEUS CONTAINS
17 PROTONS AND
17± NEUTRONS

CHLORINE

conservation of mass states that there will be an equal number of atoms on either side of the equation because mass can be neither created nor destroyed.

Looking at an example we can see how representation as a formula is a method to help us to interpret the results that are seen during experimentation. Let us write down the reaction of carbon disulfide, CS_2, with chlorine, Cl_2, to produce halon 104 or carbon tetrachloride, CCl_4. The additional product in this reaction would be disulfur dichloride, S_2Cl_2:

$$CS_2 + Cl_2 \rightarrow CCl_4 + S_2Cl_2$$

carbon disulfide (CS_2)
DOT: flammable liquid, poison
BP: 46°C
FP: −30°C
VP: 297 mm Hg
VD: 2.647
SG: 1.26 @ 20°C
IDLH: 500 ppm
STEL: 10 ppm
TWA: 10 ppm
LEL: 1.3%
UEL: 50%
Solubility in water: 0.3%

chlorine (Cl_2)
DOT: nonflammable gas, poison
BP: −34.6°C
FP: N/A
VP: 6.8 atm
VD: 2.5
SG: 1.565 @ −35°C
IDLH: 30 ppm
STEL: 1 ppm
TWA: 1 ppm
LEL: N/A
UEL: N/A
Solubility in water: 7%

supports combustion

carbon tetrachloride (CCl_4)
DOT: poison
BP: 76.7°C
FP: N/A
VP: 91 mm Hg
VD: N/A
SG: 1.59 @ STP
IDLH: 200 ppm
STEL: 200 ppm
TWA: 5 ppm
LEL: N/A
UEL: N/A
Solubility in water: 0.05%

decomposition produces phosgene and hydrogen chloride

disulfur dichloride (S_2Cl_2)
DOT: corrosive
BP: 138°C
FP: 118°C
VP: 7 mm Hg
VD: 4.7
SG: 1.7
IDLH: 10 ppm
STEL: not determined
TWA: 1 ppm
LEL: N/A
UEL: N/A
Solubility in water: decomposes in water

a combustible liquid

■ **Note**

When looking at the periodic table, the symbols hold valuable information.

But this equation does not express what occurs in nature. Because of the law of conservation of mass, we must have as many atoms at the end of a reaction as we had at the beginning. The equation must balance so that equality exists between masses.

First we look at the number of carbons on the left and compare them with the number on the right. We can see that we have one carbon on both sides of the equation. We repeat the process for sulfur and then for chlorine. After counting the sulfur atoms, we see two sulfur atoms on either side. However, there is a discrepancy when it comes to the number of chlorine atoms. We can see that on the left side we have two chlorine atoms, and on the right we have six. To correct this imbalance we place a "3" in front of the chlorine reactant:

$$CS_2 + 3Cl_2 \rightarrow CCl_4 + S_2Cl_2$$

By doing this we have now balanced the equation and conserved mass. Therefore, we can see that for this reaction to take place, there must be three molecules of chlorine to every one molecule of carbon disulfide.

Another convenient way to describe the outer shell's movement and the law of conservation is by the use of Lewis dot structures. Gilbery Lewis was the first to suggest that a sharing or transfer of electrons takes place during bonding. He introduced the visual representation of atoms utilizing only the outer, or octet, shell. In a pattern around the chemical symbol, we can schematically describe the activity of the electrons. For example, the structure of formaldehyde is as shown in Figure 2–3. We can see that the carbon has eight electrons, the number of electrons that it requires to be stable. Each hydrogen has two and the oxygen has eight. By sharing electrons, all the elements have their outer shells satisfied.

The balancing of equations illustrates the fact that chemicals need certain quantities in order for a reaction to take place. If there is a limit to the reactants (the molecules on the left side of the equation), the chemical reaction may not take place completely. Another factor that may impede a reaction is the amount of energy needed for the thermodynamics of the reaction, that is, the energy required to break a bond and create a new one. In addition to this factor, the orientation of the molecule, its polarity, and an open space for bonding to occur in all play a role in reactions. Many laws enter into the occurrence of a chemical reaction; some can be controlled and others cannot. We as hazardous materials responders can predict the overall chemical outcome if given a few chemical facts by our reference section.

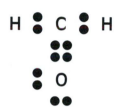

Figure 2–3 *The structure of formaldehyde.*

■ **Note**

Calcium carbide, the precursor to acetylene, has a structure that is a bit confusing and does not match the rules we have set forth. It is a calcium ion in association with two triple bonded carbons: $Ca^{++} {}^-C \equiv C^-$. Hence the formula CaC_2.

When using Lewis dot structures, the idea is to manipulate the electrons to satisfy the outer shell. Outer-shell electrons are sometimes referred to as the valence electrons. The total number of negative valence charges must equal the total number of positive valence charges. Subscripts show the number of atoms in a compound, as opposed to the number placed in front of a compound that indicates the number of molecules required to complete the reaction. For example, H_2O represents a molecule of water. The subscript 2 indicates that there are two atoms of hydrogen to one oxygen. Because the valence charge for hydrogen is $+1$ and the valence charge for oxygen is -2, to have a molecule that is electrically neutral requires two $+1$ and one -2 for an overall valence of zero. However, if we were to place a "2" in front of the molecule ($2H_2O$) we would be depicting two entire molecules of electrically neutral water. Similarly, NaCl can be written as Na^+ and Cl^- or, better yet, as Na^{+1} and Cl^{-1}. Cross valence now suggests that one atom of sodium and one atom of chlorine make the molecule sodium chloride:

$$Na^{+1}Cl^{-1}, \text{ or } Na_1Cl_1, \text{ or NaCl}$$

Problem Set 2.1   Identify the valence, cross-identification of subscript, and resultant formula for the following:

$$Mg + Cl \longrightarrow$$
$$K + S \longrightarrow$$
$$Ba + O \longrightarrow$$
$$Li + F \longrightarrow$$
$$Al + Cl \longrightarrow$$

■ Note

When we look at a combination of elements, we first place their respective valences under the symbol:

$$\begin{array}{cc} Mg & Cl \\ +2 & -1 \end{array}$$

We then cross the valences in order to achieve the correct number of atoms in combination:

$$\begin{array}{cc} Mg & Cl \\ +2 & -1 \end{array}$$

$$\frac{1}{+2} \qquad \frac{2}{-2}$$

When we multiply the valences with the number of atoms, we should get a balancing of valence where plus charges equal negative charges:

OR

$$MgCl_2$$

■ Note

Valences of common elements: Mg $-$ +2 Cl $-$ $-$1; K $-$ +1 S $-$ $-$2; Ba $-$ +2 O $-$ $-$2; Ca $-$ +2 C $-$ $-$4; Li $-$ +1 F $-$ $-$1. Find these elements on the periodic table. Do you agree with the valences cited?

● Caution

With a few exceptions, water in contact with some metal salts can produce an alkali solution. This corrosive solution is a chemical reaction that produces a hydroxide of the base metal.

BINARY COMPOUNDS

Binary Ionic Compounds I and II
M + NM   No Oxygen

Metal Salts

Chemistry Quick Reference Card

Family Class	Naming	Hazards
Metal salts M + NM ø O	-ide	General hazards, with four being water reactive

When a metal and a nonmetal combine, the result is a salt. Metal salts include only those compounds composed of a metal and a nonmetal but no oxygen. In this category, the name of the nonmetal is formed with the suffix -ide. Some examples are:

>   Potassium sulfide (K_2S), which is flammable, may spontaneously ignite, and explosive in dust form
>
>   Aluminum chloride ($AlCl_3$), which may have a violent reaction with water and evolves hydrogen chloride gas
>
>   Cupric chloride ($CuCl_2$), which is toxic by ingestion or inhalation
>
>   Calcium nitride (Ca_3N_2), which evolves ammonia gas when in contact with water

The hazards of these compounds depend on the metal ion of attachment. Four of the metal salts are water reactive and extreme care should be taken with them. They are carbides, which produce acetylene gas; hydrides, from which can evolve hydrogen gas; nitrides, which give off ammonia gas; and phosphides, which produce phosphine gas.

■ Note

In an acetylene gas cylinder, acetone is placed within a honeycomb ceramic to stabilize the acetylene. Therefore, you never place an acetylene tank on its side.

This family's members are called salts and they comprise several subfamilies. Each of these subfamilies is ionically bonded, but each has its own intrinsic health and reactivity uncertainties. The first group of salts, metal salts (traditional chemistry textbooks may call this group binary ionic compounds I), includes compounds resulting from the combination of a metal and a nonmetal. Here the respective cation combines with an anion. If the overall negative charge is 1, then the respective positive charge for the combination is 1. However, in the case of calcium nitride, calcium has an overall +2 charge, and nitrogen has a −3 charge. When these two elements combine, it takes three calcium atoms to combine with one diatomic molecule of nitrogen to make calcium nitride:

$$3Ca^{+2} + N_2^{-3} \longrightarrow Ca_3N_2 \text{ (calcium nitride)}$$

Problem Set 2.2 Identify the valences and the resulting compounds, and name each compound:

Mg + S $\longrightarrow$

Al + Cl $\longrightarrow$

Ra + Cl$_2$ $\longrightarrow$

Ca + N $\longrightarrow$

Na + Cl $\longrightarrow$

Na + F $\longrightarrow$

Ba$_3$ + N$_2$ $\longrightarrow$

Al + P $\longrightarrow$

A second group of compounds exists within this same family. This group is sometimes referred to as binary ionic compounds II. The difference between binary ionic compounds I and II is that the metal component has the potential to have different charges. Because of the possibility of having different charges, two methods have evolved in naming these compounds (see Table 2–1). As you can see, the Latin names are listed. Latin or Greek names are sometimes used in chemistry and are referred to as the alternate naming system (alternate to the systematic system, which uses Roman numerals). In the table you can see that the suffix -ous represents the lower-charged cation, and -ic, the higher-charged

Table 2–1 *Common cations.*

+1	+2	+3	+4	Alternate Latin Name (old version)	Systematic Name
Cu				Cuprous	Copper I
	Cu			Cupric	Copper II
Hg				Mercurous	Mercury I
	Hg			Mercuric	Mercury II
Au				Aurous	Gold I
		Au		Auric	Gold III
	Fe			Ferrous	Iron II
		Fe		Ferric	Iron III
	Sn			Stannous	Tin II
			Sn	Stannic	Tin IV
	Co			Cobaltous	Cobalt II
		Co		Cobaltic	Cobalt III
	Cr			Chromous	Chromium II
		Cr		Chromic	Chromium III
	Mn			Manganous	Manganese II
		Mn		Manganic	Manganese III
	Pb			Plumbous	Lead II
			Pb	Plumbic	Lead IV
	Zn			Zincic	Zinc II

cation. This configuration is common in the naming of popular chemicals. For example:

Cupric chloride or copper II chloride ($CuCl_2$), which can react violently with K and Na and is a poison

Cobaltous chloride or cobalt II chloride ($CoCl_2$), which is incompatible with metals K and Na and is a poison

Ferric chloride or iron III chloride ($FeCl_3$), which is a corrosive material that releases toxic chlorine gas

---

Types of Fires

Class A: ordinary combustibles
Class B: flammable liquids
Class C: electrical
Class D: combustible metals

■ Note

Some symbols are derived from the Latin names of elements: sodium = natrium = Na; tungsten = wolfrium = W; potassium = kalium = K; iron = ferium = Fe.

■ Note

The DOT hazard class that most of this group represents is the flammable solids. The alkali metals react violently when in contact with water; the alkaline earth metals, although not as reactive as group I, do not have the tendency to exist in nature as a pure metal. For example, this group (II) has the tendency to form an oxide around the outer surface of the metal. This oxide coating usually interrupts possible reactions. Magnesium is an example of an exception. Finely turned shavings of this metal can produce a dramatic and difficult-to-extinguish fire when heated. This dramatic pyrophoric condition has to do with the increased surface area the shavings provide. Some of the carbides, nitrides, chlorides, and phosphides have reactive qualities.

❗Safety

Metal or binary oxides all produce a corrosive solution, are water reactive, and will produce varying degrees of heat during reaction depending on the base metal. If ordinary combustibles are present, the heat of reaction may cause a Class A fire. Some are classified as Class 4 DOT hazards; however, in solution, they may be in a different hazard class.

Problem Set 2.3   Write the correct name (both alternate and systematic) and identify the valences for the following:

$$Cu + Cl_2 \longrightarrow$$
$$Fe + Cl_2 \longrightarrow$$
$$Pb + Br_2 \longrightarrow$$
$$Pb_3 + N_2 \longrightarrow$$
$$Sn_3 + P_4 \longrightarrow$$
$$Fe + Cl_3 \longrightarrow$$
$$Cu_2 + S \longrightarrow$$

M + Oxygen

Metal Oxides

Chemistry Quick Reference Card

Family Class	Naming	Hazards
Metal salts M + NM ø O	-ide	General hazards, with four being water reactive
Metal oxides **M + oxygen**	-oxide	Water reactivity, corrosive solutions, and oxidation

Another group of binary compounds has a metal and oxygen in the attachment. The names of elements in this group end with -oxide. These are called the metal oxides and they tend to be water reactive, which in turn produces heat once the reaction takes place. The resulting solutions are corrosive, the degree of which depends on the base metal that was attached to the oxygen atom. If the oxygen were attached to an alkali metal, the corrosivity would increase. Some of these compounds may even act like an oxidizer. Examples are:

Aluminum oxide (Al_2O_3), which is used for electrical insulators and heat-resistant fibers, and produces a toxic dust

Arsenic oxide (As_2O_5), which is used in insecticides, herbicides, and colored glass

Beryllium oxide (BeO), which is toxic by inhalation, and is used in reactor systems and electron tubes

Calcium oxide (CaO), which evolves heat when exposed to water

---

● Caution

Peroxides have an unstable oxygen–oxygen bond leading to strong oxidative potential.

Problem Set 2.4   Name the following:

Be + O ⟶

Na + O ⟶

Mg + O ⟶

Fe + O ⟶

Al + O ⟶

As + O ⟶

Ca + O ⟶

M + O$_2$

Inorganic Peroxides

Chemistry Quick Reference Card

Family Class	Naming	Hazards
Metal salts M + NM ø O	-ide	General hazards, with four being water reactive
Metal oxides M + oxygen	-oxide	Water reactivity, corrosive solutions, and oxidation
Inorganic peroxides **M + O$_2$**	Peroxide	Water reactivity, corrosiveness, and very strong oxidizers

Sometimes called the metal peroxides, inorganic peroxides are a metal in combination with a peroxide ion, a complex anion (O$_2$). The "per" with "oxide" means that there is one more oxygen atom than the normal state of one (in the naming, dioxide is occasionally used instead of peroxide). They tend to be strong oxidizers, which generate heat on reaction. This heat can ignite ordinary combustibles in the general area. As a family, most of them are water reactive. Their degree of corrosiveness depends on the base metal in the compound. Examples are:

> Barium peroxide (BaO$_2$), which can produce fire on contact with organic materials

> Calcium peroxide (CaO$_2$), which can produce fire on contact with organic materials

> Cesium peroxide (Cs$_2$O$_4$), which can produce fire on contact with organic materials

> Hydrogen peroxide (H$_2$O$_2$), which presents a severe fire hazard

The diatomic molecule of oxygen in these compounds is a good example of covalent bonding. Each oxygen has six electrons in its outer shell. However, when two oxygen molecules are in close proximity, the two oxygen atoms share the number of electrons in order to satisfy each atom's requirement of stability. As shown in Figure 2–4, the complete state of stability for the entire molecule is not achieved until the peroxide radical (with a valence of −2) combines with another compound or atom.

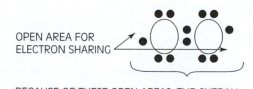

OPEN AREA FOR
ELECTRON SHARING

BECAUSE OF THESE OPEN AREAS, THE OVERALL
VALENCE FOR THIS POLYATOMIC ION IS −2

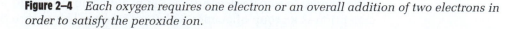

Figure 2–4   *Each oxygen requires one electron or an overall addition of two electrons in order to satisfy the peroxide ion.*

!Safety

●Inorganic peroxides, sometimes referred to as peroxide salts or metal peroxide salts, are in DOT hazard Classes 4 or 5.1, depending on the compound. As oxidizers, they can be extremely reactive.

■ Note

Two or more nonmetals that are covalently bonded behave as a single unit during reaction. Because of the excess oxygen present, these compounds release the oxygen radical.

Problem Set 2.5   Name the following:

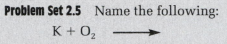

$$K + O_2 \longrightarrow$$
$$Mn + O_2 \longrightarrow$$
$$Na + O_2 \longrightarrow$$
$$Mg + O_2 \longrightarrow$$
$$Ba + O_2 \longrightarrow$$
$$Ca + O_2 \longrightarrow$$
$$Cs + O_2 \longrightarrow$$

COVALENT BONDING

A covalent bond is created by the sharing of electrons between atoms to make a molecule. Each covalent bond is a pair of electrons that are in opposite spin directions, creating a cloud about the nucleus of the respective atom. In this configuration, the octet limit is achieved. We can see that the upper-right side of the periodic table (nonmetals) houses most of the elements that make up the discipline of organic chemistry. We can also see that these elements have high ionization potentials. In other words, they do not lose their electrons easily, but they will share electrons.

■ Note

When a covalent bond is broken, there is a corresponding release of energy.

   With the ability to share, we can have single, double, and even triple bonds. However, we must remember that energy was required to formulate these bonds. If broken, the energy required to make the bond will be released. (This condition is relative to the potential energy that the molecule possesses.) Looking at a molecular reaction, we see that if one bond is broken, energy is released to the adjacent molecule, which in turn releases energy to two or more molecules, thus producing a chain reaction.

   A group of atoms can bond together covalently, but have a positive or negative charge on the group collectively. This quality is seen with individual atoms called ions. In covalent bonding, the same basic principle is taking place; the exception is a group of elements that act as one. The charge that this group possesses is called the formal charge, or valence. For example, the ammonia molecule has three hydrogen atoms attached to the nitrogen. The nitrogen has a shell that is open, with two electrons that are available:

$$H{:}N{:}H + H^+ \rightarrow (NH_4)^+$$
$$::$$
$$H$$

If a positive ion is in close proximity to, say, a hydrogen atom that has been stripped of its electron, the available electron pair will fill the hydrogen's requirements for its needed electrons. The total complex, or ion, will have an overall positive charge. If another negatively charged ion is available, the unit will act as one molecule. The negative ion will combine with the positive ion to form another compound, which is electronically stable.

PROPERTIES OF IONIC AND COVALENT COMPOUNDS

In ionic compounds, electrostatic forces create an attachment between all the molecules. These forces are so strong that in order to melt an ionic compound, a great amount of thermal energy must be supplied. The energy that is thus supplied breaks the intermolecular attachment. Because of this intermolecular attraction, ionic-bonded compounds have high melting and boiling points, are hard, and are very soluble in polar solvents.

Nonpolar, covalently bonded molecules are held together by what are called Van der Waals forces. These forces are weak and are caused by the movement of the electrons about the nucleus and its position relative to the protons. Sometimes a nonpolar compound has an area that is negative one moment and then positive the next. This constant moving of positive and negative fields creates the electrostatic attraction needed to hold the compound together. Because they are weak (due to the slight movement of the electron in relation to the proton), these compounds are extremely soft and have low melting points. If such a compound is a liquid (in predominately water, hydrogen bonding also occurs and adds to the strength of the molecule—see Figure 4–14), it may be extremely volatile.

Polar covalent bonds possess characteristics somewhere between the polar ionic and the nonpolar covalent bond. The polar, covalently bonded molecules have higher melting and boiling points (although this depends on the size of the molecule). Each molecule acts as a unit in a chemical reaction. Just as in individual atoms, molecules also strive to reach a comfortable state (the ground state, or octet satisfaction) in order to become stable. These polar, covalently bonded

---

■ Note
In a colvalently bonded molecule, a relatively strong electronegative pull is created. This pull shifts the density of the electron cloud toward the stronger electronegative atom. This shift creates polarity—a positive end and a negative end—in the molecule. A nonsymmetrical cloud distribution is referred to as being polar.

■ Note
Van der Waals forces are a type of magnetic attraction. If a nonpolar, covalently bonded compound is a liquid, it can be volatile. Hydrogen bonding is a strong type of dipole interaction. Van der Waals and dipole forces are electrostatic.

■ Note
Thio- designates sulfur groups.

■ Note
Nonsalts are mostly covalently bonded nonmetals: the hydrocarbons.

Table 2–2 *Binary compounds of two nonmetals forming polyatomic ions that act like singular entities.*

Anions		Cations	
Amide	NH_2^{-1}	Ammonium	NH_4^{+1}
Cyanide	CN^{-1}	Diammine silver (I)	$Ag(NH_3)_2^{+1}$
Carbonate	CO_3^{-2}	Mercury amide	$Hg(NH_2)_2^{+1}$
Oxalate	$C_2O_4^{-2}$		
Chromate	CrO_4^{-2}		
Phosphate	PO_4^{-3}		
Dichromate	$Cr_2O_7^{-2}$		
Molybdate	MoO_4^{-2}		
Thiocyanate	SCN^{-1}		
Thiosulfate	$S_2O_3^{-2}$		
Chlorate	ClO_4^{-1}		
Iodate	IO_4^{-1}		
Nitrate	NO_3^{-1}		

molecules are sometimes referred to as nonmetal binary compounds or nonsalts (the hydrocarbons). Table 2–2 lists binary compounds that form to make a polyatomic ion that acts like a singular entity.

● Caution

Nonsalts:

☠ Are primarily liquids or gases   ☠ Are nonconductors of electricity   ☠ Are often insoluble in water

☠ Are burnable   ☠ Are toxic

Oxysalts
M + Oxy ion

Oxygenated Inorganic Compounds

Chemistry Quick Reference Card

Family Class	Naming	Hazards
Metal salts M + NM ø O	-ide	General hazards, with four being water reactive
Metal oxides M + oxygen	-oxide	Water reactivity, corrosive solutions, and oxidation
Inorganic peroxides M + O$_2$	Peroxide	Water reactivity, corrosiveness, and very strong oxidizers
Oxygenated inorganic compounds M + oxy ion	Per—ate -ate -ite Hypo—ite	Strong oxidizers can generate heat if wet

By definition, an oxygenated salt, or oxysalt, is a metal in combination with a complex oxygenated ion. Nomenclature is based on the normal state of the oxygenated ion, for which the name starts with the metal and ends with the oxygenated compound as -ate. The prefixes per- and hypo- aid in the designation of the level of oxygen that the compound contains. If the oxygenated compound contains one more oxygen than the normal state, the metal and the oxygenated compound are identified, in which case the oxygenated compound takes the prefix per- and the suffix -ate. If the compound contains one oxygen less than normal, then the name ends with -ite. For two less oxygen than normal, the metal again is identified and the oxygenated compound radical takes the prefix hypo- and the suffix -ite. (Tables 2–3, 2–4, and 2–5 summarize this information.)

■ **Note**

Sometimes referred to as the oxysalts or inorganic oxygen compounds, the DOT classification for many of the oxygenated compounds is Class 5.

For example, as an oxygenated molecule, the chlorine molecule has three oxygens attached to the chlorine for the normal state. One more oxygen would

Table 2–3 *Normal states.*

−1	−2	−3
ClO_3	CO_3	PO_4
BrO_3	SO_4	BO_3
IO_3		AsO_4
NO_3		
MnO_3		

Table 2–4 *Naming configuration.*

State	Prefix/Suffix
More than 1 oxygen	Per—ate
Normal	-ate
1 less oxygen	-ite
2 less oxygen	Hypo—ite

Table 2–5 *Base naming conventions for the oxions.*

−1		−2		−3	
ClO_3	Chlorate	CO_3	Carbonate	PO_4	Phosphate
BrO_3	Bromate	SO_4	Sulfate	BO_3	Borate
IO_3	Iodate			AsO_4	Arsenate
NO_3	Nitrate				
MnO_3	Manganate				

give a per—ate state; one less, an -ite state; and two less, a hypo—ite state. So, the radicals would look like this:

-ClO_4 Perchlorate

-ClO_3 Chlorate

-ClO_2 Chlorite

-ClO Hypochlorite

The metal that is attached would be placed in front of the oxygenated ion. These are extremely strong oxidizers. If they become wet, heat is generated, which, in turn, can ignite combustibles. If mixed with fuels, the resulting compound becomes violently explosive.

A typical hazard within this group is the solubility of some of these compounds. They tend to dissolve in water. The water–oxysalt solution can become absorbed into packaging or, worse yet, firefighters' bunker gear. Once the water evaporates, the salt is left behind. If this material is exposed to heat or fire, it will burn in an accelerated manner and explode in some circumstances. Here are some examples:

Sodium perchlorate ($NaClO_4$), which is an oxidizer and can generate fire on contact with organic materials

Sodium chlorate ($NaClO_3$), which is an oxidizer and can generate fire on contact with organic materials

Sodium chlorite ($NaClO_2$), which is a flammable oxidizer and an irritant to tissue

Sodium hypochlorite ($NaClO$), which is an oxidizer and an irritant, and can generate fire on contact with organic materials

Remember that the ClO_x ion has an overall valence of −1 and sodium has a +1 formal charge. If sodium were attached to the sulfate ion, because SO_4 has an overall valence of −2, you would need two sodium atoms.

In looking at this group, it is the relationship of oxygen to the base non-metal in the natural state that identifies the valence. The overall valence does not change when attached to the nonmetal.

Problem Set 2.6   Write out the four configurations of each radical listed in Tables 2–3, 2–4, and 2–5 using sodium as the base metal.

		Per—ate	-ate	-ite	Hypo—ite
ClO_3	Chlorate				
BrO_3	Bromate				
IO_3	Iodate				
NO_3	Nitrate	N/A			
MnO_3	Manganate				
CO_3	Carbonate	N/A			
SO_4	Sulfate	N/A			
PO_4	Phosphate	N/A			
BO_3	Borate				
AsO_4	Arsenate	N/A			

N/A = not applicable; these compounds do not have a per—ate state.

ACID–BASE REACTIONS

During the Middle Ages, chemists (then known as alchemists) placed substances into categories based on their acidity as determined by taste. This testing procedure did not last very long, for obvious reasons. However, during the twentieth century, fire departments became involved with hazardous materials, and this obsolete identification procedure came into use. Because of the resulting injuries and deaths, and the awareness of the potential hazards due to our increase in chemical knowledge, this testing procedure once again has been put to rest. But looking at this technique gives us a cognitive understanding of how we relate to acids and bases. If a certain substance tasted sour or bitter, it was classed as an acid. Bases have a variety of tastes (aromatics were classed under this heading for the fragrances a material possessed), but mostly they created a soapy film. Although this film also causes tissue damage, this was the classification method used at the time (and the method of identification fire departments used for much too long).

We can define acids and bases in terms of qualitative numbers. These numbers can be utilized in deciding what form of protective gear an entry-team member must use. The numbers are calculated from the chemicals' ability to produce the hydronium ion, H_3O^+, or the hydroxide ion, OH^-.

● Caution

Both organic and inorganic acids can be water reactive, combustible, corrosive, and explosive by nature.

■ Note

Acid–base reactions are ionic solutions in an equilibrium state. It is possible to have an acid stronger than a pH of 0 or a base greater than a pH of 14. These substances are called superacids and superbases, respectively; pH paper will not react to show their acidity or alkalinity.

An acid is a compound that is capable of transferring a hydrogen ion (proton) in solution. A base is a group of atoms that contain or can generate one or more hydroxyl groups in solution. The base is referred to as the proton acceptor. It is this acceptance and donation that produces the acid–base reaction. Acids and bases exist because of their ability to dissociate. It is an equilibrium between the hydronium ion and the hydroxide ion that creates a neutral solution in water. The following equation is a general expression for the equilibrium constant for the ionization of water:

$$K_w = \frac{[H_3O^+][OH^-]}{[H_2O]^2}$$

It is this relationship in water that gives rise to the acidity or alkalinity; in other words, hydronium or hydroxide production.

Acids and bases are classified according to their particular ionization in water. It is this ionization or degree of cation or anion state that we term as strength. The pH scale for measuring acidity and alkalinity ranges from 0 to 14. Zero to 6.9 is termed acidic, 7.1 to 14 is basic, and 7 is considered neutral. This scale measures the relationship of cations and anions within solution. Generally, compounds that form acids are covalently bonded. Compounds that form bases are ionically bonded.

● **Caution**

Because of the equilibrium reaction, using water to neutralize an acid leak is not recommended. All that will result is a larger acid spill. One gallon of an acid at a pH of 0 requires more than 1 million gallons of water to dilute it to a pH of 7. In addition, you should never add water to an acid because a violent reaction may ensue.

■ **Note**

Acid–base is a relationship of equilibrium.

■ **Note**

Technically the "p" in pH means the negative logarithm, and in this case refers to the negative log of hydrogen concentration.

While some reference sources use the pH scale (pH is an abbreviation for positive hydronium ions), others use a numerical system based on the number of hydronium ions in comparison to the number of hydroxide ions in a different format. The total reaction is expressed and measured in K, where K is the ionization constant (usually measured at 25°C). The subscripts a and b identify the constant for an acid or a base, respectively (K_a, K_b). It is this expression that gives the true relationship between acids and bases. See Table 2–6 for a comparison of the two systems.

A neutral solution has the hydronium ion and the hydroxide ion in equilibrium. When the solution contains a greater concentration of hydronium ion with respect to the hydroxide ion, the solution is said to be acidic. Conversely, if the hydroxide ion is in greater proportion than the hydronium ion, the solution is said to be basic. In general, the larger the K value, the stronger the acid or the base:

Extremely strong	K value greater than 10^3
Strong	K value between 10^3 and 10^{-2}
Weak	K value between 10^{-2} and 10^{-7}
Extremely weak	K value less than 10^{-7}

● **Caution**

An anhydrous (without water) acid solution will not give a meaningful reading using pH paper.

■ **Note**

The acidity or alkalinity of a solution can be expressed using the hydronium or hydroxide ion concentration: $pH = -\log[H^+]$ or $pOH = -\log[OH^-]$, where the autopyrolysis of water is $pK_w = pH + pOH = 14.00$.

Table 2–6   *Sample comparison of acidity measures.*

Acid/Base	K	pH
Sulfuric acid	Large	0.3
Acetic acid	1.8×10^{-5}	2.4
Hydrocyanic acid	6.2×10^{-10}	5.1
Ammonia	1.8×10^{-5}	11.6
Calcium hydroxide	3.74×10^{-3}	12.4

The strength of an acid depends on its ability to give up protons. Weak acids give up their protons with great difficulty and will react slowly, whereas a strong acid will react rapidly. In comparing acids (and bases) it is necessary to compare each as equivalent concentrations. These concentrations are obtained as volumes of solution that contain the same weight of available hydrogen (proton donor). The degree of electronegativity and bond polarity has a bearing on the donation of the proton. In general, the strength of an acid is inversely proportional to the strength of the bond. With a weak acid, the hydrogen has a strong bond; the strong acid has a weakly bonded hydrogen.

Acids and alkalis can produce devastating injuries depending on their concentrations. Acid burns, while serious, are typically self-limiting due to the effect acids have on the protein within the skin. In contrast, alkali burns cause a change in the fatty substances in the skin and will cause liquefaction. This reaction makes an alkali burn twenty times more destructive than an acid of corresponding strength and concentration.

We can utilize strength and concentration information to determine the amount of base needed to neutralize an acid spill. Say, for example that 1 gallon of 70% nitric acid has spilled and we have to figure out how much soda ash is needed to neutralize it. First we need to know the weight of the acid. From the material safety data sheet (MSDS) we find that 1 gallon of 70% nitric acid weighs 8.77 pounds. Then we need the conversion factor for the appropriate neutralizing agent, which in this case is 0.844. Multiplying these two figures (8.77 pounds $\times$ 0.841) tells us that 7.4 pounds of soda ash would be required. (See Chapter 6— page 231 for a detailed explanation.)

M + OH

Metal Hydroxides (Bases)

Chemistry Quick Reference Card

Family Class	Naming	Hazards
Metal salts M + NM ø O	-ide	General hazards, with four being water reactive
Metal oxides M + oxygen	-oxide	Water reactivity, corrosive solutions, and oxidation
Inorganic peroxides M + O_2	Peroxide	Water reactivity, corrosiveness, and very strong oxidizers
Oxygenated inorganic compounds M + oxy ion	Per—ate -ate -ite Hypo—ite	Strong oxidizers can generate heat if wet
Metal hydroxides **M + OH**	Hydroxide	Corrosiveness and heat production during reaction

■ Note

The metal hydroxides are DOT classified as corrosives and are in Class 8. Sometimes referred to as hydroxide salts, they are a combination of a nonmetal and the hydroxide radical. The addition of water to metal oxides produces a caustic hydroxide or base solution.

A hydroxide is a metal combined with the hydroxide radical ($-OH$). It is caustic in the solid state (the valence of the hydroxide radical is -1). If mixed with water, the result is a caustic solution. All hydroxides are classified as corrosives and produce heat during reaction. The level of corrosiveness depends on the base metal of attachment. The names of substances in this category end in "hydroxide." Examples of this type of salt include sodium hydroxide, potassium hydroxide, and calcium hydroxide, as well as the following:

Aluminum hydroxide ($Al(OH)_3$), which is used for flame retardants, mattress batting, and cosmetics

Beryllium hydroxide ($Be(OH)_2$), which decomposes to an oxide and is extremely toxic

Cesium hydroxide ($CsOH$), which is extremely toxic

Potassium hydroxide (KOH), which is toxic and corrosive

Problem Set 2.7   Name the following:

$NaOH \longrightarrow$

$LiOH \longrightarrow$

$Ca(OH)_2 \longrightarrow$

$KOH \longrightarrow$

$Pb(OH)_4 \longrightarrow$

$Mg(OH)_2 \longrightarrow$

$Fe(OH)_3 \longrightarrow$

$NH_4OH \longrightarrow$

H + poly/NM

Inorganic Acids (Acids)

Chemistry Quick Reference Card

Family Class	Naming	Hazards
Metal salts M + NM ø O	-ide	General hazards, with four being water reactive
Metal oxides M + oxygen	-oxide	Water reactivity, corrosive solutions, and oxidation
Inorganic peroxides M + O_2	Peroxide	Water reactivity, corrosiveness, and very strong oxidizers
Oxygenated inorganic compounds M + oxy ion	Per—ate -ate -ite Hypo—ite	Strong oxidizers can generate heat if wet
Metal hydroxides M + OH	Hydroxide	Corrosiveness and heat production during reaction
Inorganic acids **H + polyNM**	Per—ic -ic -ous Hypo—ous	Corrosiveness

Sometimes called oxyacids, inorganic acids are formed when a nonmetal combines with a complex oxygen ion. The ion complex is the base for the name, which uses the prefixes and suffixes per—ic, -ic, -ous, and hypo—ous and ends with the word "acid"; for example, perchloric acid ($HClO_4$), chloric acid ($HClO_3$), chlorous acid ($HClO_2$), and hypochlorous acid ($HClO$). (The latter should not be confused with HCl, or hydrochloric acid, which is, however, easily recognized as not having a polyatomic ion and is a binary acid.)

The normal base state must be identified. In the examples, the chlorate ion is the base or normal state. The -ate suffix is changed to -ic. For the most part, and for our level of discussion, inorganic acids will have a hydrogen connected to the **polyatomic** nonmetal ion.

polyatomic ion

an ion having more than one constituent atom

Substances in this group are both strong oxides and corrosives:

Hypochlorous acid ($HClO$), which is moderately toxic and can emit fumes that can cause pulmonary edema

Boric acid (H_3BO_3), which is a poison, a mutagen, and a human skin irritant that is incompatible with K

Sulfuric acid (H_2SO_4), which has fuming and mist forms in industry that are poisonous and pose a variety of health problems

Nitric acid (HNO_3), which presents variable fire and explosion hazards

Problem Set 2.8   Name the following:

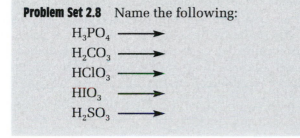

$$H_3PO_4 \longrightarrow$$
$$H_2CO_3 \longrightarrow$$
$$HClO_3 \longrightarrow$$
$$HIO_3 \longrightarrow$$
$$H_2SO_3 \longrightarrow$$

■ **Note**

As we observed for oxygenated compounds, a group of acids can also be formed by the oxygenated inorganic compounds. This group has strong oxidative and corrosive qualities.

BINARY ACIDS

H + NM

Hydrogen Halides

Chemistry Quick Reference Card

Family Class	Naming	Hazards
Metal salts M + NM ø O	-ide	General hazards, with four being water reactive
Metal oxides M + oxygen	-oxide	Water reactivity, corrosive solutions, and oxidation

(continues)

Chemistry Quick Reference Card (Continued)

Family Class	Naming	Hazards
Inorganic peroxides $M + O_2$	Peroxide	Water reactivity, corrosiveness, and very strong oxidizers
Oxygenated inorganic compounds M + oxy ion	Per—ate -ate -ite Hypo—ite	Strong oxidizers can generate heat if wet
Metal hydroxides M + OH	Hydroxide	Corrosiveness and heat production during reaction
Inorganic acids H + polyNM	Per—ic -ic -ous Hypo—ous	Corrosiveness
Binary acids **H + NM** **Hydrogen halides**	Hydro- -ic	Corrosiveness

This category comprises a specific number and association of hydrogen with a nonmetal from group VIIB elements. In pure form, hydrogen halides are gases. When they are in a water solution (aqueous [aq is the chemical abbreviation for aqueous]), the prefix hydro- and the suffix -ic are used, with the word "acid" added. For example, HF is hydrogen fluoride, which becomes hydrofluoric acid in an aqueous solution: hydrogen is changed to hydro, the fluoride becomes fluoric, and "acid" is added at the end. Here are other examples:

Hydrobromic acid (HBr), which is a toxic and irritant, faintly yellow, and used as a catalyst

Hydrofluoric acid (HF), which is extremely toxic and colorless, and used in etching glass

Hydrochloric acid (HCl), which is a toxic and irritant, colorless to slight yellow, and used in metal cleaning

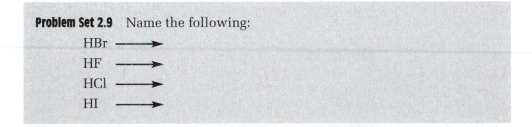

Problem Set 2.9   Name the following:

HBr $\longrightarrow$

HF $\longrightarrow$

HCl $\longrightarrow$

HI $\longrightarrow$

Salt · H_2O

Hydrates

Chemistry Quick Reference Card

Family Class	Naming	Hazards
Metal salts M + NM ø O	-ide	General hazards, with four being water reactive

(continues)

Chemistry Quick Reference Card (Continued)

Family Class	Naming	Hazards
Metal oxides M + oxygen	-oxide	Water reactivity, corrosive solutions, and oxidation
Inorganic peroxides M + O_2	Peroxide	Water reactivity, corrosiveness, and very strong oxidizers
Oxygenated inorganic compounds M + oxy ion	Per—ate -ate -ite Hypo—ite	Strong oxidizers can generate heat if wet
Metal hydroxides M + OH	Hydroxide	Corrosiveness and heat production during reaction
Inorganic acids H + polyNM	Per—ic -ic -ous Hypo—ous	Corrosiveness
Binary acids H + NM Hydrogen halides	Hydro- -ic	Corrosiveness
Hydrates **Salt • H_2O**	Name the salt then add the Greek numbering to "hydrate"	Strong affinity for water

---

● Caution

Even though hydrates are dry (appearing wet) their strong affinity for water (water seekers) makes their classification very dangerous.

Table 2–7 *Greek numbering for use in naming hydrates.*

One	Mono
Two	Di
Three	Tri
Four	Tetra
Five	Penta
Six	Hexa
Seven	Hepta
Eight	Octa
Nine	Nona
Ten	Deca

Hydrates are substances that include water molecules. You would think, then, that they are wet in nature, but they are actually dry substances, even if they appear to be wet. The formula for a hydrate has a dot (sometimes described as a multiplication dot) in it. This dot is not a multiplication sign, but rather the identifying mark of the hydrate. The chemical to the left of the dot is named according to either the systematic or the alternative method, depending on the specific atomic structure. To the right of the dot is the water molecule with the appropriate Greek numbering system (see Table 2–7) incorporated in the term. So, for example, BaCl • $2H_2O$ is named from the left first, which is barium chloride. Then, to the right of the dot in this formula we see two water molecules, so we use the Greek prefix di- for two and place that in front of the hydrate to signify the number of water molecules. The compound is, thus, barium chloride dihydrate. Here are other examples:

Barium hydroxide octahydrate, $Ba(OH)_2$ • $8H_2O$, which is a white powder used in analytical chemistry

Aluminum oxide trihydrate, Al_2O_3 • $3H_2O$, which is a white powder used for rubber reinforcement

Aluminum chloride hexahydrate, $AlCl_3$ • $6H_2O$, which is a yellow-white powder used in roofing material

The term "hydrate" is sometimes used to describe the hydroxides, but hydroxides, of course, do not have a water molecule in the formula. So when we see a formula with a water molecule attached, or the description identifies a water molecule, it is a hydrate and may be named a hydroxide. When there is no water molecule and the compound follows the metal plus a hydroxide ion, the formula represents a hydroxide.

● **Caution**

A hydride is a binary or complex inorganic compound of hydrogen and another element. In general, hydrides are flammable and react violently with water and oxidizing agents.

● **Caution**

The term "hydrate" is sometimes used to describe the hydroxides. The main difference is that the hydroxides as we have described them will not have a water molecule in the formula.

Problem Set 2.10   Name the following:

$Na_2SO_4 \cdot 10H_2O \longrightarrow$

$MgSO_4 \cdot 9H_2O \longrightarrow$

$Na_2SO_4 \cdot 7H_2O \longrightarrow$

$BaCl_2 \cdot 2H_2O \longrightarrow$

$Na_2SO_4 \cdot H_2O \longrightarrow$

$CuSO_4 \cdot 5H_2O \longrightarrow$

NM + NM, no Carbon

Binary Nonsalts

Chemistry Quick Reference Card

Family Class	Naming	Hazards
Metal salts M + NM ø O	-ide	General hazards, with four being water reactive
Metal oxides M + oxygen	-oxide	Water reactivity, corrosive solutions, and oxidation
Inorganic peroxides M + O_2	Peroxide	Water reactivity, corrosiveness, and very strong oxidizers
Oxygenated inorganic compounds M + oxy ion	Per—ate -ate -ite Hypo—ite	Strong oxidizers can generate heat if wet
Metal hydroxides M + OH	Hydroxide	Corrosiveness and heat production during reaction
Inorganic acids H + polyNM	Per—ic -ic -ous Hypo—ous	Corrosiveness
Binary acids H + NM Hydrogen halides	Hydro- -ic	Corrosiveness
Hydrates Salt • H_2O	Name the salt then add the Greek numbering to "hydrate"	Strong affinity for water
Binary nonsalts **NM + NM ø C**	Name the first NM, then the number of the second NM, and change ending to -ide	Flammability, can react violently with water and oxidizers, irritation

The category of binary compounds can become confusing. Technically, the binaries are compounds made of two different elements. Our discussion so far has identified the binaries having a metal and nonmetal attachment. When two nonmetals, neither of which contains carbon or a polyatomic ion, combine, the result is a binary nonsalt. For these compounds, we name the atom that has the fewest number and combine this with the atom(s) that are contributing the most like atoms. The second name ending takes the suffix -ide, just like we did for the metal salts, along with the number utilizing the Greek method to identify the number of the second nonmetal's atoms. For example, NH_3 is nitrogen (the first nonmetal atom, which has the fewest numbers) and hydrogen (the second atom, which contributes the most number of atoms to the complex and must take the ending -ide). Because there are three hydrogens, we use the Greek nomenclature for three, which is tri-. Thus the compound is called nitrogen trihydride, usually referred to as ammonia. Most of the naming in this category follows common names, and the systematic naming system is not usually used. Other examples are:

> Iodine pentafluoride (IF_5), which is used in fluorination and as an incendiary agent
>
> Bromine pentafluoride (BrF_5), which is an oxidizer used in liquid rocket propellants
>
> Phosphorus heptasulfide (P_4S_7), which is flammable

● Caution
Hydrides release gas, making them potentially flammable and usually toxic.

These hydrides release gas, making them potentially flammable and often toxic. Depending on the configuration, water reactivity and extreme toxic values are also potential hazards of this classification.

Problem Set 2.11   Name the following:

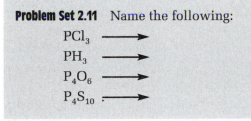

$PCl_3 \longrightarrow$

$PH_3 \longrightarrow$

$P_4O_6 \longrightarrow$

$P_4S_{10} \longrightarrow$

NM + O$_2$

Nonmetal Oxides

Chemistry Quick Reference Card

Family Class	Naming	Hazards
Metal salts M + NM ø O	-ide	General hazards, with four being water reactive
Metal oxides M + oxygen	-oxide	Water reactivity, corrosive solutions, and oxidation
Inorganic peroxides M + O$_2$	Peroxide	Water reactivity, corrosiveness, and very strong oxidizers

(continues)

Chemistry Quick Reference Card (Continued)

Family Class	Naming	Hazards
Oxygenated inorganic compounds M + oxy ion	Per—ate -ate -ite Hypo—ite	Strong oxidizers can generate heat if wet
Metal hydroxides M + OH	Hydroxide	Corrosiveness and heat production during reaction
Inorganic acids H + polyNM	Per—ic -ic -ous Hypo—ous	Corrosiveness
Binary acids H + NM Hydrogen halides	Hydro- -ic	Corrosiveness
Hydrates Salt • H_2O	Name the salt then add the Greek numbering to "hydrate"	Strong affinity for water
Binary nonsalts NM + NM Ø C	Name the first NM, then the number of the second NM, and change ending to -ide	Flammability, can react violently with water and oxidizers, irritation
Nonmetal oxides **NM + O$_x$**	Name the NM, then the oxide, using the Greek nomenclature for the number of each	Toxic gases mostly produced by fire

The nonmetal oxides are gases mostly produced by combustion and, as such, are referred to as the "fire gases." They include asphyxiants, irritants, and carcinogens. The bases for these gases come from the materials used in building construction and in technology-based industry, and from the natural fibers utilized in the home. They contain a nonmetal with oxygen. The name starts with the principle atom (not oxygen) and takes the suffix -oxide, to which is added the Greek number; for example:

Sulfur dioxide (SO_2), which is toxic by inhalation

Sulfur trioxide (SO_3), which forms sulfuric acid when in contact with moisture, such as humidity in air

The lethal effects of smoke have been recognized and utilized as far back as the first century. Romans used smoke of green wood to execute prisoners. Today, both fire and EMS responders frequently encounter victims of smoke inhalation. The chemicals that make up the fire gases (see Table 2–8) are the oldest hazardous materials emergency.

Table 2–8   *Some hazardous gases produced by the burning of various materials.*

Combustible Source	Gas Produced
Wool, nylon	Hydrogen cyanide
PVC, some fire-retardant materials	Hydrogen chloride, phosphorous pentaoxide
Resins, some fire-retardant materials, wood	Halogen acid gases, dioxin
Wool, silk, nylon	Ammonia
Cotton fabrics, cellulose	Nitrogen dioxide, dioxin
Polyolefins	Acrolin

Problem Set 2.12   Name the following:

$$N_2O_4 \longrightarrow$$
$$CO \longrightarrow$$
$$NO_2 \longrightarrow$$
$$CO_2 \longrightarrow$$

OTHER HAZARDOUS INORGANIC COMPOUNDS

We have purposely skated around ammonia, cyanide, and the azide compounds, all of which can be found in a variety of industrial processes. Ammonia is sometimes attached to another compound and called ammonium plus the name of the compound, as in ammonium bicarbonate, ammonium bifluoride, ammonium bisulfite, and ammonium bromide. Organic compounds utilizing the ammonia ion as a base are named similarly; for example, ammonium formate, ammonium oxalate, and ammonium acetate. In general, ammonium compounds are readily absorbed through the skin, mucous membranes, and respiratory tree. Once in the bloodstream, they combine with the hemoglobin to create a non-oxygen-carrying compound called methemoglobin. Additionally, the nitrogen component has the ability to relax the smooth muscle in the circulatory system causing vasodilatation, and thus lowered blood pressure.

The ammonium, cyanide, and azo compounds have distinct but similar toxicological effects. Ammonium compounds cause methemoglobinemia, which limits the transport of oxygen in the bloodstream, and the azo and cyanide compounds act similarly towards cellular metabolism. The process of metabolism in the cell is a very precise and organized production of energy, water, and heat. Each has to be produced in order for the cell to live. The process of electron transfer between the enzymes is called electron transport and is extremely important for cellular respiration. Cyanide and hydrogen sulfide attack electron transport by combining with the cytochrome oxidase and thus stopping cellular respiration. Azo compounds are thought to interfere with this process as well.

■ Note

The iron in the hemoglobin molecule has a valence of +2 (Fe^{++}). The nitrogen compound converts this Fe^{+2} into Fe^{+3}, or ferrous to ferric iron. This conversion of valences changes the hemoglobin's ability to carry oxygen, making it incapable of carrying oxygen. This condition is called methemoglobinemia.

■ Note

The cyanide antidote kit can be used to cause methemoglobinemia to pull the cyanide away from the cytochrome oxidase and maintain cellular respiration. There is no evidence that the cyanide antidote kit should be used for the azo and azide compounds.

■ Note

Amides, also belonging to this general group, are extremely reactive.

Cyanide utilizes organic chemistry nomenclature and the common name of prussic acid. Typically, cyanide has a compound attached, which is then named for the ion of cyanide (HCN = hydrogen cyanide). In some forms of cyanide, cyano- is the common reference to the cyanide ion.

Terrorist activity has been on the increase globally. The three United States strikes that received high media coverage were the bombings of the World Trade Center in New York in 1993, the Murrah Federal building in Oklahoma City in 1995, and the World Trade Center and the Pentagon in 2001. Each incident utilized a "homemade" version of an explosive device. At the World Trade Center, nitrourea was used in conjunction with hydrogen. Cyanide was placed in the bomb for dispersion and further effect. Nitrourea is a white powder that has high explosive potential and is approximately 34% more powerful than TNT, kilogram for kilogram. The cyanide was consumed in the reaction.

The Oklahoma City bomb was a mixture of ammonium nitrate and fuel oil (ANFO). Over-the-counter ammonium nitrate was mixed with ammonium sulfate or calcium carbonate, which decreases the explosive potential of the ammonium nitrate. During the decomposition reaction, the ammonium nitrate melts, especially at high temperature, releasing nitric acid vapor and water vapor. Within a closed container, the result is explosive.

In 2001, the device was the infrastructure of transportation. The use of fuel-laden airplanes served as the mechanism of delivery, causing catastrophic building failure.

Whereas both ammonium and cyanide are polyatomic ions that can satisfy the octet of several compounds, azides are a group of compounds that have inorganic molecule attachments, and in some cases, organic attachment. The azide group of compounds can have any metal, hydrogen, or halogen attached to the ammonium radical: $R(N)_x$, where R is the group attached to three atoms of nitrogen (or sometimes two, $-N=N-$, in which case the name then begins with azo-). The attachment may also have an organic compound such as a methyl, phenyl, or nitrophenol. These compounds are discussed in more detail in the next few chapters. This is a broad classification of compounds not found in nature. They are specifically man-made structures, primarily used in the manufacturing of dyes, although the plastics industry utilizes these compounds in other chemical operations, for example, as a blowing agent. They can also be used as a carrier gas for fumigants, for curing unsaturated polyester resins, to control polymerization reactions, or as propellants.

Being manufactured structures, they do not necessarily follow a recognizable pattern for naming. For example, Acetamine is Dupont's trademark for an azo dye. Acetate dye, acid dye, and amaranth dye are also examples of nontraditional naming of the azo (azide) compounds. However, hydrogen azide, lead azide, azoxybenzene, and azoxyethane are recognized as belonging to this particular group of dangerous compounds and are named utilizing conventional nomenclature.

The structures of these azide compounds are complicated and are beyond the scope of this book; however, their frequency of use in industry and the dangers that these chemicals pose warrants this brief discussion. The toxic effects of these compounds are similar to those of cyanide, and that the same process occurs. Both cyanide and azide compounds inhibit cytochrome oxidase within the cell, thus collapsing cellular metabolism. Additionally, some are light- or shock-sensitive, and under appropriate conditions such as fire, they exhibit explosive qualities. Heavy- and light-metal azides, hydrogen azide, and many of the organic azides can be light-, shock-, or heat-sensitive. (Sodium azide is used as the propellant in automobile air bags.)

COMBUSTION

combustion

the rapid oxidation of a material in the presence of an oxidizer

In Chapter 1 we introduced some basic terminology and concepts related to **combustion**. The fire cycle was briefly introduced in connection with the definitions and chemistry presented there. Now, because of the discussion of the salts (and future discussion of the organics), we can revisit the ideas of combustion with the knowledge we have attained thus far.

> ■ **Note**
>
> Flammable solids are of three types: wetted explosives, self-reactive substances, and readily combustible substances. DOT regulations also specify three categories: "flammable solid," "spontaneously combustible," and "dangerous when wet." "Dangerous when wet" is a description used for material that becomes spontaneously flammable or that evolves flammable or toxic vapors on interacting with water. "Spontaneously combustible" material is a self-heating or pyrophoric substance. Pyrophoric material is a solid or liquid that may ignite without an external ignition source within 5 minutes of coming into contact with the air, even in small quantities.

combustible

an ignitable and free-burning material

During the 1700s it was commonly believed that all material contained a substance called phlogiston. This weightless substance explained why there was a weight change when a **combustible** burned, or why fire would go out in an enclosed space and animals would die in a burning room (because the phlogiston would saturate the air). Although technically incorrect, it was a plausible theory for the time. As silly as it sounds, this combustion theory was years ahead of its time, and, in fact, the chemist who introduced these ideas was executed for them: the father of modern chemistry, Antoine Lavoisier.

> ■ **Note**
>
> As the son of a prominant lawyer in Paris during the late 1700s, Antoine Lavoisier was educated in the finest schools. His studies were broad but inclined toward the sciences. In 1765 he published a paper on how to improve the street lighting around Paris. By 1775 he was appointed to the National Gunpowder Commission. In 1794, because of his liberal views and ideas, he was arrested, put in prison, tried, convicted, and sentenced to execution, all in one day.
>
> ■ **Note**
>
> A combustible liquid has a flash point at or above 100°F, and a flammable liquid has a flash point below 100°F. Standards such as *NFPA 30* identify flammable and combustible liquids in terms of flash point and boiling point.
>
> ■ **Note**
>
> A flammable gas is any product that is a gas at 68°F or lower and a pressure of 14.7 psi that is ignitable at 14.7 psi when the mixture is 13% or less. It can also be the vapors of such a product possessing a flammable range of at least 12% regardless of the LEL. (Consider why ammonia is a nonflammable gas.)
>
> ■ **Note**
>
> Combustible metals—such as the alkali metals, alkaline earth metals, and transition metals, including titanium, magnesium, aluminium, zirconium, and zinc—burn in air and are denoted as Class D fires.

Lavoisier began his work on combustion in 1773, when similarities between respiration and combustion were noticed. Through a series of experiments, he noticed that proportions of air had an effect on the burning process. Additionally he observed that animals within a closed container would suffocate, thus loosing "air." Thanks to Lavoisier, the science of chemistry changed at this point from a qualitative science to a quantitative one.

oxidation

the chemical process or reaction in which there is a loss of electrons, or an increase in the **oxidation number**

oxidation number

a number assigned to an element's atom in a compound or the valence number in an ionic compound

Although the subject of combustion and **oxidation** is a field unto itself, a basic understanding is essential in the hazardous materials industry and especially for emergency responders. It is the knowledge of these concepts that enables identification of the level and type of mitigation (e.g., foam application, halon use, Met-L-X) at the scene of an emergency. In the field of fire protection engineering, this subject has received much study in order to make buildings safer and to make materials used in the home noncombustible.

In Chapter 1 we discussed activation energies, or the minimum amount of energy that is required for a chemical collision to take place. As it applies here, heat production from the reaction fuels the reaction to continue, while at the same time the proportional amount of oxygen is entering into the reaction (either from the environment or from additional chemicals placed in the reaction vessel). Both of these concepts relate to the velocity between the molecules. As these molecules bombard each other, free radicals are formed. Reactions between these chemical species are critical for many reactions. If inhibited, the reaction stops; if enhanced, as with a catalyst, the reaction becomes dynamic. When the hill is large, as with an inhibitor, the reaction slows over the hill, and with a small hill, the amount of energy required is minimal and the reaction occurs quickly.

Conditions of Combustion

We normally think of combustion as a heat-producing reaction, or fire. Sometimes we call this reaction oxidation. Both "combustion" and "oxidation" are used to describe an **exothermic** reaction (as opposed to **endothermic**, which absorbs heat). Technically there is a subtle difference between the two terms that depends on the chemical(s) undergoing the reaction. For example, we typically think in terms of a fire that is being fed by oxygen from an outside source, that is, the surrounding air. In this case, the chemical is undergoing a reaction in which two or more substances are uniting chemically. Heat is being liberated faster than the chemical can absorb it, which results in an increase in heat and the liberation of light and energy.

Oxidation is a reaction occurring slowly between a chemical and oxygen that releases heat, while at the same time absorbing that heat. The rusting of metal is an example of oxidation, or what is termed *slow oxidation*. The reason for the distinction between these two terms is a matter of semantics that has occurred in the emergency field. Rapid oxidation is combustion, whereas slow oxidation is rusting.

Combustion does not need to occur in the presence of oxygen. Mixtures of certain chemicals can produce extremely violent combustion in even an inert atmosphere under fire conditions. Chemicals such as the peroxides and oxy-compounds can give the reaction enough oxygen to sustain the combustion process because they have oxygen in the molecules. Other materials such as ozone, peroxides, the halogens, nitrates and nitrites, oxides of heavy metals, and a few of the transition metals have the ability to support combustion through inherent oxidative potentials in the chemical structure of their molecules.

Combustion is dependent on the substance's *state of matter*. Solids, for example, can absorb heat, making them difficult to ignite. Liquids also absorb heat, to a point. Beyond that point, they give off vapors. It is the gas phase that is the most significant. If the substance is a gas, depending on the appropriate level of mixing with oxygen, it may produce a **flammable** environment, especially if the substance has flammable qualities.

exothermic

a reaction that evolves heat

endothermic

a reaction that absorbs heat

flammable

capable of being easily ignited and susceptible to burning intensely

■ **Note**

Direct-reduced iron sponge (created by hot carbon monoxide treatment of iron ore briquettes offshore) becomes a catalyst for a hydrogen shift reaction in the presence of water. There have been at least five marine incidents involving this reaction in which a hot hold was identified (actually due to rain water leaking into a loose hold cover) and was subsequently flooded with seawater, resulting in a runaway reaction. The first two burned through the hull and resulted in sinking. The third, at Baltimore, was saved, as well as another in Spain a few months later, and a fifth in Delaware Bay three years later. The technique used was to immerse the ship in as much water as possible, including the space between the hold affected and the hull, without compromising the ability to stay afloat and to hot-unload the contents at a steel plant (heat was absorbed by surrounding material, which exceeded the heat generation) or to starve the system for water. The same reaction applies to "stump fires" and to many barn fires in which dirt or lack of air circulation starves the system for air. Green wood and wet hay produce biological decomposition with heat and carbon monoxide generation. Any contained dirt that includes iron filings becomes the catalyst, thus beginning the process. Starving these systems of water or quickly exposing them to open air will stop the reaction. Attempts to extinguish them using foam or direct water application only feeds the reaction. A "hydrogen shift reaction" requires carbon monoxide and water with a metallic catalyst of iron or nickel. Direct-reduced iron is actually an iron sponge with the microscopic voids filled with CO. If liquid water is introduced, a slow reaction begins converting the CO to CO_2 plus H_2. If the water is in the form of steam, the reaction is very fast. In both cases the equilibrium temperature is about 800°C, sufficient to flare off the hydrogen. This is why the Baltimore incident was nicknamed the "world's largest floating barbecue grill." In the case of the *Federal Rhine*, after stepping down the gangplank it was possible to look inside and see 9400 metric tons of glowing briquettes with blue flames licking through it.

Pyrolysis comes from the Greek words meaning "fire breakdown," and is the dividing of covalent bonds found in hydrocarbons. It occurs near the surface of the material but not directly on it. Combustion occurs at a distance at which the flammable range is reached through the heating of the material and mixing with air. This breaking of bonds is also called cracking, and is a term used by the oil refinery industry.

A solid has the ability to absorb heat. If the material is combustible, heat is absorbed until its ignition temperature is reached. At this point the material's molecules break down and produce flammable gases. These gases combine with oxygen to create a flammable range of the gas. For the reaction to continue, heat must be generated to feed the breakdown of more molecules, and thus more gas. If at some point the material is able to absorb heat, the gas will not be liberated and the fire will go out. However, if the heat from the source and the heat from the reaction sustain the burning process, the fire will continue to burn until all the material (fuel) has been consumed.

Achieving combustion for a liquid or gas is simpler. The liquid absorbs heat; the heating enables it to liberate vapor. When this vapor mixes with the appropriate concentration of oxygen and an ignition source is available, it will ignite. Obviously, then, for a gas all that has to occur is the appropriate mixture of product vapor and oxygen. A cloud of gas will extend from its source and produce a flashback if an ignition source is encountered.

Whatever the state of the matter, *surface area* plays an important role in the combustion of the materials. The faster a substance can absorb heat without dissipation, the quicker it reaches its ignition temperature. Therefore, the more finely divided the particles, the less heat is absorbed, and the more quickly heat increases toward the ignition temperature.

The heat energy can be a flame, spark, friction, sunlight, or internal movement of the molecules. This movement of the molecules requires collisions of the appropriate orientation (see "Rate of Reaction" under "Chemical Bonding" in Chapter 1). Orientation, collision frequency, and air-to-material ratio (flammable limits) must all be at the appropriate levels for a fuel to ignite. Once the gas ignites, continued reaction is sustained by the increased frequency of collisions between molecules. As a simple example, consider vapors from a flammable liquid, which for the purpose of this discussion are hydrocarbons (fuel) and oxygen in the air. Remember that air is made up of several gases: 78% nitrogen, 21% oxygen, and 1% inert gases. When the fuel (the hydrocarbon vapors) is released into the environment, molecular collisions occur between the hydrocarbon and nitrogen, the hydrocarbon and the inert gases, and the hydrocarbon and oxygen under flame combustion. The increasing temperature of the surrounding environment increases the movement of the air (oxygen), which thus increases the frequency of collision.

In addition to the temperature, oxygen potentials, and surface area (gases have a tremendous "surface area"), the thermal expansion of the gases must be considered. A gas will "grow" as it is heated. This concept also applies to solids and liquids to some degree. As material heats up, its strength diminishes, allowing it to expand due to convection, conduction, and radiant heat. For a hazardous materials incident, flame impingement can (1) fatigue the container, (2) cause the material in the container to heat and expand, and (3) cause the chemicals in the container to enter into reaction. Catastrophic failure of the container will ensue. Even without flame impingement, an internal chemical reaction can result in failure of the container due to the reaction causing material to expand. Conversely, an endothermic reaction (absorbing heat) can result in temperatures well below the intended operational strength of the container, causing implosion or collapse. It can also cause condensation (water accumulation), increased viscosity, or solidification. Another hazardous result may then be that a material that once was a liquid may be shock sensitive as a solid.

Halting Combustion

The combustion process requires temperatures above the ignition point in combination with chemical reactions. The keys to mitigating or extinguishing combustion are to reduce the input heat (cool the material), isolate the fuel (reduce the vapors produced), limit the oxygen (or the supportive chemicals of oxidation), and maintain the vapor and air ratio below the LEL (vapor limitation or dissipation) so that the reaction is stopped. For most materials (hydrocarbons), these results can be accomplished through the use of foam, carbon dioxide, dry chemical agents, halogenated agents, or wetting agents.

foam
a frothy firefighting–extinguishing agent that forms a barrier to limit the vapors being produced

Foam is a frothy firefighting–extinguishing agent that forms a barrier to limit the vapors being produced. There are two types of fire foams: chemical and mechanical. Chemical foam is produced from a mixture of chemicals. It is the older of the two types and generally not used in the fire service today. An example is bicarbonate of soda with aluminum sulphate, which was used during 1920–1950. That mixture forms carbon dioxide gas in the bubbles of aluminum hydrate, and protein hydolyzates.

Mechanical foam has the advantage of being a single substance as opposed to the two or three substances that would have to be mixed to produce chemical foam. Mechanical foam was used as early as 1904, but it was not widely accepted until the rotary foam pump was developed in 1929.

Mechanical foam has several classifications in use today. Aqueous film-forming foam (AFFF) releases a thin aqueous solution, forming a film barrier on the surface of the fuel. The foam has low viscosity and surface tension that allows it to spread over the fuel. Flashback resistance depends on the additives in the fuel. Polar solvents in the fuel have a tendency to break down the foam layer fairly rapidly. Film-forming fluoroprotein (FFFP) has the characteristics of AFFF and retains water in the foam. It is compatible with dry chemical usage, increasing its usefulness. Alcohol foams are used on polar solvent fuels. The most common is an alcohol-resistant polymeric AFFF. Alcohol foams are frequently called universal foams. On hydrocarbons without polar additives, a 3% rate is adequate; with polar solvents or additives to the fuel, a 6% rate is suggested. These foams can also be used to suppress vapor from flammable liquid spills to prevent ignition. Newer foams just on the market have flow concentration of 1–3%.

High-expansion foams have special application for suppression of a fire in a confined area. The expansion rate varies with percentage and atmospheric conditions; however, 1% and 3% with velocity through the foam generator are commonly used. The water drainage with this foam is extremely rapid, necessitating frequent applications. Application of this foam has been successful for shelter in-place applications. Foam applied to the fire room can protect the occupants of the building. It has also shown usefulness for insulating the shock wave from small explosive devices in a room.

The effectiveness of any foam is related to the chemistry of the fuel that is on fire. All foams have water drainage problems (water draining out of a foam film blanket eliminates effectiveness); the question is, will that drainage affect mitigation efforts? All foam works by removing the air space above the fuel, thus reducing the possibility of flammable limits being produced. Due to the application of water within a "soap" solution, cooling of the material occurs, but at a very slow rate. Evaluate your primary hazards before purchasing foam to ensure compatibility.

■ Note

Water drainage time is the amount of time it takes for water to drain out of a foam film blanket, eliminating the effectiveness of the foam application.

The extinguishing principle behind the use of carbon dioxide (CO_2) is to reduce the atmospheric oxygen below 15%, which is required for flame production. Carbon dioxide is colorless, odorless, inert, and approximately 50% heavier than air. It will not conduct electricity. The white cloud that is seen as CO_2 comes out of the extinguisher is due to the temperature difference between the air moisture and the temperature of the CO_2 ($-110°F$). Once discharged, the small particles of CO_2 evaporate extremely rapidly, thus giving CO_2 a minimal cooling effect on the material being extinguished. Incompatibilities are chemicals that are pyrophoric or are oxidizing agents, and reactive metals such as sodium, potassium, magnesium, titanium, zirconium, and the metal hydrides.

A **halogenated agent** is a halogen used in combination with methane or ethane to produce a nonflammable colorless gas that can be used as an extinguishing agent, or refrigerant. These agents are either liquids that vaporize or liquefied gases. Pound for pound, they are the most effective method of extinguishment. They work by interrupting free radical production during a fire.

The disadvantages of the halogenated agents are the displacement of air and confinement of the extinguishing agent such that the appropriate level of concentration is maintained. Also, some of the agents are toxic to humans and have an environmental impact. The halons are being replaced with more environmentally safe extinguishing agents called clean extinguishing agents (CEAs).

All **dry chemical** extinguishment works on basically the same principle: inhibition of the production of free radicals and heat during a fire. The action of a dry chemical depends on the velocity of the agent applied, which is why there

---

halogenated agent
halogen used in combination with methane or ethane to produce a nonflammable colorless gas that can be used as an extinguishing agent, or refrigerant

dry chemical
an extinguishing agent that works by inhibiting the production of free radicals and heat during a fire and blocks oxygen from the fire

are several different dry powder combinations; for example, potassium chloride, monoammonium phosphate, sodium bicarbonate, potassium bicarbonate, and Met-L-X, a foam-compatible dry chemical. Just as when using the foams and halogens, dry chemicals must blanket the entire area. Having the appropriate amount will determine the success of the operation.

Wetting agents are liquid "soap" concentrates that, when mixed with water, reduce the surface tension of the water, thus allowing for deeper penetration into the material. Water to which a wetting agent has been added is sometimes referred to as wet-water. As a detergent foam, its actions are similar to water in that it reduces the heat, stopping the pyrolysis of the material. Wetting agents are used for Class A and B materials, with the limiting factor on Class B being the polar solvents in the fuel.

wetting agents

liquid "soap" concentrates that, when mixed with water, reduce the surface tension of the water, thus allowing for deeper penetration into the material

RADIATION

Emergency responders are exposed to a variety of hazards, not the least of which is radiation. Radiation is all around us; the common forms seem not so threatening: sunshine and diagnostic medical procedures such as X-rays and medicines, for example. However, the topic of nuclear power strikes fear into the hearts of emergency responders and the public alike. But it's basically a fear of the unknown. In fact, radiological accidents are few. Other hazard classifications are responsible for larger and more frequent releases. As with anything else, it is increased understanding that reduces fears and provides the capability for handling these types of releases, while also educating us about when we should be concerned.

> ■ **Note**
> In addition to their radioactive qualities, metals at the bottom of the periodic table can undergo the same reactions that we have been discussing.

We bring up radiation at this point because most of the fuels and isotopes that are radioactive fall into the salt category. Most are metal salts, but a variety of compounds are used in the nuclear power industry. Even though the following discussion focuses on the radiological hazards, never forget that these compounds can undergo other chemical reactions as well and pose other hazards in addition to radiation.

The constituents of atomic components give rise to the subject and understanding of radiation. Recall that the atomic number found on the periodic table represents the number of protons, which in turn represents the number of electrons. Adding the number of protons and the number of neutrons gives the mass number (A). For example, uranium can be shown as $^{238}_{92}U$, or U-238, where the atomic number is 92 and the mass number is 238 (i.e., Z = 92 and A = 238).

> ■ **Note**
> Radioactivity is not changed if temperature or pressure is applied to the element under decay.

An element consists of a designated number of protons, neutrons, and electrons that make up the atom of that particular element. Atoms in their natural state try to achieve a level of stability. Due to electron configuration and the satisfaction of the octet rule, this level of stability can be achieved, so why does radiation, the releasing of electrons and nuclei, exist?

Table 2–9 *An example of the radioactive decay process toward stability.*

Isotope	Emission
U-238	Alpha
Th-234	Beta and gamma
Pa-234	Beta and gamma
U-234	Alpha
Th-230	Alpha
Ra-226	Alpha
Rn-222	Alpha
Po-218	Alpha
Pb-214	Beta and gamma
Bi-214	Beta and gamma
Po-214	Alpha
Pb-210	Alpha and beta
Bi-210	Alpha
Po-210	Alpha
Pb-206	Stable

beta particle

an electron (negatively charged) or positron (positively charged) emitted from a radioactive isotope during beta decay

alpha particle

the radiation particle consisting of two neutrons and two protons that is emitted from certain radioisotope decay

radioactive decay

the process by which an unstable nucleus releases particles or energy in order to become stable; sometimes referred to as radioactivity or spontaneous transmutation

half-life

the length of time for half the atoms in a radioactive substance to decay

curie

the outdated unit of measurement of radio-activity as compared to 1 gram of radium whereby 1 Ci = 3.700 × 10^{10} decays per second

becquerel

the unit of measurement for radioactivity equivalent to one disintegration of a nucleus per second

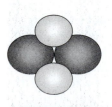

Figure 2–5 *An alpha particle contains two neutrons and two protons, hence the name helium nucleus.*

All elements have isotopes that have different numbers of neutrons and differing relationships to the number of protons in the nucleus. Hydrogen, for example, has three isotopes differing by the number of neutrons in the nucleus. These isotopes also try to achieve a more stable state. In doing so, they release an electron (**beta particle**) or a particle consisting of two neutrons and two protons (**alpha particle**), which we call a helium particle because helium has two protons and two neutrons when its electrons are stripped. In addition to the release of particles, energy is also released as a wave or package called a *photon*.

This whole process encompasses the terms *radioactivity, radioactive decay*, and *spontaneous transmutation*. The unstable nucleus releasing these items is said to be radioactive and the process is **radioactive decay** (see Table 2–9), which determines a substance's **half-life**. This disintegration is not a chemical or physical property. By losing neutrons, protons, and electrons to achieve a stable state, a reduction of the atom's components results, which produces a different element. The radioactivity of substances was formerly measured in **curies** (Ci) and is now measured in **becquerels** (Bq).

Alpha Particles

The nucleus of a radioactive element has excessive repulsion in it. In order to reduce this "pressure" from within the nucleus, a helium nucleus, or alpha particle consisting of two neutrons and two protons (see Figure 2–5), is released. This occurrence in effect reduces the atomic number by two and the mass number by four, and the element transforms or transmutates into another, usually radioactive, element.

Alpha particles can travel only approximately 4 inches from their source. These particles can be stopped by this page of paper and, therefore, cannot penetrate even light clothing. If an alpha particle came in contact with the skin, it could penetrate, but only the top layer. However, wind and air turbulence can keep alpha particles airborne, and thus they can be inhaled.

Figure 2–6 *The beta particle can be negatively or positively charged, and thus called the negatron or positron, respectively.*

Beta Particles

Beta particles can be positively or negatively charged (referred to as a negatron or positron) depending on the type of radiological emitter and the needs of the releasing and accepting atom (see Figure 2–6). These particles travel at higher speeds than the alpha particle and are of smaller configuration (basically, they are electrons). Most beta particles can travel about 30 feet to 100 feet from the source, depending on the air turbulence and environment. Penetration can occur up to half an inch into the skin, but is blocked by dense material or a thin layer of metal.

The beta particle is actually a neutron that has been released by the nucleus when the neutron-to-proton ratio is too great. The release of this neutron from an unstable nucleus transforms the nucleus into a proton and an electron. The proton is incorporated into the nucleus and the electron is released. This occurrence describes a negative beta particle release. With this type of beta release, the atomic number increases by one and the mass number remains essentially unchanged. The converse occurrence is the release of a positively charged electron or positron. Here the neutron-to-proton ratio is too small, and a proton converts into a neutron and a positive electron. With the positive beta release, the mass number remains the same but the atomic number is decreased by one. A third process that may occur is when the unstable nucleus of the element that is releasing beta radiation captures an electron from the environment. The extra-nuclear electron is then incorporated in a proton, thus forming a neutron. The atomic number decreases by one and the mass number remains the same. This process occurs because of the neutron-to-proton ratio being too small.

Gamma Radiation

gamma radiation
electromagnetic radiation of energy emitted from radioisotopes

roentgen (R)
the measurement of radioactivity for gamma radiation; it is the amount of ionization per cubic centimeter of air; it is a measurement of exposure and represents the amount of gamma radiation that produces two billion ion parts in dry air

Gamma radiation is not the result of a particle release, but rather of an emission of energy (ionization). Like X-rays, ultraviolet, infrared, and microwave radiations are electromagnetic energy (see Figure 2–7). Ultraviolet and infrared radiation have long wavelengths as compared to X-rays and gamma radiation, which have shorter wavelengths. Gamma rays are photons, released from a nucleus of a high level of energy falling to a lower energy. The atomic number or mass of the atom remains unchanged. Measured in **roentgens** (R), the energy is extremely strong and can penetrate most material.

During a nuclear fission reaction, a neutron is sent into the nucleus of uranium-235, which captures the neutron and becomes U-236. This compound splits apart into two fission fragments: two or three neutrons and gamma rays. The energy that is released is in the form of kinetic energy and is used in an electrical reactor to increase the temperature of water to create steam to produce electricity by means of a turbine.

However, in a nuclear age and with the threat of terrorism, governments are concerned about the growing level of nuclear waste. Fission fragments are produced in a fission reaction, and all will be radioactive. These fragments are typically cesium-137, strontium-90, and iodine-131. Enriched uranium has a certain percentage of U-238 and U-235. Uranium-238 does not go through the fission reaction, but U-235 does. When U-235 is used in the compound, the material is spent but radioactive. However, if U-238 captures a neutron, it is transformed into plutonium-239. Thus, nuclear waste has nonusable U-238, P-239, Cs-137, Sr-90, and I-131, to name a few. The problem is that the P-239 waste can be refined and the plutonium removed. This would shorten the half-life of the

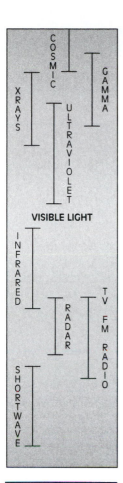

Figure 2–7
*Electromagnetic
energy spectra.*

RAD (radiation absorbed dose)
a unit, slightly greater than 1 roentgen, that is used to measure dosage

REM (roentgen equivalent man)
a measurement of biological effect, it represents the amount of absorbed radiation of any type that produces the same effect on the human body as 1 roentgen of gamma radiation

waste, but would concentrate the plutonium. Once this process is complete, it would not take much more work to make weapons-grade plutonium. This is one of the reasons for the purchase of radiological waste by the United States from other countries.

Radiation Hazard

The particles or energy released from a radioactive isotope have the potential to interfere with biological systems. This disruptive action can impair cell function in the human body, resulting in a variety of illnesses. The outward result of injury has to do with the amount of the radiation (measured in **RAD**s and **REM**s) absorbed during the time of exposure. For example, an acute exposure of 350 RADs kills 50% of the exposed population within 30 days (LD_{50-30}).

> **■ Note**
>
> A situation described as "ALARA" (as low as is reasonably achievable) means that every reasonable effort has been taken to limit radiation exposure to below the dose limit.

Each type of radiation has a different effect on biological tissue. What occurs is that a particle, such as an alpha or beta particle, changes the atoms or molecular biological compounds into ions, or the biological chemicals have a displacement of the normal atomic nuclei (gamma ionizes or breaks the chemical bonds). Additionally, the energy of motion of beta and gamma radiation heats the cells, causing the injury.

There are four basic safety factors in avoiding harm from radiation:

1. Time: Limit exposure. The shorter, the better.
2. Distance: Stay as far away as possible from the source. Remember that by doubling the distance from the source, exposure is decreased by a factor of four (inverse square law).
3. Quantity: Limit the amount of the exposed radioactive material.
4. Shielding: Put almost anything between yourself and the source.

> **■ Note**
>
> The inverse square law looks at a source of energy that spreads out in all directions equally. From a geometric standpoint it is the intensity at any given radius from the source. In other words, energy that is twice as far from the source will spread over an area that is four times the original area, or reflect ¼ the intensity at distance. We use the following to identify the radiation at distance, where R is the rad measurement and d is the distance:
>
> $$R/d^2 = \text{radiation at distance}$$
>
> so a package at 20 ft is expressed as $100 \ R/(20 \ ft)^2 = 0.25 \ R$
>
> This equation assumes that the first responder will walk up to the package and retrieve that information in order to make the calculation. However, a first responder can use this law to look at the radiation at distance and reverse the logic to get the radiation at the package when at a distance.
>
> The reverse inverse square law states that
>
> $$D^2 \times \text{Rad} = \text{rad at package,}$$
>
> so at 20 ft there is 0.25 R or $(20 \ ft)^2 \times 0.25 \ R = 100 \ R$.

Summary

Chemistry Quick Reference Card

Family Class	Naming	Hazards
Metal salts M + NM ø O	-ide	General hazards, with four being water reactive
Metal oxides M + oxygen	-oxide	Water reactivity, corrosive solutions, and oxidation
Inorganic peroxides M + O_2	Peroxide	Water reactivity, corrosiveness, and very strong oxidizers
Oxygenated inorganic compounds M + oxy ion	Per—ate -ate -ite Hypo—ite	Strong oxidizers can generate heat if wet
Metal hydroxides M + OH	Hydroxide	Corrosiveness and heat production during reaction
Inorganic acids H + polyNM	Per—ic -ic -ous Hypo—-ous	Corrosiveness
Binary acids H + NM Hydrogen halides	Hydro- -ic	Corrosiveness
Hydrates Salt • H_2O	Name the salt then add the Greek numbering to "hydrate"	Strong affinity for water
Binary nonsalt NM + NM ø C	Name the first NM, then the number of the second NM, and change ending to -ide	Flammability, can react violently with water and oxidizers, irritation
Nonmetal oxides NM + O_x	Name the NM, then the oxide, using the Greek nomenclature for the number of each	Toxic gases mostly produced by fire

Review Questions

1. Cross-valence the following compounds, determining their formulas:

NaCl

CoN

NiI

KS

MgO

AlO

SnO

PbO

CoI

HCl

FeO

CuBr

CaF

CuS

PbCl

AlS

MgCl

SnCl

BaO

HI

SnBr

2. Name the following -1 anions:

A. H

B. F

C. Cl

D. Br

E. I

3. Name the following -2 anions:

A. O

B. S

C. Se

4. Name the following -3 anion:

A. N

5. Name the following metal salts:

A. KBr

B. $MgBr_2$

C. RbI

D. NaBr

E. Al_2S_3

F. NaI

G. Cs_2S

H. Na_2S

I. K_2S

J. $AlCl_3$

K. Li_2S

L. MgF_2

M. SrS

N. Sr_3P_3

O. Ba_3N_2

P. AlP

Q. MgF_2

R. $BeBr_2$

S. KCl

6. Identify the type of ion (positive or negative) and the name of the following:

A. mercurous

B. cuprous

C. manganic

D. ferrous

E. stannic

F. cupric

G. mercuric

H. ferric

I. stannous

J. manganous

7. Name the following using both the Latin and systematic naming systems:

A. FeS

B. Fe_2S_3

C. K_2S

D. Cu_3N

E. Na_2S

F. $SnCl_2$

G. Cu_3P_2

H. CuI_2

I. SnS_2

J. CuF

K. $PbBr_2$

L. $SnCl_4$

M. $CuCl_2$

N. $PbBr_4$

O. Pb_3N_2

8. Give the chemical formulas for the following:

A. iron III sulfide

B. tin II chloride

C. potassium chloride

D. cuprous fluoride

E. calcium bromide

F. barium chloride

G. copper II chloride

H. ferrous chloride

I. lead IV bromide

J. gold III chloride

K. stannic phosphide

L. ferric sulfide

M. mercurous fluoride

N. plumbous nitride

O. copper II phosphide

P. sodium fluoride

Q. potassium nitride

R. cesium sulfide

S. potassium sulfide

T. tin IV chloride

U. calcium chloride

V. copper I nitride

W. lithium sulfide

X. strontium fluoride

Y. magnesium fluoride

9. Name the following oxides:

A. MgO

B. HgO

C. SnO

D. K_2O

E. PbO

F. Fe_2O_3

G. Na_2O

H. Al_2O_3

I. CoO

J. PbO_2

K. CaO

L. CsO

10. Give the chemical formulas for the following:

A. ferrous oxide

B. plumbic oxide

C. stannous oxide

D. aluminum oxide

E. mercury II oxide

F. cobaltous oxide

G. tin II oxide

H. lead oxide

I. iron III oxide

J. mercuric oxide

11. Name the following:

A. MgO_2

B. H_2O_2

C. Hg_2O_2

D. K_2O_2

12. A covalent bond between two atoms can be best described as:

A. An atom of low electronegativity and high ionization

B. An atom of high electronegativity and low ionization

C. An atom of low electronegativity and low ionization

D. An atom of high electronegativity and high ionization

13. Name the following inorganic -1 ions:

A. NO_2

B. ClO_2

14. Name the following -2 inorganic ion:

A. SO_3

15. Name the following oxygenated -1 ions:

A. NO_3

B. ClO_3

16. Name the following oxygenated -2 ions:

A. SO_4

B. CrO_4

C. CO_3

17. Name the following -3 oxygenated ions:

A. PO_4

B. BO_3

18. Name the following -1 ions:

A. ClO_4

B. ClO_3

C. ClO_2

D. ClO

19. Name the following -2 ions:

A. CO_4

B. CO_3

C. CO_2

D. CO

20. Name the following -3 radicals:

A. PO_5

B. PO_4

C. PO_3

D. PO_2

21. Name the following compounds and write out the complete formulas:

	Iodate	Chlorate	Bromate	Carbonate	Phosphate
Sodium					
Calcium					
Copper II					
Lead					
Potassium					
Mercury II					

22. For the oxygenated inorganic compounds in Question 21, write out all four configurations with all the possible naming conventions.

23. The acid–base reaction can be described as which three of the following molecules in solution?

A. water, H_2O

B. hydronium ion, H_3O^+

C. hydroxide ion, OH^-

D. hydrogen ion, H

24. What are the ranges of pH?

A. 1–7 acidity; 7.1–14 alkalinity

B. 0–6.9 acidity; 7.1–14 alkalinity

C. 0–7 acidity; 7–14 alkalinity

D. 0.6.8 acidity; 7.2–14 alkalinity

25. Name the following:

A. $Pb(OH)_4$

B. $Hg_2(OH)_2$

C. AgOH

D. $Mg(OH)_2$

E. KOH

F. $Ba(OH)_2$

G. $Ca(OH)_2$

H. NaOH

I. $Al(OH)_3$

J. $Cu(OH)_2$

K. $Sr(OH)_2$

L. LiOH

26. Write the formula that would identify the following compounds:

A. cupric hydroxide

B. strontium hydroxide

C. mercuric hydroxide

D. magnesium hydroxide

E. sodium hydroxide

F. potassium hydroxide

G. lithium hydroxide

H. lead IV hydroxide

27. Write out the four configurations of the following ions utilizing the inorganic acid nomenclature:

A. chlorate

B. carbonate

C. sulfate

D. borate

28. Name the following:

A. HF

B. HBr

C. HCl

D. HI

29. Name the following acids:

A. H_3PO_4

B. HNO_2

C. HIO_3

D. H_2CO_3

E. HCN

F. H_3PO_3

G. $HClO_3$

H. HI

30. Give the chemical formulas of the following:

A. nitrous acid

B. perchloric acid

C. carbonic acid

D. sulfurous acid

E. hydrocyanic acid

F. hydrobromic acid

31. Name the following compounds and identify the hazard group to which they belong:

A. $CuSO_4 \cdot 5H_2O$
B. BrF_5
C. PH_3
D. P_4S_7
E. NO
F. N_2O_5
G. $BaCl_2 \cdot 2H_2O$
H. IF_5
I. PCl_3
J. CO_2
K. ClO_2
L. P_2O_5
M. $MgSO_4 \cdot 9H_2O$

32. Identify the following compounds by writing out the formula and the hazard group to which they belong:

A. cuprous nitride
B. sulfurous acid

C. iron III oxide
D. potassium sulfide
E. lead peroxide
F. nitrous acid
G. sodium oxide
H. sodium hydride
I. lead IV bromide
J. mercuric oxide
K. barium hydroxide
L. barium carbonite
M. calcium nitride
N. sodium hypochorite
O. cupric chloride
P. diphosphorous pentoxide
Q. hydrochloric acid
R. sodium iodide
S. calcium periodate
T. copper II hypoborite

Problem Set Answers

Problem Set 2.1

$MgCl_2$	K_2S	BaO	LiF	$AlCl_3$

Problem Set 2.2

$Mg^{+2} + S^{-2}$	MgS	magnesium sulfide	$Al^{+3} + Cl^{-1}$	$AlCl_3$	aluminum chloride
$Ra^{+2} + Cl^{-1}$	$RaCl_2$	radium chloride	$Ca^{+2} + N^{-3}$	Ca_3N_2	calcium nitride
$Na^{+1} + Cl^{-1}$	NaCl	sodium chloride	$Na^{+1} + F^{-1}$	NaF	sodium fluoride
$Ba^{+2} + N^{-3}$	Ba_3N_2	barium nitride	$Al^{+3} + P^{-3}$	AlP	aluminum phosphide

Problem Set 2.3

$Cu^{+2} + Cl_2^{-1}$	$CuCl_2$	copper II chloride	cupric chloride
$Fe^{+2} + Cl_2^{-1}$	$FeCl_2$	iron II chloride	ferrous chloride
$Pb^{+2} + Br_2^{-1}$	$PbBr_2$	lead II bromide	plumbous bromide
$Pb_3^{+2} + N_2^{-3}$	Pb_3N_2	lead II nitride	plumbous nitride
$Sn_3^{+4} + P_4^{-3}$	Sn_3P_4	tin IV phosphide	stannic phosphide
$Fe^{+3} + Cl_3^{-1}$	$FeCl_3$	iron III chloride	ferric chloride
$Cu_2^{+1} + S^{-2}$	Cu_2S	copper I sulfide	cuprous sulfide

Problem Set 2.4

Be + O	BeO	beryllium oxide	Na + O	Na_2O	sodium oxide
Mg + O	MgO	magnesium oxide	Fe + O	FeO	iron oxide
Al + O	Al_2O_3	aluminum oxide	As + O	As_2O_5	arsenic oxide
Ca + O	CaO	calcium oxide			

Problem Set 2.5

$K + O_2$	K_2O_2	potassium peroxide	$Mn + O_2$	MnO_2	manganese peroxide
$Na + O_2$	Na_2O_2	sodium peroxide	$Mg + O_2$	MgO_2	magnesium peroxide
$Ba + O_2$	BaO_2	barium peroxide	$Ca + O_2$	CaO_2	calcium peroxide
$Cs + O_2$	Cs_2O_4	cesium peroxide			

Problem Set 2.6

Ion	Ion Name	One More Per—ate	Normal State -ate	One Less -ite	Two Less Hypo—ite
ClO_3	Chlorate	ClO_4	ClO_3	ClO_2	ClO
BrO_3	Bromate	BrO_4	BrO_3	BrO_2	BrO
IO_3	Iodate	IO_4	IO_3	IO_2	IO
NO_3	Nitrate	XXXXXXX	NO_3	NO_2	NO
MnO_3	Manganate	MnO_4	MnO_3	MnO_2	MnO
CO_3	Carbonate	XXXXXXX	CO_3	CO_2	CO
SO_4	Sulfate	XXXXXXX	SO_4	SO_3	SO_2
PO_4	Phosphate	XXXXXXX	PO_4	PO_3	PO_2
BO_3	Borate	BO_4	BO_3	BO_2	BO
AsO_4	Arsenate	XXXXXXX	AsO_4	AsO_3	AsO_2

Problem Set 2.7

sodium hydroxide lithium hydroxide	calcium hydroxide potassium hydroxide
lead IV hydroxide (plumbic hydroxide)	magnesium hydroxide
iron III hydroxide (ferric hydroxide)	ammonium hydroxide

Problem Set 2.8

phosphoric acid	carbonic acid	chloric acid	iodic acid	sulfurous acid

Problem Set 2.9

hydrobromic acid	hydrofluoric acid	hydrochloric acid	hydroiodic acid

Problem Set 2.10

sodium sulfate decahydrate	magnesium sulfate nonahydrate
sodium sulfate heptahydrate	barium chloride dihydrate
sodium sulfate monohydrate	copper II sulfate petahydrate

Problem Set 2.11

phosphorous trichloride	phosphorous trihydrate
phosphorous decasulfide	phosphorous hexaoxide

Problem Set 2.12

nitrogen tetroxide (dinitrogen teroxide)	nitrogen dioxide
carbon monoxide	carbon dioxide

Chapter 3

Organic Compounds I
Organic Structure

Learning Objectives

Upon completion of this chapter, you should be able to:

- Explain the differences in bond types.
- Explain the basic principles of bonding and the application to nomenclature.
- Duplicate the alkanes nomenclature.
- Duplicate the alkenes nomenclature.
- Duplicate the alkynes nomenclature.
- Explain the isometric nomenclature.
- Describe the *cis-* and *trans-* configurations.
- Indicate the difference between cyclo compounds and benzene.
- Identify the gas laws and the relationship to critical temperature and pressure.
- Reproduce the effects that branching has on a compound.

Figure 3–1 *The ability of the outer orbital of carbon to utilize the electrons in half-full or half-empty configurations leads to the uniqueness of organic chemistry.*

Organic chemistry, the largest branch of chemistry, deals with over 6 million compounds that contain carbon. Most of the chemicals we interact with on a daily basis are organic compounds. This vast number of compounds is possible because of carbon's ability to bond with other carbon atoms and a variety of other nonmetals. The carbon atom has four valence electrons. It is this configuration of a half-full or half-empty outer orbital that gives carbon the unique characteristics that we are about to explore (see Figure 3–1).

> **■ Note**
>
> The qualifier "organic" in organic chemistry comes from the word "organism." In the early 1800s scientists were convinced that the synthesis of organic compounds from inorganic compounds was impossible. It was believed that living organisms were necessary to produce organic compounds, because up to this time all known organics came only from organisms or their remains.

PRINCIPLES OF BONDING

Hydrocarbons are the branch of organic chemistry in which the chemistry of fire and hazmat incidents is best represented, for it is these materials that are the most commonly encountered in situations involving flammability, combustibility, and accidental releases.

In hydrocarbons, the carbon–carbon bond exists with hydrogen attached around the center carbon or along the chain of carbons. Inorganic compounds tend to transfer electrons in order to manage an ionic bond, and even when covalent bonds occur, the covalent complex has overall charges that function in an ionic fashion. Organic compounds encompass a large variety of covalently bonded complexes, which have the unique capacity to bond with the nonmetals. In the organics, a carbon–carbon configuration can form ever-larger molecules (see Figure 3–2). The size, polarity shift, and weight of these molecules have distinct effects on chemical and physical properties such as melting points, boiling points, and vapor pressure.

> **● Caution**
>
> Covalently bonded compounds may have one or more of the following qualities: ☠ flammability, ☠ corrosiveness, and ☠ explosiveness. In addition, they may be ☠ irritants, ☠ oxidizers, and ☠ poisonous.

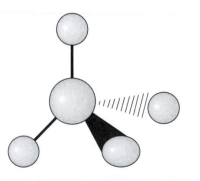

Figure 3–2 *The geometry of the covalent structure involving carbon is a tetrahedron, with each hydrogen or bond at the corners of such a tetrahedron. This configuration enables the structure to react with other compounds, thus forming new structures all with the basic configuration. The orientation of the molecule and the structure of the bonding can influence the reactivity of the molecule due to the polarity shift, weight of the molecule, and overall size.*

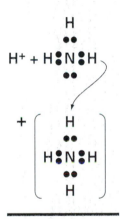

Figure 3–3
Hydrogen ions and ammonia react to form coordinate covalent bonds.

In the inorganics, namely in inorganic acids and bases, the solution has a water component. It is this water component that contributes to the bonding activity and thus affects the physical properties. Hydrogen bonding, which is an intermolecular attraction of polarity, causes an increase in the boiling point, density, and surface tension of many liquids, water included. However, some of the hydrocarbons cannot form this intermolecular attraction and, therefore, we see in general low boiling and melting points and densities (organic acids and alcohols have hydrogen bonding that increases the boiling point). The boiling and melting point of many compounds is a function of molecular weight: the greater the weight, the higher the boiling and melting points.

In addition to covalent and some hydrogen bonding, we will see compounds that have an overall charge, but the bonds will be covalent (see Figure 3–3). It is the ability of some compounds to produce a coordinate covalent bond that results in an overall charge to the complex by adding or subtracting a proton (hydrogen atom with its electrons stripped away) as in acids and bases. Coordinate covalent bonds cannot be distinguished from ordinary covalent bonds; however, they do explain why some compounds have the ability to form without any apparent contribution to the bond. It is this bond type that explains the fact that the hydronium ion is produced when an acid dissolves in water.

Polar covalent bonds occur more frequently in the organics than in the inorganics (see Figure 3–4). The molecule has a positive end and negative end and acts very much like a tiny magnet. These bonds are created by the size of each respective atom in the molecule and the distribution of electrons within the entire complex. In other words, the electron cloud about the nucleus lies, on the average,

Figure 3–4 *It is the orientation in space that gives the chemical configuration a positive and negative end. In this example, ammonia has a negative area due to the unshared pair of electrons. A polar covalent bond is formed.*

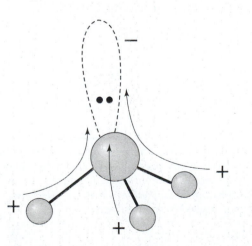

to one end, creating a predominantly negative charge and exposing the other side of the molecule, which in turn becomes predominantly positive.

Problem Set 3.1　Identify the following compounds as coordinate covalent, polar covalent, or resonance structures. (If necessary, refer back to Chapter 1 concerning resonance structures.) Draw each structure.

H_2O

CH_3

NH_3

CO_3

STRUCTURE

The structure of molecules in organic chemistry is very precise. The orientation in space in the molecule is important in connection with reactions. This orientation and the movement of the molecule in space will affect reactions, and thus the outcomes at the scene of a hazardous materials spill. Molecular orientation and the activation energies become more important when exposure to a chemical occurs. The chemical sometimes changes the body's metabolism, producing a variety of potential outcomes. The prediction is difficult at best; however, one should realize that the structure of the compound in reaction produces the exposure outcome.

■ Note

When referencing chemicals, it is important to notice naming nomenclature, such as in the enantiomers.

■ Note

The compounds making up an enantiomer can each have different boiling points, melting points, densities, and overall chemical characteristics. They are identical in structure, but they cannot be superimposed due to their mirror-image quality.

---

enantiomers
a pair of chemical compounds having molecular structures that are mirror images of each other

　　　　The exact orientation and movement of a molecule has an effect on the refraction of light waves. Therefore, we can test this orientation by using light. Consider, for example, **enantiomers** (see Figure 3–5). Under certain optically active conditions, the molecule will rotate to the right or to the left. If a molecule moves to the right, or in a clockwise direction, the + sign is used, and if the rotation is to the left, or in a counterclockwise direction, the − sign is used. If we have a mixture

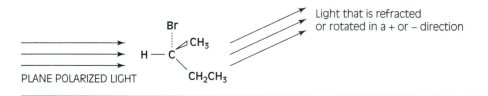

Figure 3–5　*The enantiomer 2-bromobutane moves the light clockwise or counterclockwise. The clockwise direction is prefixed by the + sign and is referred to as dextrorotatory. The counterclockwise direction is prefixed by the − sign and is referred to as levorotatory. Nomenclature normally will utilize the − or + sign, but may also describe direction by the respective names.*

racemic mixture

an optically inactive mixture of exactly equal amounts of two enantiomers

of both the right and the left enantiomer, we call the mixture a **racemic mixture** and use the $\pm$ sign. The name designates the optical shift by using $-$, $+$, or $\pm$.

It is not necessary to understand the physics of the optically active compound, but rather the reason for the designation of right or left. These designations can influence the decision-making process. Without adequate understanding, important clues in researching these chemicals may be overlooked.

ORGANIC NOMENCLATURE

saturated hydrocarbons

hydrocarbons with only single bonds between the carbons:

Alkanes: C_nH_{2n+2}
Alkenes: C_nH_{2n}
Alkynes: C_nH_{2n-2}
Aromatics: C_nH_n

unsaturated hydrocarbons

molecules that have a carbon chain with at least one double or triple bond (some unsaturated hydrocarbons may polymerize)

If we have a chain of carbons with single bonds between the carbons, we call these compounds alkanes, or **saturated hydrocarbons**. They are referred to as "saturated" because at every available point where a hydrogen can bond there is a carbon–hydrogen attachment; thus, the compound is saturated with hydrogen. There are cases when not all of the points of attachment involve a hydrogen molecule. It is the number of carbons multiplied by two, plus two, that gives the number of hydrogen atoms that can attach to the molecule (C_nH_{2n+2}).

At times, a double bond exists in the carbon–carbon chain. Compounds that have at least one double bond between the carbons are called alkenes, or **unsaturated hydrocarbons**. In other words, a full saturation of hydrogen is not possible with these compounds. The two electrons that would normally share with the hydrogen are bound with another bond to the attached carbon. Compounds having at least one triple bond in the carbon–carbon chain are called alkynes, and they are also unsaturated hydrocarbons. Saturation is dictated by the carbon–carbon bonds and their attachment to hydrogen, because these bonds are responsible for the single hydrogen–carbon single bond.

> ■ **Note**
> In hydrocarbon nomenclature, the suffix indicates whether it contains a single, double, or triple bond; the prefix denotes the number of carbons in the chain.

Therefore, if a chemical has the suffix -ane, we know it is a single-bonded hydrocarbon. If the suffix is -ene, it contains at least one double bond, and if -yne, a triple bond. In the nomenclature of organic chemistry, the number of carbon–carbon bonds is also important. Each carbon has a name when in a chain. This name or prefix designates the number of carbons in the compound. For every attached carbon, a different prefix is used, as follows:

Number of Carbons	Corresponding Prefix
One	methyl-
Two	ethyl-
Three	propyl-
Four	butyl-
Five	pentyl-
Six	hexyl-
Seven	heptyl-
Eight	octyl-
Nine	nonyl-
Ten	decyl-
Eleven	undecyl-
Twelve	dodecyl-

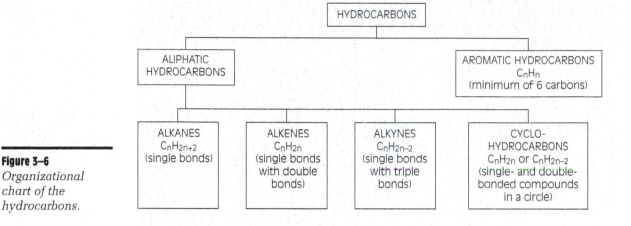

Figure 3–6

Organizational chart of the hydrocarbons.

Figure 3–6 provides an overview of the hydrocarbon groups.

To figure out which class of organics you may have encountered, use the following formulas: alkanes (C_nH_{2n+2})—if you have a four-carbon chain, $2 \times 4 + 2 = 10$ or C_4H_{10}; alkenes (C_nH_{2n})—if you have a five-carbon chain, $2 \times 5 = 10$, or C_5H_{10}; alkynes (C_nH_{2n-2})—if you have a six-carbon chain, $2 \times 6 - 2 = 10$, or C_6H_{10}; aromatics (C_nH_n)—if you have a six-carbon group, $n = 6$, so C_6H_6.

Alkanes

Chemistry Quick Reference Card

Family Class	Naming	Hazards
Alkanes C_nH_{2n+2}	Single bond, prefix depends on number of carbons, suffix -ane	Flammability Boiling point of the molecule increases 20–30 degrees for each carbon added to the chain Nonpolar

aliphatic hydrocarbons
hydrocarbons whose molecules are *not* composed of benzene or benzene-like structures

alkane
a hydrocarbon composed of carbon–carbon single bonds

The simplest of the **aliphatic hydrocarbons** are the **alkanes**. The names of all such compounds end with -ane. Each respective compound adds a carbon to the chain for the next respective compound. Because each carbon can have four single bonds (four electrons in the outer orbital), each carbon added represents one carbon and two hydrogens also added to the molecule in order to satisfy the octet rule. The first four members of this group utilize the common name rather than the Greek prefix because these names are an accepted naming configuration; that is, methane should be methyl hydride, ethane should be dimethyl, propane should be dimethyl methane, and butane should be dimethyl ethane (see Table 3–1). However, the common name is utilized. We can call this series an analogous event, in that each consecutive compound differs only by a $-CH_2$. This series is sometimes referred to as the paraffin series because the first solid is paraffin, or candle wax.

The first ten compounds in the alkane series represent the majority of the flammable gases and liquids encountered at hazmat incidents (see Table 3–2). Therefore, this discussion focuses on these lower molecular-weighted compounds and does not cover anything above ten carbons. The general formula for the alkanes is C_nH_{2n+2}, where the n stands for the number of carbons in the compound. Utilizing this formula you can see that the number of corresponding hydrogens becomes twice the number of carbon atoms, resulting in a saturated compound. If a hydrogen is pulled from these compounds, we end up with a

Table 3–1 *Alkanes.*

Name	Formula	Structure Formula
Methane	CH_4	CH_4
Ethane	C_2H_6	CH_3CH_3
Propane	C_3H_8	$CH_3CH_2CH_3$
Butane	C_4H_{10}	$CH_3CH_2CH_2CH_3$
Pentane	C_5H_{12}	$CH_3CH_2CH_2CH_2CH_3$
Hexane	C_6H_{14}	$CH_3CH_2CH_2CH_2CH_2CH_3$
Heptane	C_7H_{16}	$CH_3CH_2CH_2CH_2CH_2CH_2CH_3$
Octane	C_8H_{18}	$CH_3CH_2CH_2CH_2CH_2CH_2CH_2CH_3$
Nonane	C_9H_{20}	$CH_3CH_2CH_2CH_2CH_2CH_2CH_2CH_2CH_3$
Decane	$C_{10}H_{22}$	$CH_3CH_2CH_2CH_2CH_2CH_2CH_2CH_2CH_2CH_3$
Undecane	$C_{11}H_{24}$	$CH_3CH_2CH_2CH_2CH_2CH_2CH_2CH_2CH_2CH_2CH_3$
Dodecane	$C_{12}H_{26}$	$CH_3CH_2CH_2CH_2CH_2CH_2CH_2CH_2CH_2CH_2CH_2CH_3$

Table 3–2 *Representative alkanes.*

Name	Formula	Condensed Formula	Molecular Weight	Melting Point (°C)	Boiling Point (°C)
Methane	CH_4	CH_4	16.04	−182.5	−164.0
Ethane	C_2H_6	CH_3CH_3	30.07	−183.0	−88.6
Propane	C_3H_8	$CH_3CH_2CH_3$	44.09	−187.7	−42.5
Butane	C_4H_{10}	$CH_3(CH_2)_2CH_3$	58.12	−138.0	−0.5
Pentane	C_5H_{12}	$CH_3(CH_2)_3CH_3$	72.15	−129.7	36.1
Hexane	C_6H_{14}	$CH_3(CH_2)_4CH_3$	86.17	−95.3	68.7
Heptane	C_7H_{16}	$CH_3(CH_2)_5CH_3$	100.20	−90.7	98.4
Octane	C_8H_{18}	$CH_3(CH_2)_6CH_3$	114.22	−56.8	125.6
Nonane	C_9H_{20}	$CH_3(CH_2)_7CH_3$	128.30	−53.3	151.0

alkyl group

the group of atoms remaining after a hydrogen atom is removed from a molecular structure such as an alkane

corresponding **alkyl group**. For example, CH_3- is an alkyl group, methyl, CH_3CH_2-, is an ethyl alkyl group, and so forth.

Problem Set 3.2   Name the following alkyl groups:

CH_3-

$CH_3(CH_2)_5CH_2$-

C_7H_{15}-

CH_3CH_2-

C_5H_{11}-

$CH_3CH_2CH_2CH_2CH_2CH_2CH_2$-

$CH_3(CH_2)_4CH_2$

C_8H_{17}-

$CH_3(CH_2)_2CH_2$-

$C_{10}H_{21}$-

$CH_3CH_2CH_2CH_2CH_2CH_2CH_2CH_2CH_2$-

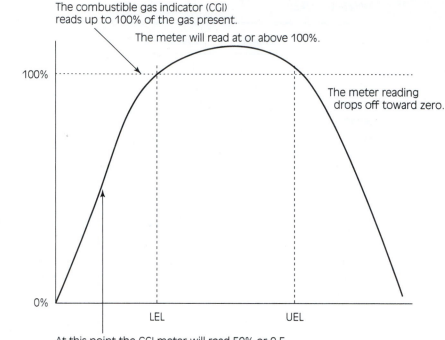

The combustible gas indicator (CGI)
reads up to 100% of the gas present.

The meter will read at or above 100%.

100%

The meter reading
drops off toward zero.

0%

LEL   UEL

Figure 3–7 *Due to the physical chemistry and the limitations of the meter, understanding how a specific meter responds in certain environments is critical.*

At this point the CGI meter will read 50% or 0.5.
If the LEL is 5%, the meter will read 2.5 or 0.025
depending on the meter's reading ability.

Problem Set 3.3   Draw the following compounds:

Heptane

Pentane

Butane

Undecane

Propane

Hexane

Ethane

Octane

Nonane

Methane

■ Note
The airborne concentrations 1300 rule gives us the ability to calculate a theoretical concentration that might occur above the product or within a confined space. These theoretical calculations use vapor pressure at a standard temperature of 68°F. Multiplying the vapor pressure of the compound in question by 1300 gives the theoretical vapor concentration in ppm this only works for vaporizing liquids.

To organize and simplify reading and writing about organic compounds, the carbons are classified as 1, 2, 3, 4, and so on, and as primary, secondary, tertiary, and quaternary carbons (see Figure 3–8). The primary carbon is defined as a carbon atom that has only one bond to another carbon atom, and the secondary carbon has two bonds each to another carbon atom. (We will see shortly that the

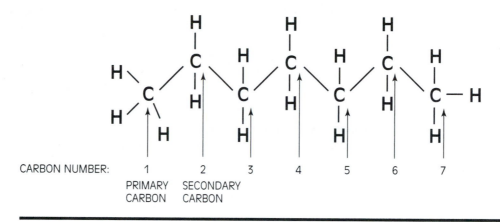

CARBON NUMBER: 1   2   3   4   5   6   7

PRIMARY   SECONDARY
CARBON   CARBON

Figure 3–8   *The naming of organic compounds starts with the carbon having the lowest carbon number in the chain. This will become a rule as we progress through the organic compounds. Each carbon has a number that identifies the prefix utilized to describe the particular hydrocarbon. The use of primary, secondary, tertiary, and quaternary is reserved for isomers.*

tertiary carbon is attached to three other carbon atoms and a quaternary carbon is attached to four other carbon atoms.) Figure 3–8 shows the carbons offset because of the bonding that is occurring with carbon–carbon bonds. Each bond is at a specific angle, allowing the hydrogen (and other elements) to attach themselves to the carbon chain. Boiling and melting points, as well as other physical properties, are affected by the configuration that each consecutive carbon (alkyl group) has in relation to the carbon of attachment.

■ Note

Wastewater treatment plants have learned to use the off-gassing of light-weight alkanes (predominantly methane) to run small generators and pumps. In these facilities, alkane, which is predominately methane, is captured and stored for future use. Depending on the level of methane produced through the water treatment process, excess gas is burned off through flaring.

Problem Set 3.4   Identify the primary and secondary carbons in the following compounds:

Heptane

Pentane

Butane

Undecane

Propane

Hexane

Ethane

Octane

Nonane

The progression of boiling points and melting points is fairly predictable. The first four alkanes are gases, the next twenty are liquids, and at the twenty-first the alkanes become solids at room temperature. The boiling point increases along the series, as does the melting point, as molecular weight increases. The melting points for the odd-numbered alkanes are slightly lower than for the even-numbered ones. If plotted on graph paper, this difference is shown to be due to the zigzag configuration illustrated in Figure 3–8.

The alkanes are insoluble in water due to the inability to form hydrogen bonds. The liquid alkanes are soluble with each other and with low-polarity compounds such as benzene and carbon tetrachloride.

Isomers. In organic chemistry, many different compounds exist in the same basic chemical family. Each compound has different chemical and physical properties but has the same number of carbons and hydrogens. These compounds having the same molecular formula but different structure and properties are called isomers, or **structural isomers**.

Butane provides the first isomer possibility of the alkane series: isobutane. In Figure 3–9 we can see two structures that have the same number of atoms but are different in structural configuration. Even though they have the same molecular weights, the difference in configuration results in different melting and boiling points and ignition temperatures (see Table 3–3).

structural isomers

compounds with the same molecular formula but different structure, chemical properties, and physical properties

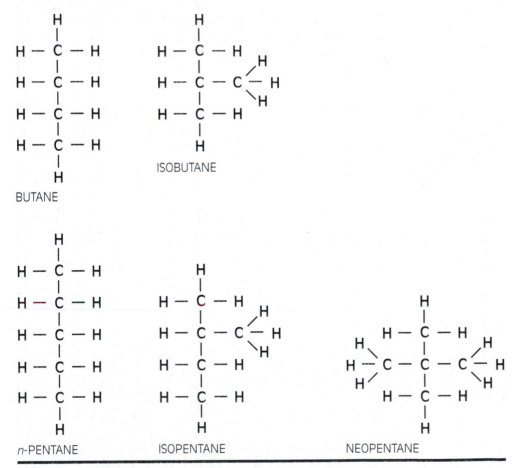

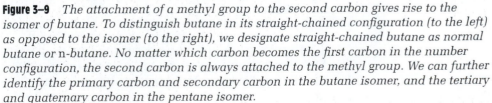

Figure 3–9   *The attachment of a methyl group to the second carbon gives rise to the isomer of butane. To distinguish butane in its straight-chained configuration (to the left) as opposed to the isomer (to the right), we designate straight-chained butane as normal butane or n-butane. No matter which carbon becomes the first carbon in the number configuration, the second carbon is always attached to the methyl group. We can further identify the primary carbon and secondary carbon in the butane isomer, and the tertiary and quaternary carbon in the pentane isomer.*

Table 3–3   *Comparison of butane and its isomer, isobutane.*

	Butane	Isobutane
Formula	C_4H_{10}	C_4H_{10}
Molecular weight	58	58
Boiling point (°F)	31.1	10.4
Melting point (°F)	−216.4	−254.2
Ignition temperature (°F)	550	860

For the purpose of this text, we mention only the first two possible isomers. Beyond these, it becomes impractical to use this form of nomenclature. However, you should realize that each analogous series has a set of structural isomers. In the first ten alkanes, butane is the first possible isomer, and pentane is next with three isomers: pentane, isopentane, and neopentane (see Figure 3–9).

■ **Note**

When we have isomers, the straight-chained hydrocarbon of an isomer is called normal or designated as *n-*.

■ **Note**

The short-chained alkenes are sometimes referred to as the olefins.

Alkenes

Chemistry Quick Reference Card

Family Class	Naming	Hazards
Alkanes C_nH_{2n+2}	Single bond, prefix depends on number of carbons, suffix -ane	Flammability. Boiling point of the molecule increases 20–30 degrees for each carbon added to the chain. Nonpolar.
Alkenes C_nH_{2n}	Double bonds in the structure, prefix depends on the number of carbons, suffix -ene Several double bonds are identified with the Greek numbering system	Flammability and reactivity. Boiling point increases as in alkanes; in both the alkanes and alkenes, branching lowers boiling point. Only weakly polar. Dienes and trienes are very reactive; the simplest diene is propadiene, or referred to allene, $H_2C=C=H_2$ ((when a double bond occurs next to each other it is called an Allene) Structure change $H_2C=C=CH_2$).

alkene

a hydrocarbon composed of at least one double bond in the carbon–carbon chain

The second analogous series of hydrocarbons is the **alkenes** series, in which the names of all the compounds end with -ene. As in the alkanes, each respective alkene compound adds a carbon to the chain for the building of the next respective compound. However, the structural formula is significantly different from that of the alkanes. It is an unsaturated hydrocarbon with single bonds and at least one double bond. The corresponding names are similar to the alkanes with the addition of the double bond configuration and the naming of the position of the double bond.

> ■ **Note**
>
> During the naming process, the appropriate prefix is used, with either -ylene or -ene as the suffix. The use of -ylene is most common.

The number of hydrogen atoms for molecules in this group compared to the carbon atoms is reduced by two, because one pair of electrons in a pair of carbon atoms shares a double bond. Overall, this configuration reduces the number of hydrogen atoms required to satisfy the octet rule, hence the unsaturated alkene. The general formula is C_nH_{2n}.

Note that the alkene family starts with the ethyl alkyl group because of the requirement of one double bond. Because hydrogen can never form more than one covalent bond **and one carbon,** as seen in the methyl group, the first compound has to be an **ethyl group.**

The naming configuration is the same as for the alkanes, utilizing the Greek prefix numbering convention and a suffix of -ene. The identification of the double bond is done by a number at the lowest carbon (see Figure 3–10).

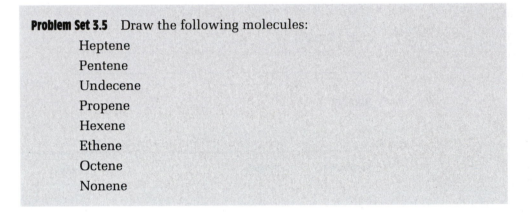

Problem Set 3.5 Draw the following molecules:

  Heptene

  Pentene

  Undecene

  Propene

  Hexene

  Ethene

  Octene

  Nonene

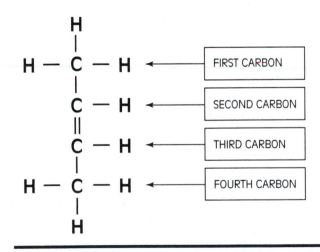

Figure 3–10 *It does not matter in this case which side we start from. Either way we come up with the double bond at the second and third carbon. The naming configuration is such that the prefix identifies the four-carbon chain, the suffix identifies the double bond, and the number shows the carbon that the double bond starts at: 2-butene or 2-butylene.*

> **■ Note**
> The alkenes and alkynes utilize the Greek numbering system to identify the number of multiple bonds: two = di, three = tri, four = tetra, five = penta, six = hexa, seven = hepta, and eight = octa.
>
> The radical of CH_2CHCH_2CH- is called a vinyl radical. See Chapter 4 concerning this naming configuration.
>
> ```
>        H   H
>        |   |
>  H — C = C —
> ```

The inclusion of a double bond in the configuration of the alkene molecule has profound effects on the chemical and physical properties of this family. When comparing the alkanes with the alkenes, the presence of the double bond greatly increases the overall chemical activity. For example, compare ethane with ethylene. Ethane will burn strong and steady, whereas ethylene polymerizes (an even stronger burn) due to the presence of the double bond.

Diene compounds are a subgroup of the alkenes. They are compounds that have two double bonds. Compounds with three double bonds are called trienes, with four double bonds, tetraenes, and so on. The position of these multiple bonds is denoted by a number in front of the name. Each number represents a double bond; so, for example, 1,3-butadiene is $CH_2=CH-CH=CH_2$. (We will see that this same configuration presents itself again in the alkynes.)

Problem Set 3.6   Draw the following compounds:

Propylene

1-Butene

2-Methyl-2-butene

2,3-Dimethyl-2-butene

2-Butene

1,2-Butadiene

1,5-Hexadiene

1,3-Pentadiene

1,4-Pentadiene

Isobutylene

Isomers. As with the alkanes, isomers of alkenes exist. More so than in the alkanes, alkene isomers create different configurations. Not only is there branching, but also the location of the double bond can be placed between different carbon groups in different compounds. Here the naming of the carbon chain is inclusive of the double bond, with all other branching as additions.

Sometimes compounds only differ in the orientation of their respective atoms in space. These isomers are called **stereoisomers**. Consider, for example, dichloroethylene, $C_2H_2Cl_2$, which has three possible isomers that all have a different set of properties, but which look basically the same. The only difference among them is the chlorine atom's orientation to the double-bonded carbons, as illustrated in Figure 3–11.

In dichloroethylene (1), the chlorine atoms are opposite each other at one end of the molecule. In dichloroethylene (2), the chlorine is at the top. (If it were

stereoisomers

compounds with the same molecular formula and connectivity, but different arrangement of atoms in space

Figure 3–11

Examples of the chlorine atom's orientation to double-bonded carbon in three isomers of dichloroethylene.

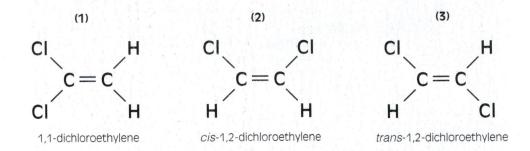

on the bottom, it would be the same molecule. Likewise in (1), if the chlorine atoms were both on the right as opposed to the left, it would be the same compound.) In dichloroethylene (3), we see the opposite ends of chlorine placement. To denote the differences of each compound, we call (1) dichloroethylene, (2) *cis*-dichloroethylene, and (3) *trans*-dichloroethylene. The *cis* and *trans* designations describe the orientation of the added groups on the parent chemical structure. Alternatively, the compounds can be named 1,1-dichloroethylene, *cis*-1,2-dichloroethylene, and *trans*-1,2-dichloroethylene, respectively.

Problem Set 3.7   Draw all three isomers of dibromoethylene. Draw all three isomers of difluoroethylene.

Alkynes

Chemistry Quick Reference Card

Family Class	Naming	Hazards
Alkanes C_nH_{2n+2}	Single bond, prefix depends on number of carbons, suffix -ane	Flammability. Boiling point of the molecule increases 20–30 degrees for each carbon added to the chain. Nonpolar.
Alkenes C_nH_{2n}	Double bonds in the structure, prefix depends on the number of carbons, suffix -ene Several double bonds are identified with the Greek numbering system	Flammability and reactivity. Boiling point increases as in alkanes; in both the alkanes and alkenes, branching lowers boiling point. Only weakly polar. Dienes and trienes are very reactive; the simplest diene is propadiene, or allene, $H_2C=C=H_2$ ((when a double bond occurs next to each other it is called an Allene) Structure change $H_2C=C=CH_2$).
Alkynes C_nH_{2n-2}	Triple bonds in the structure, prefix depends on the number of carbons, suffix -yne	Flammability. Insoluble in water. Boiling points are unusual in that they increase with increasing carbon attachment. Branching has the same effect as in the alkanes and alkenes.

Acetylene is commonly known as a very reactive substance. Actually it is this compound that is the beginning of the next analogous series of hydrocarbons,

alkyne

a hydrocarbon composed of at least one triple bond in the carbon–carbon chain

the **alkynes**. These compounds contain triple bonds and are very reactive because of the energy contained in the bonds.

Single bonds exist at angles that are conducive to reactions and bonding. Double bonds maintain that geometry, but the bonds are strained. Triple bonds also have that basic geometry, but these bonds are under a high degree of internal "pressure," which makes alkynes extremely active chemically. A slight amount of external pressure, movement (shock), or heat will release the energy that is contained in the bonds.

As shown with the first two series, the names are derived from the number of carbons in the chain. Alkyne compounds end with -yne. As observed with the other series, the respective compound adds a carbon to the chain for the building of the next compound. Although the corresponding names are similar to that of the alkanes and alkenes, the structural formula is considerably different. Because the alkynes are unsaturated hydrocarbons, the general formula is C_nH_{2n-2}. As with the alkenes, this series starts with an ethyl alkyl group.

■ Note

Acetylene, an alkyne, is actually named ethyne. Because of the -ene ending in the common name, it could be confused as an alkene.

■ Note

Many organics have common names that are used in industry rather than the IUPAC nomenclature. Memorize common names along with the IUPAC system.

Problem Set 3.8   Draw the following compounds:

 2-butyne

 1,3-butadiyne

 1-butyne

 3-hexyne

 5-methyl-2-hexyne

General IUPAC Naming

As new chemicals were produced, a system for naming was needed. The International Union of Pure and Applied Chemistry (IUPAC) has established rules for the naming of compounds to ensure uniformity of nomenclature so that everyone can understand the basic structure of the compound and the chemistry can be immediately identified. These rules follow a logical progression of application. Here we present the rules pertaining to alkanes, alkenes, and alkynes. As we move through the families of organic compounds, we will present additional rules.

1. The molecule is looked at in respect to the longest chain. If the chain is found to have a continuous single bond, then the suffix -ane is used. When the chain has one or more double bonds, the suffix -ene is used. If a triple bond is seen in the chain, the suffix -yne is used. For example:

   $CH_3–CH_2–CH_2–CH_2–CH_3$ Pentane

   $CH_3–CH_2–CH_2–CH=CH_2$ Pentene (see Rule 2)

   $CH_3–CH_2–CH_2–CH\equiv CH$ Pentyne (see Rule 2)

1a. If there is a halogen coming off the branch, we name the compound utilizing Rule 1 and the number of the carbon to which the halogen is attached. For example:

$$
\begin{array}{ccccc}
1 & 2 & 3 & 4 & 5 \\
CH_3-CH-CH_2-CH_2-CH_3 & & & & \\
\quad\; | & & & & \\
\quad\; Cl & & & &
\end{array}
$$
2-Chloropentane

2. The number of carbons is then derived by counting the carbon–carbon bonds and selecting the longest continuous chain. When double or triple bonds are noted, the carbon at the right or left of the double or triple bond is denoted as the number one carbon. The appropriate suffix and prefix are applied to derive the name, and the number is incorporated into the name so as to identify the double or triple bond. For example:

$$
\begin{array}{ccccccc}
CH_3-CH_2-CH_2-CH_2-CH_2-CH=CH_2 \\
\;\;7 & \;6 & \;5 & \;4 & \;3 & \;2 & \;1
\end{array}
$$
Heptene or 1-heptene (1 is used to identify the placement of the double bond; if no number is used, 1 is understood)

$$
\begin{array}{ccccc}
CH_3-CH_2-CH=CH-CH_3 \\
\;5 & \;4 & \;3 & \;2 & \;1
\end{array}
$$
2-Pentene or pent-2-ene

$$
\begin{array}{ccccc}
CH_3-CH_2-C{\equiv}C-CH_3 \\
\;5 & \;4 & 3 & 2 & 1
\end{array}
$$
2-Pentyne or pent-2-yne

$$
\begin{array}{cccccc}
CH_3-CH_2-C{\equiv}C-CH_2-CH_3 \\
\;1 & \;2 & 3 & 4 & \;5 & \;6
\end{array}
$$
3-Hexyne or hex-3-yne

$$
\begin{array}{cccccc}
CH_3-CH=CH-CH=CH-CH_3 \\
\;1 & \;2 & \;3 & \;4 & \;5 & \;6
\end{array}
$$
2,4-Hexadiene or hexa-2,4-diene

3. When an alkyl group comes off of the chain, the appropriate group name is applied and incorporated into the naming process. The carbon closest to the end of the chain with respect to the alkyl group determines how the carbon chain is derived or named. Each group is placed in alphabetical order or increasing size, and is preceded by the number of orientation. For example:

$$
\begin{array}{ccccccc}
1 & 2 & 3 & 4 & 5 & 6 & 7 \\
CH_3-CH-CH_2-CH-CH_2-CH_2-CH_3 \\
\quad\; | & & \quad\; | & & & & \\
\quad CH_3 & & CH_2CH_3 & & & &
\end{array}
$$
The chain is heptane; a methyl alkyl group is off the second carbon and an ethyl alkyl group is off of carbon 5: 2-methyl-5-ethylheptane

---

Problem Set 3.9   Draw the following compounds:

Methane

Heptene

2-Methyl-2-butene

2,3-Dimethyl-2-butene

2-Butene

1,2-Butadiene

1,5-Hexadiene

2-Methyl-5-ethyl-heptane

(continues)

Problem Set 3.9   *(Continued)*

Butane

Hexyne

Octane

Chloromethane

Undecane

1-Butyne

3-Hexyne

5-Methyl-2-hexyne

Isobutane

n-Pentane

4-Fluoro-2-heptene

1,2-Butadiene

Dichloroethylene

Neopentane

An Aside: Petroleum Derivatives

Most of the common fuels that we use come from petroleum that is removed from the earth; the word petroleum comes from the Greek, meaning "rock oil." Petroleum itself is a complex mixture of hydrocarbon chains, numbering from one to sixty. Because it is a mixture of compounds, and each of those compounds has a specific boiling point, we can use the process of **distillation** to separate these compounds. In particular, **fractional distillation** is used to remove different compounds at particular temperatures by boiling off specifically weighted hydrocarbons.

distillation

the physical process of converting a liquid to a vapor and condensing it back into a liquid

fractional distillation

a process of evaporation and recondensation used for separating a mixture into its various constituents based on the individual boiling points of those constituents

---

■ Note

The general formula for the cycloalkanes is C_nH_{2n}, and for cycloalkene, C_nH_{2n-2}.

● Caution

The cyclo compounds have a formula similar to the alkanes, alkenes, and alkynes. Both possibilities exist until one can be positively ruled out.

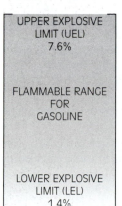

UPPER EXPLOSIVE
LIMIT (UEL)
7.6%

FLAMMABLE RANGE
FOR
GASOLINE

LOWER EXPLOSIVE
LIMIT (LEL)
1.4%

Each fraction is distilled off at specific temperature ranges at atmospheric pressure, and retrieved in separate containers. The lightest compounds distill off first at temperatures less than 70°F, giving rise to methane, ethane, propane, and butane (natural gas and liquefied petroleum). These gases used to be burned off using a flaring method, largely due to the expense of recovery, but now are used as commodities. The next fraction, occurring between the five and nine carbons, gives us the light naphtha products: gasoline and petro ethers. These boil off between 158°F and 284°F. The group that distills off between 284°F and 392°F has chains of seven to nine carbons and is represented by the heavy naphthas. Fuel oils (1 and 2), mineral spirits, kerosene, and solvents, having nine to sixteen carbons, distill off between 347°F and 527°F. Between sixteen and twenty-five carbons, we see heavy fuel oils drop out (fuel oils 4, 5, and 6) at fractional temperatures ranging from 392°F to 698°F. Above twenty carbons, we see lubricating oils, asphalt, and tar, as examples, boiling off above 698°F.

The petroleum derivatives are simple carbon-chained compounds. When these chains of hydrocarbons are heated in the presence of a catalyst, bonds rupture, producing a variety of hydrocarbons. This process is called **cracking**. This breaking of bonds is what happens when an isoalkane is chemically moved to become an alkene, for example. Alkylation involves the combination of alkanes and alkenes to produce branched chained hydrocarbons. Each of these processes provides the petroleum industry with a variety of compounds, which can then be mixed to give products such as aviation fuel. Other chemical processes can yield a larger variety of substances, leading to products for the plastic, rubber, and synthetic fiber industries.

cracking

a chemical process in which organic molecules are broken down into smaller molecules by heating

Cyclo Compounds

Chemistry Quick Reference Card

Family Class	Naming	Hazards
Alkanes C_nH_{2n+2}	Single bond, prefix depends on number of carbons, suffix -ane	Flammability. Boiling point of the molecule increases 20–30 degrees for each carbon added to the chain. Nonpolar.
Alkenes C_nH_{2n}	Double bonds in the structure, prefix depends on the number of carbons, suffix -ene Several double bonds are identified with the Greek numbering system	Flammability and reactivity. Boiling point increases as in alkanes; in both the alkanes and alkenes, branching lowers boiling point. Only weakly polar. Dienes and trienes are very reactive; the simplest diene is propadiene, or allene, $H_2C=C=H_2$ ((when a double bond occurs next to each other it is called an Allene) Structure change $H_2C=C=CH_2$).
Alkynes C_nH_{2n-2}	Triple bonds in the structure, prefix depends on the number of carbons, suffix -yne	Flammability. Insoluble in water. Boiling points are unusual in that they increase with increasing carbon attachment. Branching has the same effect as in the alkanes and alkenes.
Cyclo Compounds C_nH_{2n} **or** C_nH_{2n-2}	Same as the parent hydrocarbon with "cyclo" or "c" in front of the name; sometimes referred to as the alicyclic hydrocarbons	The rings that are smaller or larger than cyclopentane and cyclohexane are unstable. The reaction potential depends on the geometry.

cyclo compound

a hydrocarbon chain in a ring formation

The **cyclo compound** group (or cyclic compounds) is vast and confusing. For our purposes, we can limit this discussion to the lower-weighted cyclo compounds, namely, the first four in the cycloalkanes and cycloalkenes series (see Figure 3–12).

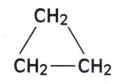

Cyclopropane or *c*-propane (C₃H₆)

Cyclopropene or *c*-propene (C₃H₄)

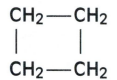

Cyclobutane or *c*-butane (C₄H₈)

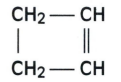

Cyclobutene or *c*-butene (C₄H₆)

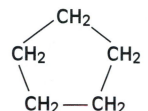

Cyclopentane or *c*-pentane (C₅H₁₀)

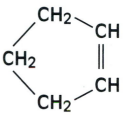

Cyclopentene or *c*-pentene (C₅H₈)

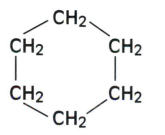

Cyclohexane or *c*-hexane (C₆H₁₂)

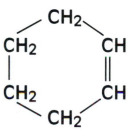

Cyclohexene or *c*-hexene (C₆H₁₀)

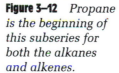

Figure 3–12 *Propane is the beginning of this subseries for both the alkanes and alkenes.*

The naming of these compounds follows the same nomenclature as for both the alkanes and alkenes. If, for example, a continuous single bond exists, then we name it as an alkane, utilizing the number of carbons as the prefix and -ane as the suffix. However, because these compounds are in a circular pattern or in a ring shape, the tail of the chain is connected to the head of the chain. So for the cycloalkane, the general formula is C_nH_{2n}, which is the same as for alkenes. If then, for example, we respond to an incident and the chemical of involvement is C_3H_6, we have two possibilities: an alkene or a cycloalkane. The package or the shipping papers should indicate "cyclo" or "*c*" to identify which compound you may have. Without this vital piece of information, both the cyclo compound and the alkene must be referenced.

● Caution

Cyclo compounds can sometimes be mistaken for straight-chained hydrocarbons. The chemistry changes with cyclo compounds. Be alert for the "*c*" in front of the name identifying each.

Problem Set 3.10 Describe the problem that can arise if alkanes, alkenes, alkynes, cycloalkanes, and cycloalkenes are present in an environment. The only information you have is from the chemist, who is also the manager, as he is running from the property: "I have aliphatic compounds with the following general formulas: C_nH_{2n+2}, C_nH_{2n}, and C_nH_{2n-2}."

Aromatics

aromatic hydrocarbons

benzene or benzene-like compounds primarily composed of carbon and hydrogen atoms in a ring structure

The aliphatic (nonaromatic) hydrocarbons—alkanes, alkenes, alkynes, and cyclo compounds—are all related chemically. The **aromatic hydrocarbons** are a bit different than those compounds covered thus far. The history of this group's discovery is quite interesting and is the reason for this chemical classification's general name: aromatic. In the early 1800s, benzene and its counterparts (the chemicals known in that day) had fragrant qualities. Most were derived from natural sources: oil from bitter almonds (benzaldehyde), resins, tolu balsam (toluene), and gum benzoin (benzyl alcohol), to name a few. It was found that this chemical benzene had an extremely low hydrogen-to-carbon ratio. Obviously this ratio could mean only one thing: that benzene is an unsaturated hydrocarbon. However, when we look at the cyclo compounds and the analogous aromatic compounds, we see a difference in the chemical and physical properties of each group.

> ● **Caution**
>
> Benzene is extremely toxic and is thought to cause cancer. Its odor threshold is 4.68 ppm, and the TLV is 1 ppm. If you can smell benzene, you are being exposed to almost five times above the permissible limit. Gasoline contains 1.5–5% benzene.

So what can explain the small but significant difference between the cyclo compounds and benzene? In organic chemistry, it is extremely important that a compound be stable. Electronic stability is what all compounds are trying to achieve. Because of this tendency toward stability, six-carbon chains are sometimes incorporated into a circle structure, which is called a benzene ring (see Figure 3–13) and has resonance bonds. A high level of stability is achieved by virtue of this ring structure due to the resonance-bonding electrons contributing to each carbon–carbon bond. The six-carbon resonant ring is more stable and stronger than the straight-chained (or cyclo-compound) counterpart.

(1)   (2)   (3)

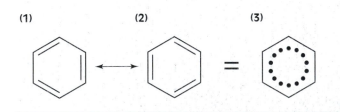

Figure 3–13 *The benzene ring is actually a resonance structure. The dash models (1 and 2) of the ring do not provide for a rational structure as compared to the experimental data. If they were, in fact, the correct model, these compounds would act like cyclo compounds, which burn well as an example. However, the (3) representation is what benzene looks like, as it follows the data quite well. Here, as you can see, the benzene structure is actually a combination of (1) and (2), not really a single bond and not really a double bond, but a resonance structure.*

Chemistry Quick Reference Card

Family Class	Naming	Hazards
Alkanes C_nH_{2n+2}	Single bond, prefix depends on number of carbons, suffix -ane	Flammability. Boiling point of the molecule increases 20–30 degrees for each carbon added to the chain. Nonpolar.
Alkenes C_nH_{2n}	Double bonds in the structure, prefix depends on the number of carbons, suffix -ene Several double bonds are identified with the Greek numbering system	Flammability and reactivity. Boiling point increases as in alkanes; in both the alkanes and alkenes, branching lowers boiling point. Only weakly polar. Dienes and trienes are very reactive; the simplest diene is propadiene, or allene, $H_2C=C=H_2$ ((when a double bond occurs next to each other it is called an Allene) Structure change $H_2C=C=CH_2$).
Alkynes C_nH_{2n-2}	Triple bonds in the structure, prefix depends on the number of carbons, suffix -yne	Flammability. Insoluble in water. Boiling points are unusual in that they increase with increasing carbon attachment. Branching has the same effect as in the alkanes and alkenes.
Cyclo Compounds C_nH_{2n} or C_nH_{2n-2}	Same as the parent hydrocarbon with "cyclo" or "*c*" in front of the name; sometimes referred to as the alicyclic hydrocarbons	The rings that are smaller or larger than cyclopentane and cyclohexane are unstable. The reaction potential depends on the geometry.
Aromatics C_nH_n **(six carbons)**	Depends on the entire structure derivative, placement on the ring, and attachments	Aromatic compounds are those compounds that resemble benzene in chemical nature, or are benzene. Fairly stable due to resonance. Multitude of chemical reactions.

The hexagonal benzene ring has a chemistry unto itself (see Table 3–4). When the electrons are arranged in a stable fashion above and below the ring, the compound is called an aromatic or benzene structure.

Table 3–4   *Comparison of properties of an aromatic and a cyclo compound.*

	Benzene	Cyclohexene
Molecular weight	78.12	82.15
Boiling point (°F)	176	181
Melting point (°F)	41.92	−154.3
Flash point (°F)	12.2	10.99
Vapor pressure (mm Hg)	100	167
Vapor density	2.7	2.8
Water solubility	0.18 g/100	Insoluble

The prefixes ortho, meta, and para are used in the nomenclature to designate the carbon group of attachment with respect to the main attachment. This nomenclature is further discussed in subsequent chapters.

The benzene ring has a naming configuration that is sometimes confusing. If the benzene ring has a functional group attached to it, the aromatic (benzene) is named using phenyl- as the prefix. The aromatic compounds are named extensively under functional groups (see Chapters 4 and 5).

KINETIC MOLECULAR THEORY

Kinetic molecular theory states that all molecules are in constant motion. This movement may be as simple as a slight vibration in a solid, or the rapid constant motion in a gas. The theory additionally states that if a material—solid, liquid, or gas—is brought down to a particular temperature, all motion in the atom, and thus in the molecule, will stop. This temperature is referred to as the absolute temperature, or 0 K.

■ Note

Absolute temperature is the temperature of absolute zero (0 K or –439.67°F). Recall the definition of STP as the temperature of 32°F (0°C or 273 K) at a pressure of 1 atm (760 torr or 760 mm Hg).

To demonstrate this theory, consider a container that has hydrogen sulfide gas in it (see Figure 3–14). The pressure is at normal pressure (14.7 psi). At normal temperature and pressure, the molecules in this container are rapidly striking the container's walls. If we increase the temperature, the molecular movement will increase. With this temperature increase and the resulting increase in molecular movement, the pressure in the container also increases. This increased

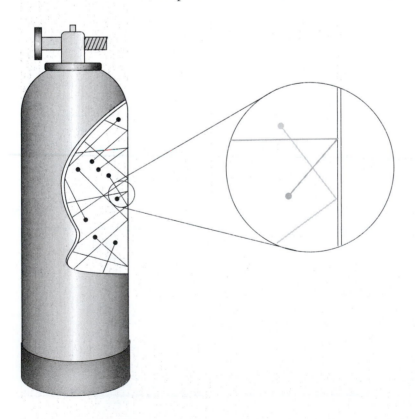

Figure 3–14

Molecular activity occurring in a compressed gas cylinder.

pressure may fatigue the container and cause a rupture, or it may damage the valve mechanism and cause a release of the gas.

If we lower the temperature, the molecules of hydrogen sulfide start to slow down. As we approach $-77°F$ ($-76.72°F$ is the boiling point of H_2S), the molecule will slow to a point at which the gas turns into a liquid. If we continue our cooling to roughly $-122°F$, the liquid hydrogen sulfide changes into a solid state. The molecular movement has been reduced. At $-459.69°F$, what was once hydrogen sulfide gas would be a solid in which all molecular movement had stopped.

There are some problems in this example. First, we are not sure if we can take a gas of this nature and place it into a solid form just by reducing the temperature (we probably also have to apply pressure). Second, freezing a chemical or liquefying a chemical depends on the temperature and the pressure that is needed to change states: the **critical temperature** and the **critical pressure**. Before we address the problem of temperature and pressure, let us look at the laws that surround the gases.

Gas Laws

The surrounding atmosphere at sea level is at what is called the standard pressure or standard **atmospheric pressure** (atm) of 14.7 psi. (Denver, Colorado, which is approximately a mile high, has an atmospheric pressure of 13.5 psi.) This pressure is the "weight" of the atmosphere in a defined space of a square inch (pounds per square inch).

Torricelli, an Italian scientist, recognized the concept of pressure in the 1600s. He measured the pressure of the atmosphere by using an evacuated tube in which mercury was placed. He then measured the height of the level of the mercury in the tube. This measurement is known as millimeters of mercury (mm Hg; at 32°F) or torr. The pressure that is exerted by the atmosphere onto a column of mercury is 760 mm Hg, or one atmosphere. Here are the various equivalent expressions:

$$1013.2 \text{ millibars} = 29.92 \text{ in} = 14.7 \text{ psi} = 760 \text{ mm Hg} = 1 \text{ atm}$$
$$= 760 \text{ torr} = 1.013 \times 10^5 \text{ pascal}$$

Boyle's Law. **Boyle's law** is a relationship between the pressure (P) of a gas and the volume (V) that it occupies at a fixed temperature, usually 77°F (25°C) and a constant number of **moles**. (When referencing a chemical, you should note at what temperature it was tested. This information is usually found under chemical and physical properties.) This law shows us that as the pressure increases, the volume decreases:

$$V = 1/P$$

where 1 can be represented by a constant known as K that is equal to the volume times the pressure. Maintaining the number of moles of the gas and temperature constant and algebraically arranging the formula, we get:

$$V_1 \times P_1 = K \quad \text{or} \quad P_1V_1 = K$$
$$V = K/P$$
$$P_1V_1 = P_2V_2$$

where $K = PV$ (a constant). So we can see that the original volume and pressure of the gas are equal to the resultant pressure and volume.

---

critical temperature
the temperature above which a gas cannot be liquefied

critical pressure
the minimum pressure required to cause a gas to liquefy at its critical temperature

atmospheric pressure
the force exerted on matter by the mass of the overlying air

Boyle's law
at constant temperature, the volume of a gas is inversely proportional to its absolute pressure

mole
the quantity of a substance that contains the same number of particles (atoms, ions, molecules) as there are in precisely 12 g of carbon 12

Charles's law

at a constant pressure, the volume of a gas is directly proportional to its absolute temperature

Charles's Law. **Charles's law** deals with the volume of a gas and its relationship to the temperature (T). The pressure (usually at 1 atm) and number of moles remain constant. Charles's law states that the volume and the temperature are directly proportional. We can express this law mathematically in the following manner, where K is equal to V/T:

$$V = KT$$
$$V/T = K \quad \text{or} \quad V_1/T_1 = V_2/T_2$$

This expression denotes that the volume of a gas will increase proportionally as the temperature increases. Jacques Charles and Joseph-Louis Gay-Lussac were interested in the hot-air balloon. By observing the eponymous law, they were able to use a hot-air balloon to ascend into the sky in the late 1700s and early 1800s.

At the scene of a hazardous materials incident, we can intuitively understand what is taking place when dealing with gases and liquids in a container:

$$P_1V_1 = P_2V_2 + V_1/T_1 = V_2/T_2$$

which is equal to

$$P_1V_1/T_1 = P_2V_2/T_2$$

If, for example, we have a propane cylinder that has flame impingement, we know that the pressure on the inside of the cylinder is increasing. In the fixed container, volume is for all practical purposes constant, and the temperature and pressure will build up, resulting in a pressurization explosion (BLEVE, or boiling liquid expanding vapor explosion). By eliminating the flame or decreasing the temperature of the liquid and cooling the cylinder, the volume and pressure will decrease.

Dalton's law

the total pressure of several gases is the sum of all the pressures of each individual gas

Dalton's Law. What pressure do we have when several gases are present? **Dalton's law** simply states that the pressures in a container that includes several gases is the sum of all the pressures of each individual gas (P_1, P_2, etc.). In mathematical form:

$$P_{Total} = P_1 + P_2 + P_3 + \cdots P_n$$

From a hazmat standpoint, we can utilize this concept to aid in the decision-making process, for example, deciding between level-A or level-B encapsulation. The conditions depend on the vapor pressures of the material. If a substance has a very low vapor pressure and the partial pressures add up to less than the atmospheric pressure, then the possibility of saturation of the vapor in the ambient air is low. This evaluation does not mean that the environment is clear of toxic gases, but simply helps to identify that the atmosphere can be entered given the appropriate level of personnel protection, not necessarily level A.

However, if the vapor pressure is high and the resulting partial pressures are also high, then the potential for a severe respiratory hazard exists. This potential has to do with the mechanism of respiration. When the thoracic cavity expands, the atmospheric pressure in the chest decreases. If the chemical already has a high vapor pressure or the partial pressures are high, a victim in this atmosphere or a team member with a breach of the protective suit is exposed to the possibility of a severe respiratory injury.

Henry's Law. What happens when a gas is placed into solution? When we open a soda bottle or a bottle of champagne, we see tiny bubbles come out of solution. These are bubbles of the gas that was forced into solution by the use of pressure

Henry's law

the amount of gas absorbed by a given volume of a liquid is directly proportional to the pressure of the gas

(the same thing occurs when a diver does not decompress: nitrogen comes out of solution). There are three factors that govern this type of reaction: the pressure of the gas surrounding the solution, the solubility constant of the gas in the fluid in question, and the temperature of that fluid. We can express this relationship between these factors, known as **Henry's law**, by using the following formula:

$$V = P \times \text{solubility constant}$$

Pressure and Temperature

Some chemicals need a certain pressure and temperature before they will move to the next state of matter, specifically referred to as critical temperature and critical pressure. With awareness of the gas laws, we can utilize these concepts.

For gases in general, the addition of pressure can force the gas into a liquid state. For certain gases, pressure is associated with a temperature before the gas will attain a different state of matter. Most of the compressed gases that we encounter as emergency responders are those that have had pressure placed on them and have thus been converted into a liquid. Sometimes the temperature has been decreased when the chemical was placed under pressure because pressure alone may not place enough restriction on the molecules. The molecules still contain enough kinetic energy to maintain the gaseous state. Once the temperature is dropped to the critical level with the associated pressure, the material then moves from a gas to a liquid.

The liquefied petroleum gases (LPGs) are often a combination of gases that are considered as one. The gas is placed into a container, and through the pressure in the tank, the gases occupy a liquid state of matter. The expansion ratio of propane, which is a major component of LPG, has an expansion ratio of 1:270. When a valve is opened on a propane tank, the pressure (equilibrium) between liquid and gas is reduced, and the propane boils to replace the equilibrium. The liquefied gas boils off to produce the vapor, thus reducing the temperature.

A phase diagram enables us to see this relationship between the states of matter and the movement from a solid to liquid and a liquid to vapor (see Figure 3–15). This movement is an equilibrium between liquids and their

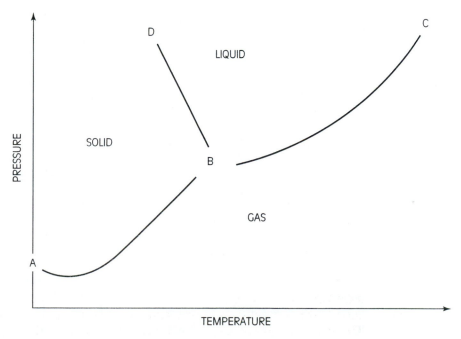

Figure 3–15 *Phase diagrams enable us to see the movement between the states of matter and the relationship with temperature and pressure. On this diagram, point C is the critical temperature and pressure.*

respective vapors (as are the equilibria for solid–liquid and solid–vapor). Many of the conditions that we have been describing thus far can be summarized and visually identified by the phase diagram.

We can see by looking at the diagram shown in Figure 3–15 that we have three basic equilibria. Along the line A to B it is the equilibrium between the solid and gas of this hypothetical compound. B to C is the equilibrium between a liquid state and gas state, and between B and D is the equilibrium between the solid and liquid. At point C, simply increasing the temperature without increasing the pressure will not convert the gas into a liquid. Moving toward the right beyond C shows that increasing only the pressure will not make the gas move into a liquid. We can see that a certain critical pressure must be attained before we can move from the gas into the liquid state.

As we move to the right along the temperature axis, states of matter change at a given pressure. As we move up the pressure axis at a particular temperature, states also change. For example, looking at the B–C portion of the graph identifies the vapor pressure–temperature of a liquid. At any point along this line, the liquid is in equilibrium with the gas phase. If this were water, liquid water and water vapor would be at a constant between the liquid and gas (vapor) state. Moving from the liquid state into the gas state is called vaporization or evaporation; from the gas to the liquid state, condensation. The equilibrium that we see along the B–C line is how **cryogenic liquids** are formed. By lowering the temperature or increasing the pressure, the gas is moved into a liquid.

The line between A and B represents the vapor pressure of the solid and identifies the equilibrium between the solid and the gas. Moving a substance from a solid to a gas is called sublimation. The compound will undergo sublimation at any temperature below the B point when the pressure is reduced below the equilibrium points. The freeze-drying of foods uses this process, for example.

B–D represents the conditions of temperature and pressure between a solid and liquid state. Movement from the solid to liquid state is called melting or fusion, and from the liquid to solid state, solidification.

Whenever we have a cryogenic liquid under the influence of temperature and pressure, a liquid-to-gas conversion will have a volume-to-volume ratio of expansion (the tendency of a liquid to evaporate into a gas) once this chemical is released. For a hazmat incident, the reference branch must be prepared to investigate this ratio and how it may affect the scene. When volume expansion ratios are present as well as a fire potential, health hazards will exist. As a rule of thumb, liquefied gases without a temperature decrease (noncryogenics) will have expansion ratios of 1 to 200–300. Cryogenic liquids have expansion ratios in the 1 to 600–1000+ range (696 for nitrogen to 1445 for neon). (See Table 3–5.)

cryogenic liquids
substances that have been cooled to extremely low temperatures

Table 3–5 *Liquefied products.*

	Boiling Point (°F)	Critical Temperature (°F)	Critical Pressure (psi)	Expansion Ratio
Propane	−43.8	206.3	617	1:270
Methane	−258.9	−115.8	672	1:693
Hydrogen	−421.6	−390	294	1:850
Nitrogen	−319.9	−230.8	485	1:696
Oxygen	−297	−180	735	1:860

The technical definition of a cryogenic is applied to those gases that have a boiling point lower than $-130°F$ at 1 atm. Cryogenics are materials that have been forced into a liquid state. Once released, this liquid will revert back to its natural state: a gas. A hazard arises when an individual is in the area of this state change. The typical injuries that are seen are traumatic and hypothermic injuries. Obviously, the cold nature of the cryogenic will produce frostbite or freezing of the tissue itself. The sudden release can cause traumatic lacerations and associated chemical exposure through absorption or injection. Once this liquid heats up to revert back to its gaseous state, the volume of the chemical multiplies. This increase of volume depends on its physical properties, such as the boiling point. The rapid expansion from the liquid state to the vapor state will also cause an increase in vapor pressure, which further increases hazard considerations of respiratory absorption during the incident.

● Caution

Be alert to expansion ratios: air mixture, 728:1; anhydrous ammonia, 855:1; argon, 842:1; carbon monoxide, 680:1; chlorine, 458:1; fluorine, 981:1; helium, 745:1; hydrogen, 850:1; krypton, 693:1; LNG, 635:1; methane, 693:1; neon, 1445:1; nitrogen, 696:1; oxygen, 860:1; propane, 270:1; xenon, 559:1.

Safety

Cryogenics have large expansion ratios. Because of the extremely low temperatures at which cryogenics exist, they can liquefy or solidify other gases, liquids, or semisolids. Between the expansion ratios and the extremely low temperatures, these products present a significant health hazard.

BRANCHING AND ITS EFFECTS ON PHYSICAL PROPERTIES

This discussion focuses on the size of a molecule and its effect on the boiling point (this also includes the branching). Recall from Chapter 1 that the lower the boiling point, the more vapor is produced. From this one property, we can gain a valuable understanding of the nature of hydrocarbons.

Three conditions primarily affect the boiling point of a substance (see Figure 3–16):

1. The weight, or molecular weight, of the substance. The lower the weight, the higher the probability that it will be a gas. Air has an atomic weight of 29. Methane, as an example, has a molecular weight of 16, is a gas, and thus rises. The first four alkanes are gases and the next twenty are liquids. In general, the low molecular-weight compounds will have low boiling points, which produce a gas or are in the gaseous state.

2. The polarity and the ability to create hydrogen bonding. For equally weighted compounds, polarity affects the boiling point. If the compound has hydrogen attached to nitrogen, oxygen, or fluorine, then hydrogen bonding is a high possibility that will thus affect the boiling point.

3. Branching. Isomers or off-branches of the hydrocarbon greatly affect its boiling point. The straight-chained hydrocarbons may form low boiling points, but isomers tend to drop the boiling point in relation to the parent hydrocarbon.

A compound with a low boiling point has a high vapor pressure and produces a gas. If this chemical has flammable qualities, the evolving gas will create

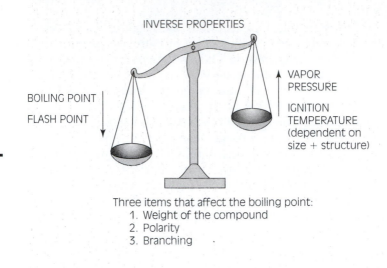

Figure 3–16 *The relationships between inverse properties and the items that affect these conditions.*

a flammable atmosphere. For example, *n*-butane has a boiling point of 0°C and a melting point of −138°C in relation to its isomer, isobutane, which has a boiling point of −12°C and a melting point of −159°C. The straight-chained parent has a low boiling point, but the isomer of the parent is even lower.

The alkanes are held together by covalent bonds. The larger the molecule, the higher the surface area of the molecule, and thus strong intermolecular forces exist. In general, as the weight of the alkane increases, so do the boiling and melting points, because of the surface area that the molecule must overcome in order to change states. The higher the molecular weight, the greater the contact or surface area with which intermolecular forces can maintain attraction. On average, the boiling point increases 20–30 degrees for each alkyl group.

The distinguishing feature of the alkene is its double bond. Here, like the alkanes, the boiling point increases with the addition of each alkyl group. Branching in this group also lowers the boiling point in relation to the parent hydrocarbon. We find very slight polar activity, which is more predominant in the isomers. In general, the *cis*- isomer has a higher boiling point when compared to its other half, the *trans*- isomer, probably due to a slightly higher polarity in the *cis*- isomer.

The alkynes have properties that are similar to the alkanes and alkenes, in that all three of these aliphatic hydrocarbons are insoluble in water but are soluble in organic solvents that possess low polarity. Again we see an increase in the boiling point as we add an alkyl group. The same effects that branching has in the alkanes and alkenes are also found in the alkynes.

Table 3–6 lists alkanes, alkenes, and alkynes with their respective boiling and melting points, and molecular weights.

Problem Set 3.11 Using Table 3–6, plot the following, where the *y*-axis is the number of carbons and the *x*-axis is the boiling point of the compounds.

1. The isomers versus the parent chain
2. The number of carbons versus the melting point of alkanes, alkenes, and alkynes
3. The number of carbons versus the boiling point of alkanes, alkenes, and alkynes

Table 3–6   *Molecular comparison between analogous groups.*

	Molecular Weight	Melting Point (°C)	Boiling Point (°C)		Molecular Weight	Melting Point (°C)	Boiling Point (°C)
			Alkanes				
Methane	16.04	−182.5	−164.0				
Ethane	30.07	−183.0	−88.6				
Propane	44.09	−187.7	−42.5				
Butane	58.12	−138.0	−0.5	iso-Butane	58.12	−159	−12
Pentane	72.15	−129.7	36.1	iso-Pentane	72.15	−160	28
				Neo-Pentane	72.15	−16.7	9.4
Hexane	86.17	−95.3	68.7	iso-Hexane	86.17	−154	60
Heptane	100.20	−90.7	98.4				
Octane	114.22	−56.8	125.6				
Nonane	128.30	−53.3	151.0				
			Alkenes				
Ethene	28.05	−169	−104				
Propene	42.08	−185	−47				
1-Butene	56.10	−130	−6	iso-Butylene	56.10	−141	−7
				cis-2-Butene	56.10		4
				trans-2-Butene	56.10		1
1-Pentene	70.13	−138	30				
3-Hexene	84.16	−99	64				
1-Heptene	98.19	−119	93				
1-Octene	112.23	−102	123				
1-Nonene	126.24	−82	146				
			Alkynes				
Ethyne	26.04	−82	−75				
Propyne	40.06	−101.5	−23				
Butyne	54.09	−122	9	2-Butyne	54.09	−24	27
Pentyne	68.12	−98	40	2-Pentyne	69.12	−101	55
Hexyne	82.14	−124	72	2-Hexyne	82.15	−92	84
Heptyne	96.17	−80	100				
Octyne	110.2	−70	126				
Nonyne	124.2	−65	151				

Branching has a naming configuration. Common names are not generally used; rather, the IUPAC name for the branch is used. The identifying number of the group(s) and the names are separated with hyphens, and are in alphabetical order (see Rule 3 on page 104). Figure 3–17 shows the compounds with the trivial names that the IUPAC nomenclature recognizes. Any other names must follow the naming rules. To that end, note the following rules:

3a. Number the carbon chain arriving at the longest continuous chain, and situate the numbering configuration such that the branch will have the lowest possible carbon number. Numbers identify the carbon that the branch arises from and the words are separated with a hyphen.

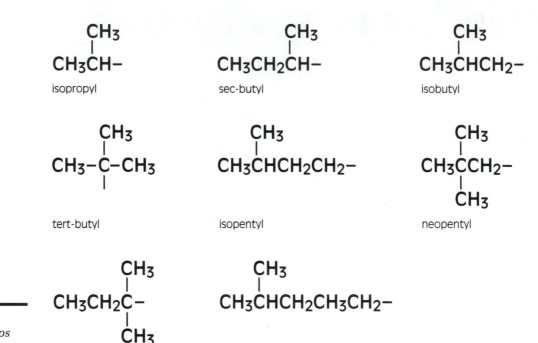

Figure 3–17
*Branched groups
used in naming
compounds.*

$$CH_3CH_2CH_2CH_2CHCH_2CH_3$$
$$|$$
$$CH_2CH_3$$

3-Ethylheptane. Whether you count from the ethyl group or from the right side of the molecule, you have seven total carbons in a continuous chain, or heptane. From right to left gives the lowest number.

3b. When there are two or more branches, each branch receives the lowest corresponding number and the groups are listed alphabetically. If two or more branches fall on the same carbon, use the number twice, with the groups in alphabetical order.

$$CH_3$$
$$|$$
$$CH_3CH_2CH_2CH_2CHCH_2CH_3 \qquad \text{3-methyl-3-ethylheptane}$$
$$|$$
$$CH_2CH_3$$

Problem Set 3.12 Draw the following compounds:

3,5-dimethylheptane
3-ethyl-2-methylpentane
3,3-dimethylpentane
2,2,4,4-tetramethylpentane
4,6-dipropylnonane
5-isopropyl-5-tert-butyldecane
4-isopropylheptane
2,5-dimethylhexane
2,3-dimethylhexene
2,2,6-trimethyl-3-heptene

Summary

If a chemical has the suffix of -ane, we know it is a single-bonded hydrocarbon. If the suffix is -ene, it contains at least one double bond, and a suffix of -yne indicates a triple bond. In the nomenclature of organic chemistry, the number of carbon–carbon bonds is also important. Each carbon has a name when in a chain. This name, or prefix, designates the number of carbons in the compound.

Chemistry Quick Reference Card

Family Class	Naming	Hazards
Alkanes C_nH_{2n+2}	Single bond, prefix depends on number of carbons, suffix -ane	Flammability. Boiling point of the molecule increases 20–30 degrees for each carbon added to the chain. Nonpolar.
Alkenes C_nH_{2n}	Double bonds in the structure, prefix depends on the number of carbons, suffix -ene Several double bonds are identified with the Greek numbering system	Flammability and reactivity. Boiling point increases as in alkanes; in both the alkanes and alkenes, branching lowers boiling point. Only weakly polar. Dienes and trienes are very reactive; the simplest diene is propadiene, or allene, $H_2C=C=H_2$ ((when a double bond occurs next to each other it is called an Allene) Structure change $H_2C=C=CH_2$).
Alkynes C_nH_{2n-2}	Triple bonds in the structure, prefix depends on the number of carbons, suffix -yne	Flammability. Insoluble in water. Boiling points are unusual in that they increase with increasing carbon attachment. Branching has the same effect as in the alkanes and alkenes.
Cyclo Compounds C_nH_{2n} or C_nH_{2n-2}	Same as the parent hydrocarbon with "cyclo" or "*c*" in front of the name; sometimes referred to as the alicyclic hydrocarbons	The rings that are smaller or larger than cyclopentane and cyclohexane are unstable. The reaction potential depends on the geometry.
Aromatics C_nH_n (minimum of six carbons)	Depends on the entire structure derivative, placement on the ring, and attachments	Aromatic compounds are those compounds that resemble benzene in chemical nature, or are benzene. Fairly stable due to resonance. Multitude of chemical reactions.

All gases can be forced into a liquid or a solid state. The critical pressure is the pressure that is required to place enough restriction on the moving molecules to change the material's state from a gas to a liquid at a temperature below the critical temperature. Above this temperature a gas cannot be condensed into a liquid. However, at temperatures below the critical temperature, the gas can be liquefied. The lower the temperature below the critical temperature, the less pressure is required for the conversion of state.

■ Note

Nonmetals and nonmetals in combination result in the following hazards:

☠ Flammability

☠ Corrosivity

☠ Irritants

☠ Oxidizer

☠ Poisonous

☠ Explosive

Appearance may be a solid, liquid, or gas. The lower the weight, the higher the probability it is in a gas form. Many will burn violently depending on the bonding configuration. Reactivity is high with other compounds and chemicals; toxicity varies.

Review Questions

1. Identify the following compounds as to whether they are coordinate covalent, polar covalent, or resonance structure. Draw each structure:

    A. H_2O

    B. CH_3

    C. NH_3

    D. CO_3

2. Compute the carbon and hydrogen for the first five compounds with each series:

    A. alkanes (C_nH_{2n+2})

    B. alkenes (C_nH_{2n})

    C. alkynes (C_nH_{2n-2})

3. List the six qualities that covalently bonded compounds can have.

4. Give the prefix for each of the numbered carbons:

    A. six

    B. eleven

    C. three

    D. seven

    E. twelve

    F. one

    G. five

    H. eight

    I. four

    J. nine

    K. two

    L. ten

5. Name the following formulas:

    A. C_4H_{10}

    B. $CH_3CH_2CH_2CH_2CH_2CH_3$

C. CH_4

D. CH_3CH_3

E. $CH_3CH_2CH_2CH_2CH_3$

F. $CH_3(CH_2)_5CH_3$

G. C_5H_{12}

H. $C_{10}H_{22}$

I. $CH_3CH_2CH_2CH_2CH_2CH_2CH_2CH_2CH_3$

J. $CH_3CH_2CH_2CH_2CH_2CH_2CH_2CH_3$

K. C_2H_6

L. $CH_3CH_2CH_2CH_3$

M. $CH_3(CH_2)_4CH_3$

N. $CH_3CH_2CH_2CH_2CH_2CH_2CH_2CH_2CH_2CH_3$

O. $CH_3(CH_2)_2CH_3$

P. C_6H_{14}

Q. C_8H_{18}

R. $CH_3CH_2CH_2CH_2CH_2CH_2CH_3$

S. C_7H_{16}

T. $CH_3(CH_2)_3CH_3$

U C_9H_{20}

V. $CH_3CH_2CH_2CH_2CH_2CH_2CH_2CH_2CH_2CH_2CH_3$

W. $CH_3CH_2CH_3$

X. $CH_3(CH_2)_6CH_3$

Y. $C_{12}H_{26}$

Z. CH_3CH_3

6. Draw the following compounds:

    A. heptane

    B. pentane

    C. butane

    D. propane

    E. hexane

F. ethane

G. octane

H. nonane

I. methane

7. Identify the primary and secondary carbons in the following compounds:

   A. heptane

   B. pentane

   C. butane

   D. undecane

   E. propane

   F. hexane

   G. ethane

   H. octane

   I. nonane

8. Draw the following molecules:

   A. heptene

   B. pentene

   C. undecene

   D. propene

   E. hexene

   F. ethene

   G. octene

   H. nonene

9. Draw the following compounds:

   A. propylene

   B. 1-butene

   C. 2-methyl-2-butene

   D. *trans*-1,2-dichloroethlene

   E. 2,3-dimethyl-2-butene

   F. 2-butene

   G. 1,2-butadiene

   H. 1,5-hexadiene

   I. *cis*-1,2-dichloroethlene

   J. 1,3-pentadiene

   K. 1,4-pentadiene

   L. isobutylene

   M. 1,1-dichloroethlene

10. Draw all the isomers of 1-bromo-1,2-dichloroethene.

11. Draw all the isomers of 1-bromo-1-chloro-propene.

12. Draw the *trans*- and *cis*- configurations of 2-butene.

13. Draw the following compounds:

    A. 2-butyne

    B. 1,3-butadiyne

    C. 1-butyne

    D. 3-hexyne

    E. 5-methyl-2-hexyne

14. Name the following compounds:

    A. $CH_3-CH_2-CH_2-CH_2-CH_3$

    B. $CH_3-CH_2-CH_2-CH_3$

    C. $CH_3-CH_2-CH_2-C{\equiv}CH$

    D. $CH_3-C{\equiv}C-CH_2-CH_3$

    E. $CH_3-CH_2-C{\equiv}C-CH_3$

    F. $CH_3-CH_2-CH_2-CH{=}CH_2$

    G. $CH_2{=}CH-CH_2-CH_2-CH_3$

    H. $CH_3-CH_2-CH{=}CH-CH_3$

    I. $CH_3-CH_2-\underset{\underset{Cl}{|}}{CH}-CH_2-CH_3$

    J. $CH_3-CH_2-CH_2-CH_2-CH_2-CH{=}CH_2$

    K. $CH_2{=}CH-CH{=}CH-CH_2Cl$

    L. $CH_3-CH_2-C{\equiv}C-CH_3$

    M. $CH_3-CH_2-C{\equiv}C-CH_3$

    N. $CH_3-CH{=}CH-CH{=}CH_2$

    O. $\underset{\underset{CH_3}{|}}{CH_2}-CH_2-CH_2-\underset{\underset{CH_2CH_3}{|}}{CH}-CH_3$

    P. $CH_3-CH_2-CH_2-\underset{\underset{CH_2-CH_2-CH_2-CH_3}{|}}{CH_2}-CH_2-CH_2-CH_2-CH_3$

15. Draw the following compounds:

    A. methane

    B. heptene

    C. 2-methyl-2-butene

    D. 2,3-dimethyl-2-butene

    E. 2-butene

    F. 1,2-butadiene

    G. 1,5-hexadiene

    H. 2-methyl-5-ethyl-heptane

    I. *c*-butane

    J. butane

    K. cyclopentene

    L. octane

    M. chloromethane

    N. undecane

    O. 1-butyne

P. c-propene

Q. 3-hexyne

R. cyclohexene

S. 5-methyl-2-hexyne

T. isobutane

U. n-pentane

V. 4-fluoro-2-heptene

W. 1,2-butadiene

X. dichloroethylene

Y. neopentane

16. What theory best describes the movement of gases?

A. Henry's law of solutions

B. Dalton's law of partial pressures

C. Kinetic molecular theory

D. Boyle's law

17. Depict the critical temperature and pressure of a hypothetical substance using a phase diagram. Discuss why temperature and pressure play a role in the movement of a gas into the liquid state.

18. In terms of pressure, what numbers can be considered as equals?

19. Describe each of the gas laws:

A. Boyle's law

B. Charles's law

C. Dalton's law

D. Henry's law

20. Describe the three factors that influence the boiling point of the aliphatic compounds.

21. Draw the following compounds:

A. 3,3-dimthylpentane

B. 4-isopropylheptane

C. 2,5-dimethylhexane

D. 4,6-dipropylnonane

E. 2,2,4,4-tetramethylpentane

F. 5-isopropyl-5-tert-butyldecane

Problem Set Answers

Problem Set 3.1

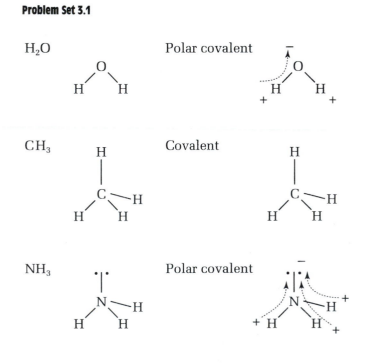

CO_3 Resonance structure

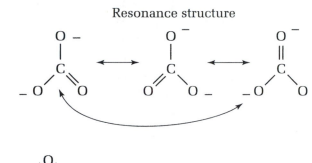

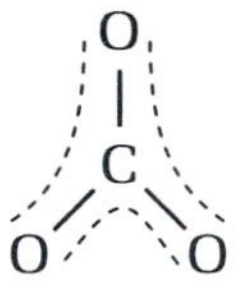

Problem Set 3.2

1. Methyl-   4. Ethyl-   7. Hexyl-   10. Decyl-
2. Heptyl-   5. Pentyl-   8. Octyl-   11. Nonyl-
3. Heptyl-   6. Heptyl-   9. Butyl-

Problem Set 3.3

1. Heptane

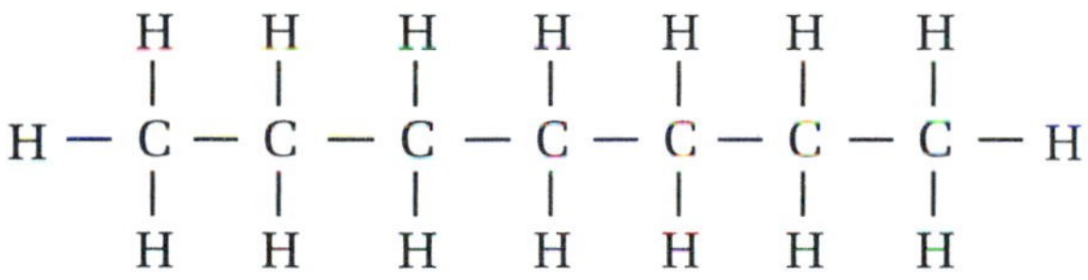

$$H - C - C - C - C - C - C - C - H$$

2. Pentane

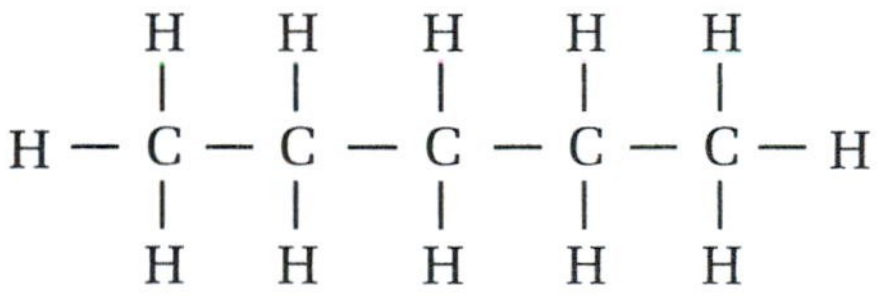

$$H - C - C - C - C - C - H$$

3. Butane

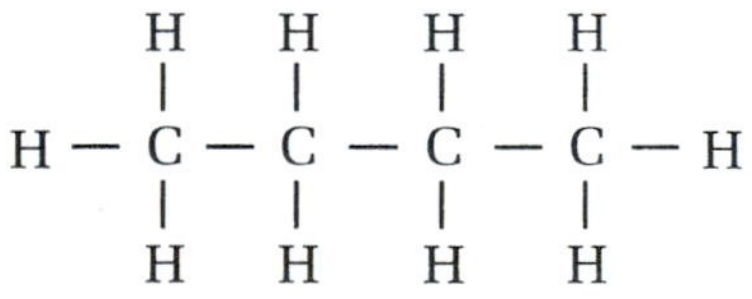

$$H - C - C - C - C - H$$

4. Undecane

$$H - C - C - C - C - C - C - C - C - C - C - C - H$$

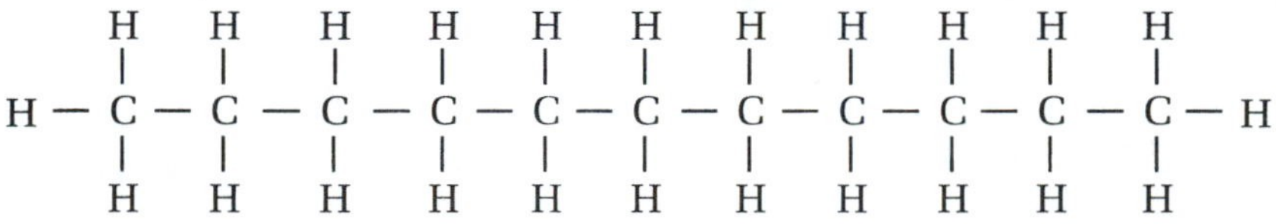

5. Propane

$$H - C - C - C - H$$

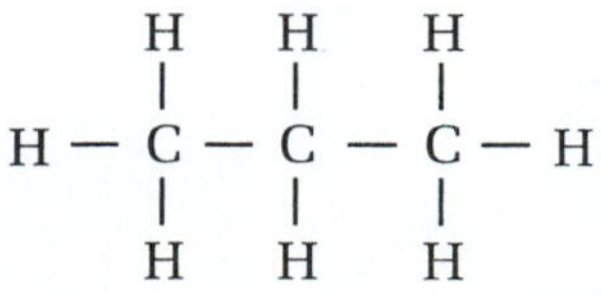

6. Hexane

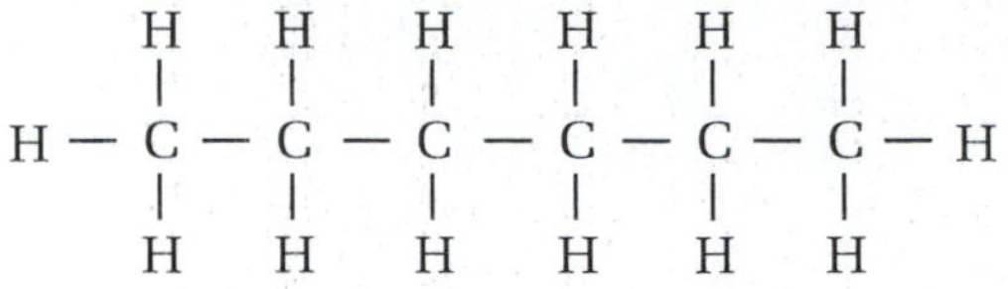

```
      H    H    H    H    H    H
      |    |    |    |    |    |
H  —  C  — C  — C  — C  — C  — C  —  H
      |    |    |    |    |    |
      H    H    H    H    H    H
```

7. Ethane

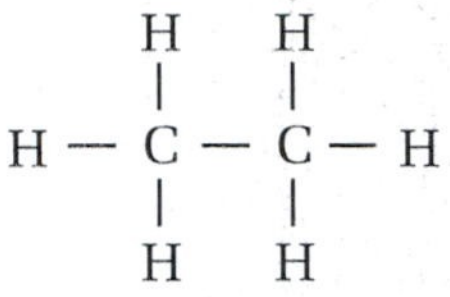

```
      H    H
      |    |
H  —  C  — C  —  H
      |    |
      H    H
```

8. Octane

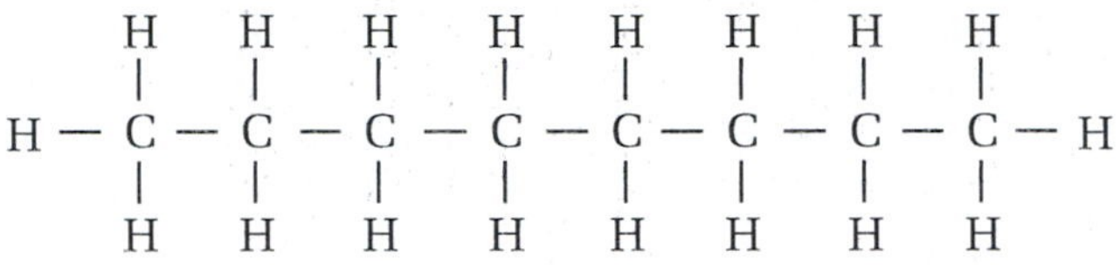

```
      H    H    H    H    H    H    H    H
      |    |    |    |    |    |    |    |
H  —  C  — C  — C  — C  — C  — C  — C  — C  —  H
      |    |    |    |    |    |    |    |
      H    H    H    H    H    H    H    H
```

9. Nonane

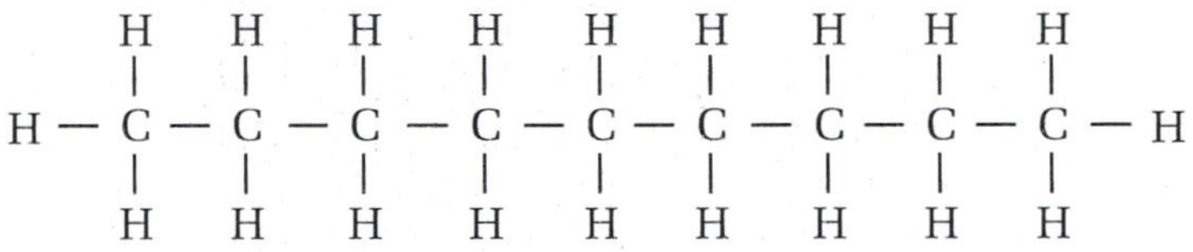

```
      H    H    H    H    H    H    H    H    H
      |    |    |    |    |    |    |    |    |
H  —  C  — C  — C  — C  — C  — C  — C  — C  — C  —  H
      |    |    |    |    |    |    |    |    |
      H    H    H    H    H    H    H    H    H
```

10. Methane

```
      H
      |
H  —  C  — H
      |
      H
```

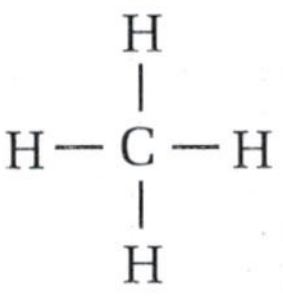

Problem Set 3.4

* = primary carbon   ' = secondary carbon

1. Heptane $C*H_3$-$C'H_2$-CH_2-CH_2-CH_2-CH_2-CH_3

2. Pentane $C*H_3$-$C'H_2$-CH_2-CH_2-CH_3

3. Butane $C*H_3$-$C'H_2$-CH_2-CH_3

4. Undecane $C*H_3$-$C'H_2$-CH_2-CH_2-CH_2-CH_2-CH_2-CH_2-CH_2-CH_2-CH_3

5. Propane $C*H_3$-$C'H_2$-CH_3

6. Hexane $C*H_3$-$C'H_2$-CH_2-CH_2-CH_2-CH_3

7. Ethane $C*H_3$-$C'H_3$

8. Octane $C*H_3$-$C'H_2$-CH_2-CH_2-CH_2-CH_2-CH_2-CH_3

9. Nonane $C*H_3$-$C'H_2$-CH_2-CH_2-CH_2-CH_2-CH_2-CH_2-CH_3

Problem Set 3.5

1. Heptene

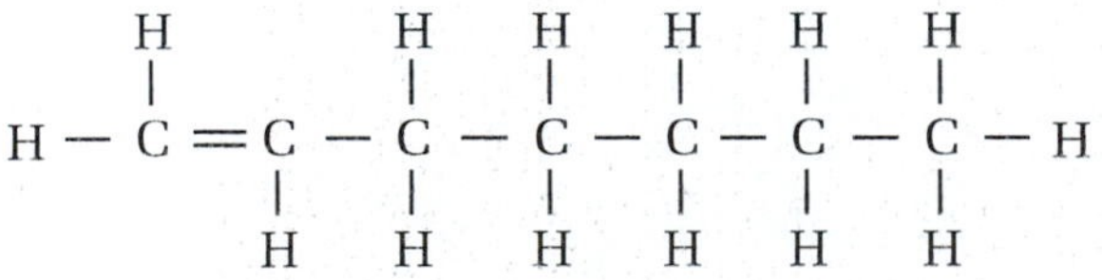

```
      H         H    H    H    H    H
      |         |    |    |    |    |
H  —  C  ═  C — C  — C  — C  — C  — C  —  H
                |    |    |    |    |
                H    H    H    H    H
```

2. Pentene

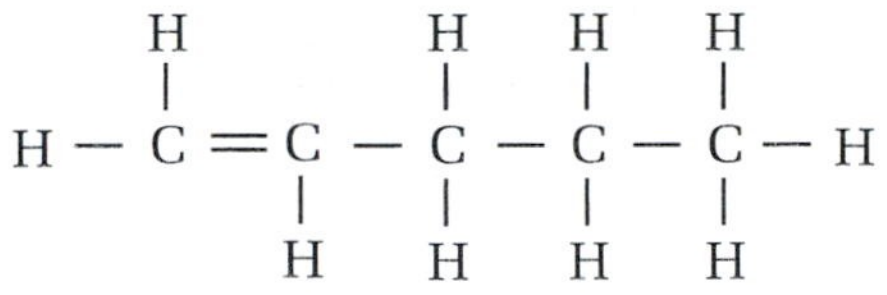

```
       H           H   H   H
       |           |   |   |
H — C = C — C — C — C — H
       |           |   |   |
       H           H   H   H
```

3. Undecene

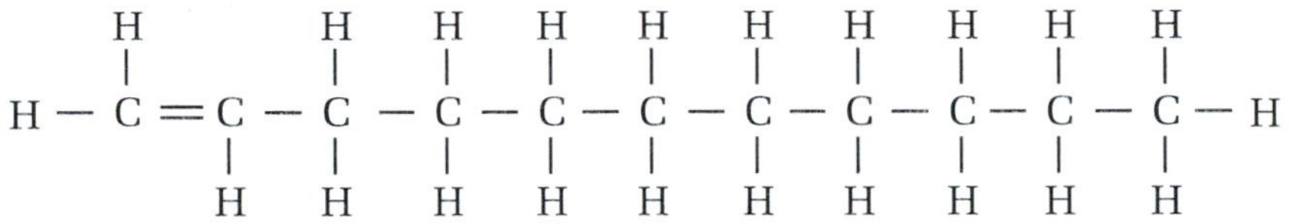

```
       H           H   H   H   H   H   H   H   H
       |           |   |   |   |   |   |   |   |
H — C = C — C — C — C — C — C — C — C — C — H
       |           |   |   |   |   |   |   |   |
       H           H   H   H   H   H   H   H   H
```

4. Propene

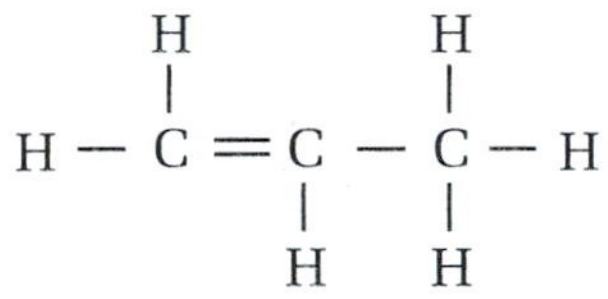

```
       H           H
       |           |
H — C = C — C — H
       |           |
       H           H
```

5. Hexene

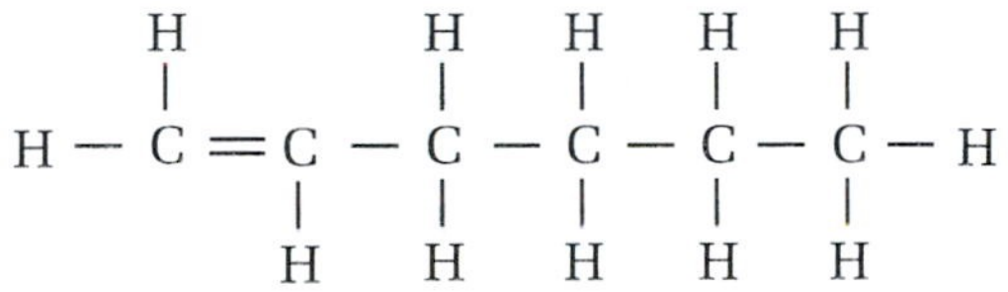

```
       H           H   H   H   H
       |           |   |   |   |
H — C = C — C — C — C — C — H
       |           |   |   |   |
       H           H   H   H   H
```

6. Ethene

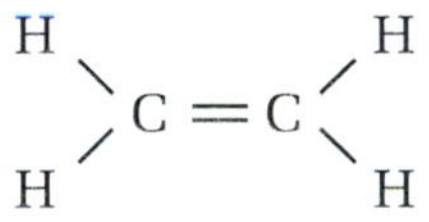

```
H             H
  \          /
   C = C
  /          \
H             H
```

7. Octene

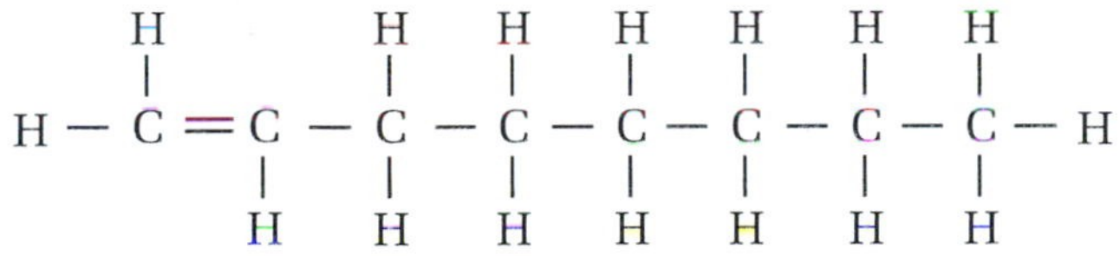

```
       H           H   H   H   H   H   H
       |           |   |   |   |   |   |
H — C = C — C — C — C — C — C — C — H
       |           |   |   |   |   |   |
       H           H   H   H   H   H   H
```

8. Nonene

```
       H           H   H   H   H   H   H   H
       |           |   |   |   |   |   |   |
H — C = C — C — C — C — C — C — C — C — H
       |           |   |   |   |   |   |   |
       H           H   H   H   H   H   H   H
```

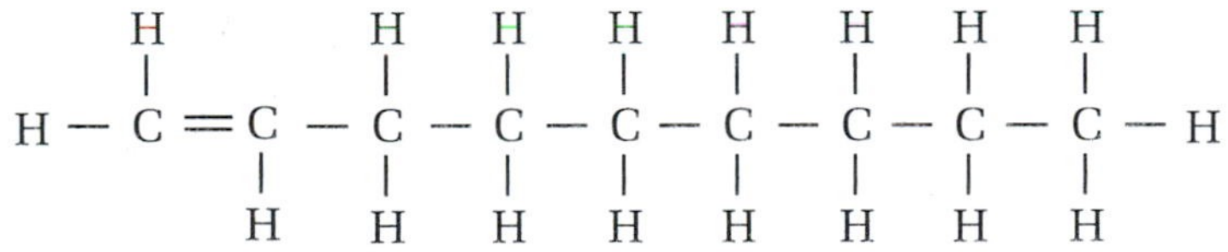

Problem Set 3.6

1. Propylene

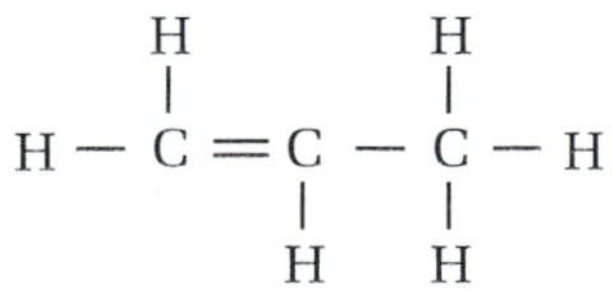

```
       H           H
       |           |
H — C = C — C — H
       |           |
       H           H
```

2. 1-Butene

```
       H           H   H
       |           |   |
H — C = C — C — C — H
       |           |   |
       H           H   H
```

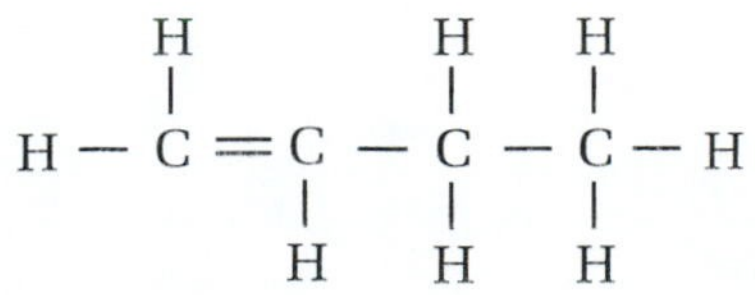

3. 2-Methyl-2-butene

4. 2,3-Dimethyl-2-butene

5. 2-Butene

6. 1,2-Butadiene

7. 1,5-Hexadiene

8. 1,3-Pentadiene

9. 1,3-Pentadiene

10. Isobutylene

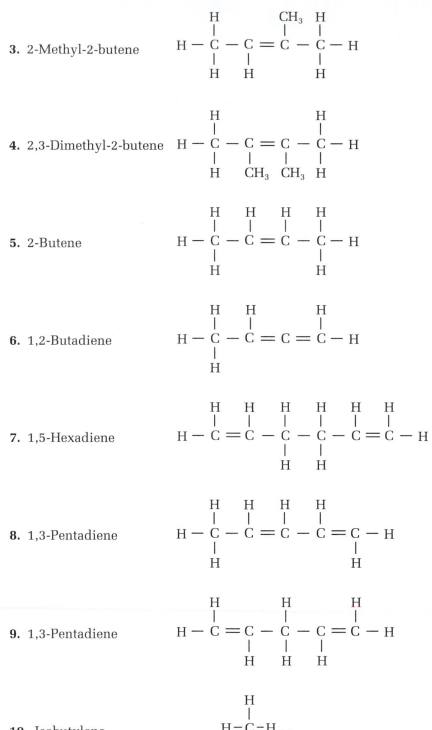

Problem Set 3.7

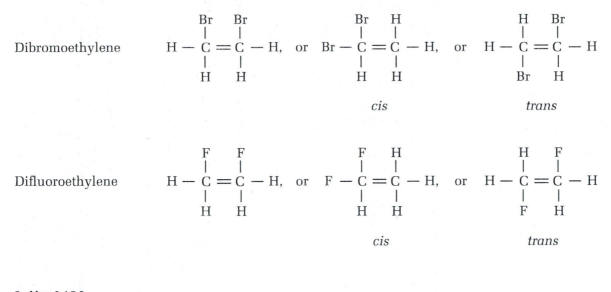

Dibromoethylene

Difluoroethylene

Problem Set 3.8

1. 2-Butyne

$$H - \overset{\overset{\displaystyle H}{|}}{\underset{\underset{\displaystyle H}{|}}{C}} - C \equiv C - \overset{\overset{\displaystyle H}{|}}{\underset{\underset{\displaystyle H}{|}}{C}} - H$$

2. 1,3-Butadiyne

$$H - C \equiv C - C \equiv C - H$$

3. 1-Butyne

$$H - C \equiv C - \overset{\overset{\displaystyle H}{|}}{\underset{\underset{\displaystyle H}{|}}{C}} - \overset{\overset{\displaystyle H}{|}}{\underset{\underset{\displaystyle H}{|}}{C}} - H$$

4. 3-Hexyne

$$H - \overset{\overset{\displaystyle H}{|}}{\underset{\underset{\displaystyle H}{|}}{C}} - \overset{\overset{\displaystyle H}{|}}{\underset{\underset{\displaystyle H}{|}}{C}} - C \equiv C - \overset{\overset{\displaystyle H}{|}}{\underset{\underset{\displaystyle H}{|}}{C}} - \overset{\overset{\displaystyle H}{|}}{\underset{\underset{\displaystyle H}{|}}{C}} - H$$

5. 5-Methyl-2-hexyne

$$H - \overset{\overset{\displaystyle H}{|}}{\underset{\underset{\displaystyle H}{|}}{C}} - C \equiv C - \overset{\overset{\displaystyle H}{|}}{\underset{\underset{\displaystyle H}{|}}{C}} - \overset{\overset{\displaystyle H}{|}}{\underset{\underset{\displaystyle CH_3}{|}}{C}} - \overset{\overset{\displaystyle H}{|}}{\underset{\underset{\displaystyle H}{|}}{C}} - H$$

Problem Set 3.9

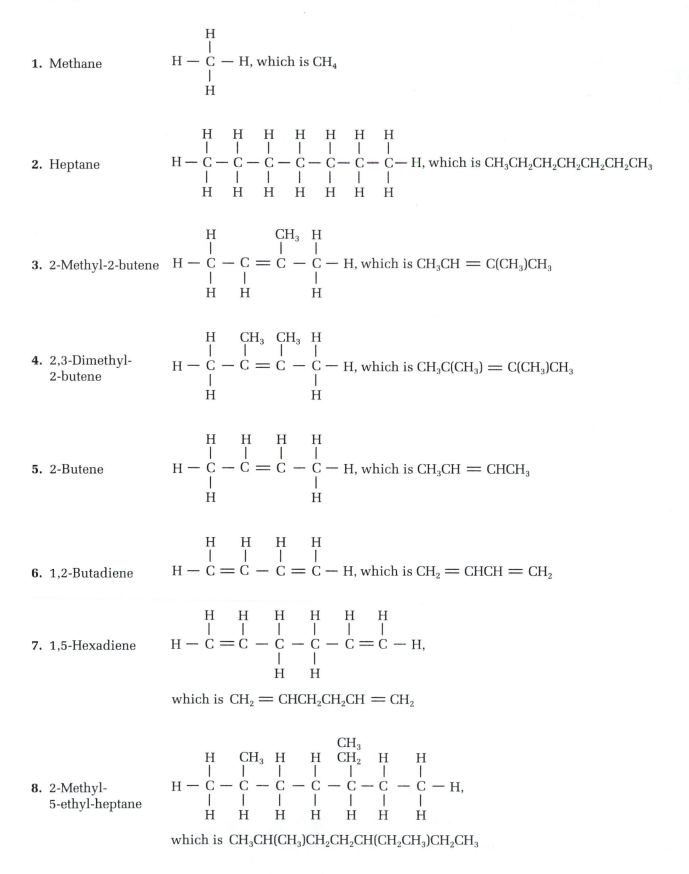

1. Methane

$$H - \overset{\overset{\displaystyle H}{|}}{\underset{\underset{\displaystyle H}{|}}{C}} - H, \text{ which is } CH_4$$

2. Heptane

$$H - \overset{\overset{\displaystyle H}{|}}{\underset{\underset{\displaystyle H}{|}}{C}} - \overset{\overset{\displaystyle H}{|}}{\underset{\underset{\displaystyle H}{|}}{C}} - \overset{\overset{\displaystyle H}{|}}{\underset{\underset{\displaystyle H}{|}}{C}} - \overset{\overset{\displaystyle H}{|}}{\underset{\underset{\displaystyle H}{|}}{C}} - \overset{\overset{\displaystyle H}{|}}{\underset{\underset{\displaystyle H}{|}}{C}} - \overset{\overset{\displaystyle H}{|}}{\underset{\underset{\displaystyle H}{|}}{C}} - \overset{\overset{\displaystyle H}{|}}{\underset{\underset{\displaystyle H}{|}}{C}} - H, \text{ which is } CH_3CH_2CH_2CH_2CH_2CH_2CH_3$$

3. 2-Methyl-2-butene

which is $CH_3CH = C(CH_3)CH_3$

4. 2,3-Dimethyl-2-butene

which is $CH_3C(CH_3) = C(CH_3)CH_3$

5. 2-Butene

which is $CH_3CH = CHCH_3$

6. 1,2-Butadiene

which is $CH_2 = CHCH = CH_2$

7. 1,5-Hexadiene

which is $CH_2 = CHCH_2CH_2CH = CH_2$

8. 2-Methyl-5-ethyl-heptane

which is $CH_3CH(CH_3)CH_2CH_2CH(CH_2CH_3)CH_2CH_3$

9. Butane

$$H - \overset{\overset{\displaystyle H}{|}}{\underset{\underset{\displaystyle H}{|}}{C}} - \overset{\overset{\displaystyle H}{|}}{\underset{\underset{\displaystyle H}{|}}{C}} - \overset{\overset{\displaystyle H}{|}}{\underset{\underset{\displaystyle H}{|}}{C}} - \overset{\overset{\displaystyle H}{|}}{\underset{\underset{\displaystyle H}{|}}{C}} - H, \text{ which is } CH_3 - CH_2CH_2 - CH_3$$

10. Hexyne

$$H - \overset{\overset{\displaystyle H}{|}}{\underset{\underset{\displaystyle H}{|}}{C}} - \overset{\overset{\displaystyle H}{|}}{\underset{\underset{\displaystyle H}{|}}{C}} - \overset{\overset{\displaystyle H}{|}}{\underset{\underset{\displaystyle H}{|}}{C}} - \overset{\overset{\displaystyle H}{|}}{\underset{\underset{\displaystyle H}{|}}{C}} - C \equiv C - H, \text{ which is } CH_3CH_2CH_2CH_2C \equiv CH$$

11. Octane

$$H - \overset{H}{\underset{H}{C}} - \overset{H}{\underset{H}{C}} - \overset{H}{\underset{H}{C}} - \overset{H}{\underset{H}{C}} - \overset{H}{\underset{H}{C}} - \overset{H}{\underset{H}{C}} - \overset{H}{\underset{H}{C}} - \overset{H}{\underset{H}{C}} - H, \text{ which is } CH_3(CH_2)_6CH_3$$

12. Chloromethane

$$Cl - \overset{\overset{\displaystyle H}{|}}{\underset{\underset{\displaystyle H}{|}}{C}} - H, \text{ which is } CH_3Cl$$

13. Undecane

$$H - \overset{H}{\underset{H}{C}} - \overset{H}{\underset{H}{C}} - \overset{H}{\underset{H}{C}} - \overset{H}{\underset{H}{C}} - \overset{H}{\underset{H}{C}} - \overset{H}{\underset{H}{C}} - \overset{H}{\underset{H}{C}} - \overset{H}{\underset{H}{C}} - \overset{H}{\underset{H}{C}} - \overset{H}{\underset{H}{C}} - \overset{H}{\underset{H}{C}} - H,$$

which is $CH_3(CH_2)_9CH_3$

14. 1-Butyne

$$H - \overset{\overset{\displaystyle H}{|}}{\underset{\underset{\displaystyle H}{|}}{C}} - \overset{\overset{\displaystyle H}{|}}{\underset{\underset{\displaystyle H}{|}}{C}} - C \equiv H - C, \text{ which is } CH_3CH_2C \equiv CH$$

15. 3-Hexyne

$$H - \overset{\overset{\displaystyle H}{|}}{\underset{\underset{\displaystyle H}{|}}{C}} - \overset{\overset{\displaystyle H}{|}}{\underset{\underset{\displaystyle H}{|}}{C}} - C \equiv C - \overset{\overset{\displaystyle H}{|}}{\underset{\underset{\displaystyle H}{|}}{C}} - \overset{\overset{\displaystyle H}{|}}{\underset{\underset{\displaystyle H}{|}}{C}} - H, \text{ which is } CH_3CH_2C \equiv CCH_2CH_3$$

16. 5-Methyl-2-hexyne

$$H - \overset{\overset{\displaystyle H}{|}}{\underset{\underset{\displaystyle H}{|}}{C}} - \overset{\overset{\displaystyle H}{|}}{\underset{\underset{\displaystyle CH_3}{|}}{C}} - \overset{\overset{\displaystyle H}{|}}{\underset{\underset{\displaystyle H}{|}}{C}} - C \equiv C - \overset{\overset{\displaystyle H}{|}}{\underset{\underset{\displaystyle H}{|}}{C}} - H,$$

which is $CH_3CH(CH_3)CHC \equiv CCH_3$

17. Isobutane

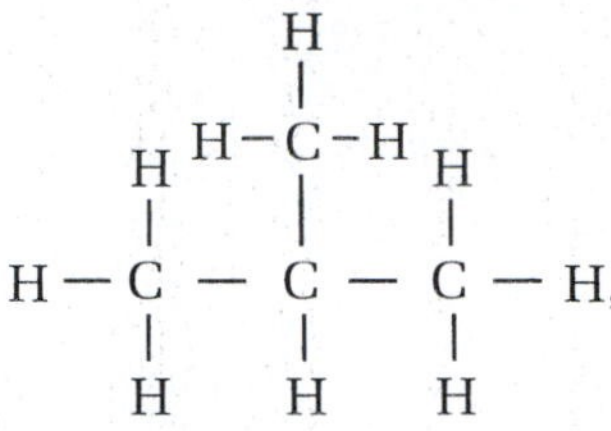

$$H-C-C-C-H, \text{ which is } CH_3CH(CH_3)_2$$

18. *n*-Pentane

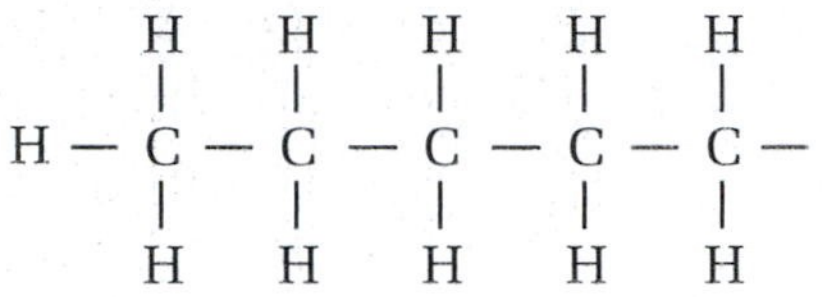

$$H-C-C-C-C-C-H, \text{ which is } CH_3CH_2CH_2CH_2CH_3$$

19. 4-Fluoro-2-heptene

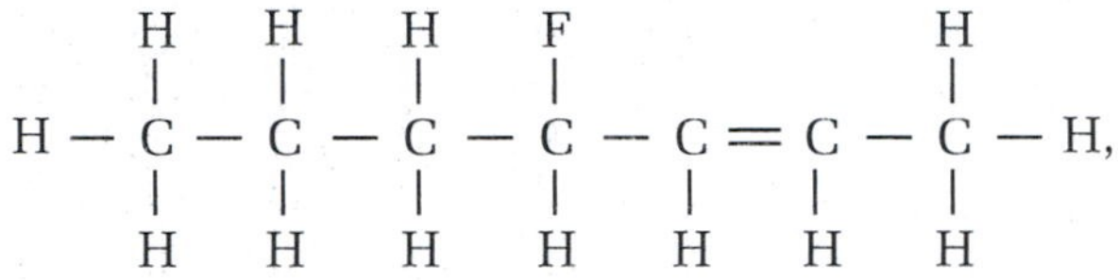

$$H-C-C-C-C-C=C-C-H,$$

which is $CH_3CH_2CH_2CH(F)CH=CHCH_3$

20. 1,2-Butadiene

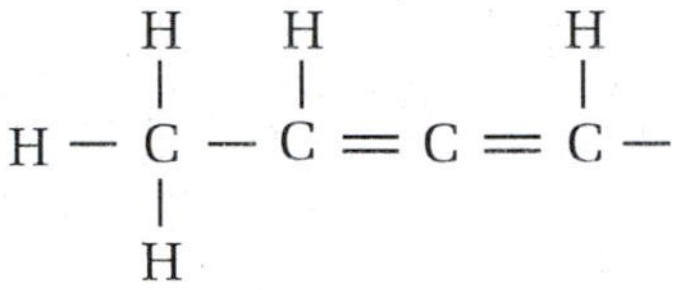

$$H-C-C=C=C-H, \text{ which is } CH_3CH=C=CH_2$$

21. Dichloroethylene

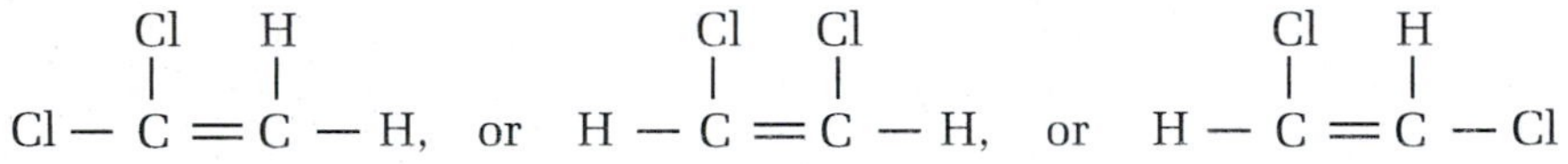

$$Cl-C=C-H, \text{ or } H-C=C-H, \text{ or } H-C=C-Cl$$

cis   *trans*

$$C(Cl_2)=CH_2 \qquad CHCl=CHCl \qquad CHCl=CHCl$$

                                cis   *trans*

22. Neopentane

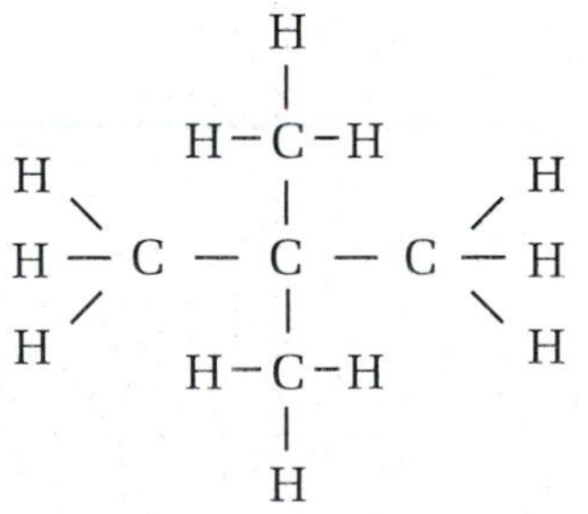

$$H-C-C-C-H$$

Problem Set 3.10

The tail of the carbon–carbon chain is connected to the head of the chain for the cycloalkane. The general formula is C_nH_{2n}, which is the same as seen in the alkene. There are two possibilities: either an alkene or a cycloalkene. Each chemical compound needs to be identified and properly researched. As seen in dichloroethylene, the isomers have different chemical and physical properties, along with different toxicological values. Additionally, the occupancy has alkanes and alkynes, each having specific chemical and toxicological qualities.

Problem Set 3.11

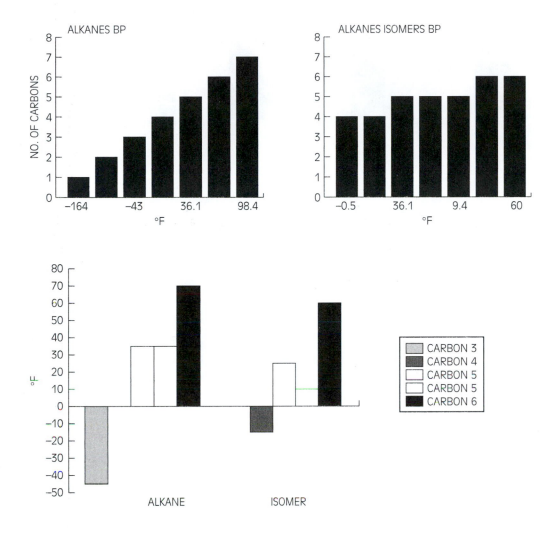

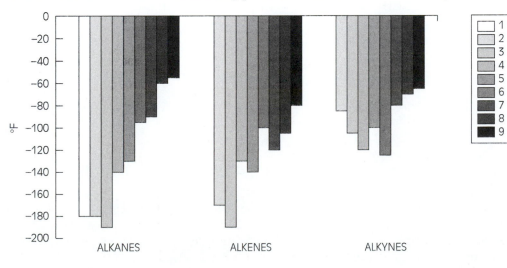

Number of carbons versus the melting point of alkanes, alkenes, and alkynes

Number of carbons versus the boiling points of alkanes, alkenes, and alkynes

200
150
100
50
0
−50
−100
−150
−200
°F
ALKANES
ALKENES
ALKYNES
1
2
3
4
5
6
7
8
9

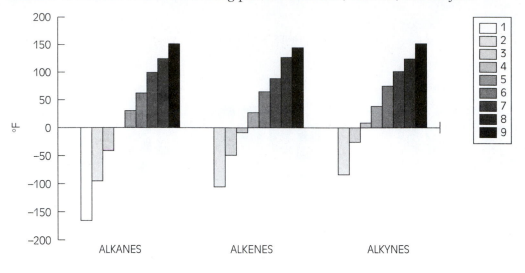

Problem Set 3.12

1. 3,5-Dimethylheptane

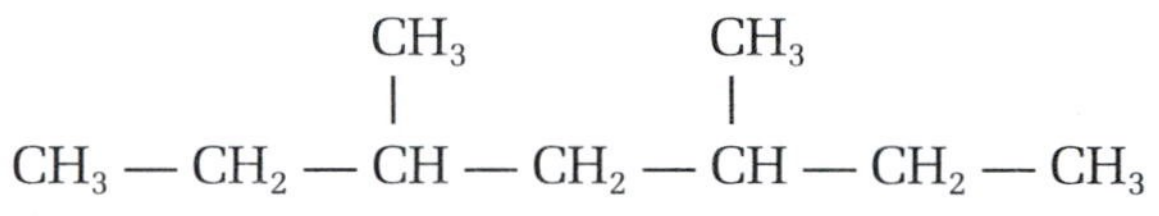

CH3 CH3
CH3 — CH2 — CH — CH2 — CH — CH2 — CH3

2. 3-Ethyl-2-methylpentane

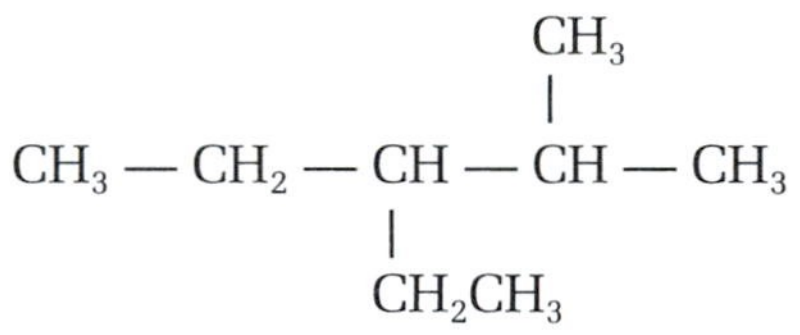

CH3
CH3 — CH2 — CH — CH — CH3
CH2CH3

3. 3,3-Dimethylpentane

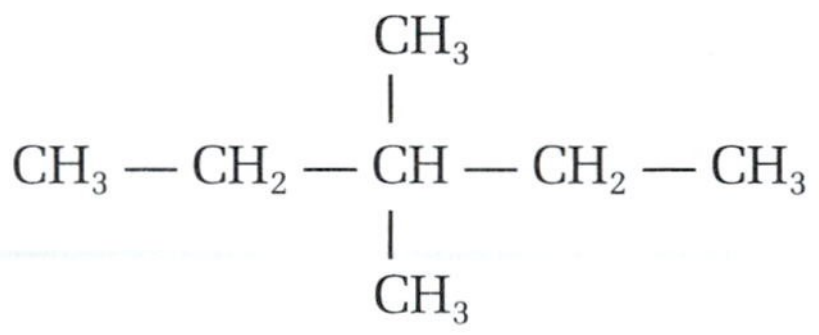

CH3
CH3 — CH2 — CH — CH2 — CH3
CH3

4. 2,2,4,4-Tetramethylpentane

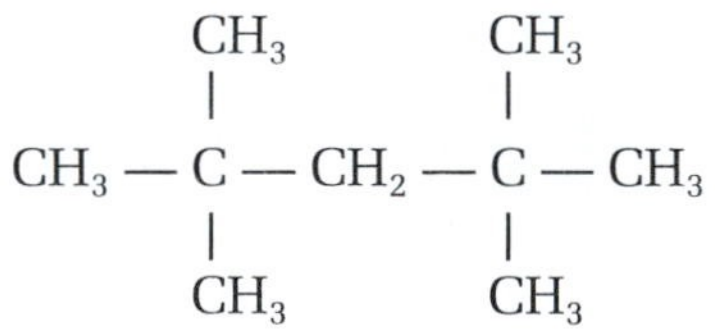

CH3 CH3
CH3 — C — CH2 — C — CH3
CH3 CH3

5. 4,6-Dipropylnonane

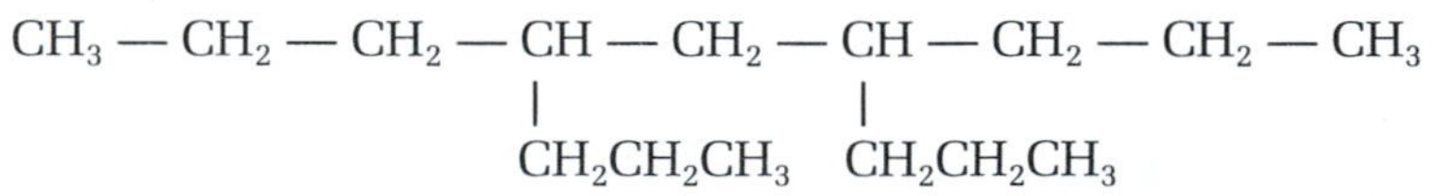

CH3 — CH2 — CH2 — CH — CH2 — CH — CH2 — CH2 — CH3
CH2CH2CH3 CH2CH2CH3

6. 5-Isopropyl-
5-tert-
butyldecane

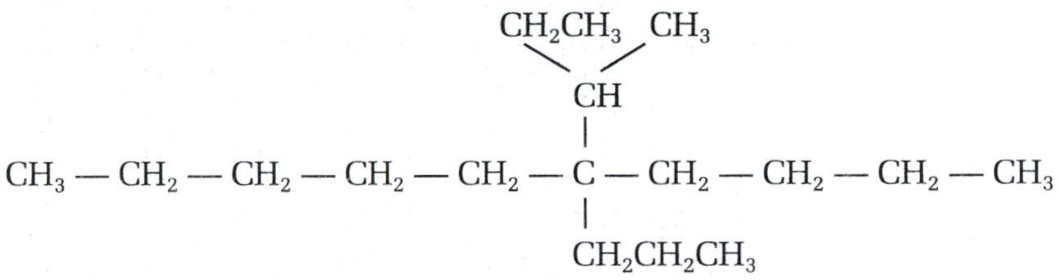

$$\overset{\displaystyle CH_2CH_3 \quad CH_3}{\underset{|}{\overset{\diagdown \quad \diagup}{\underset{CH_3-CH_2-CH_2-CH_2-CH_2-\underset{|}{\overset{|}{C}}-CH_2-CH_2-CH_2-CH_3}{\overset{CH}{}}}}}$$

$$CH_2CH_2CH_3$$

7. 4-Isopropylheptane

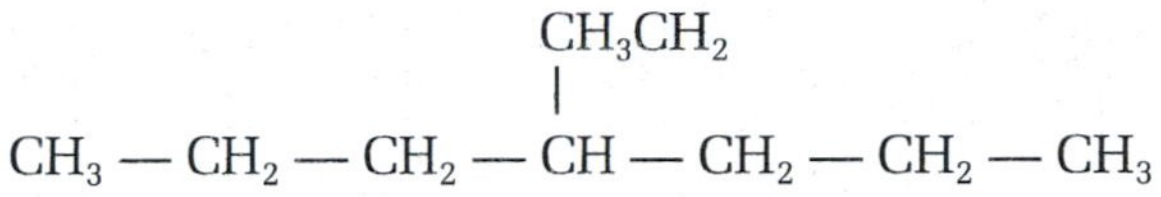

$$\overset{CH_3CH_2}{\underset{|}{CH_3-CH_2-CH_2-CH-CH_2-CH_2-CH_3}}$$

8. 2,5-Dimethylhexane

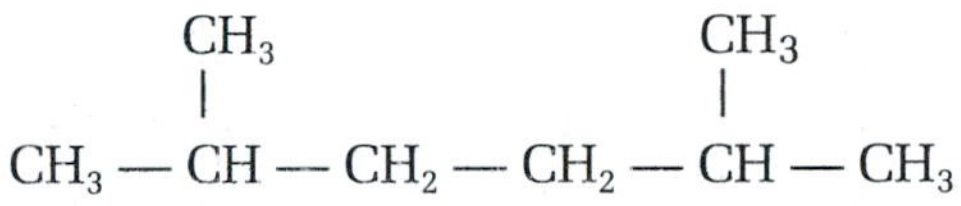

$$\overset{CH_3 \qquad\qquad CH_3}{\underset{|}{} \qquad\qquad \underset{|}{}}$$
$$CH_3-CH-CH_2-CH_2-CH-CH_3$$

9. 2,3-Dimethylhexene

$$\overset{CH_3}{\underset{|}{CH_2=C-CH-CH_2-CH_2-CH_3}}$$
$$\underset{|}{}$$
$$CH_3$$

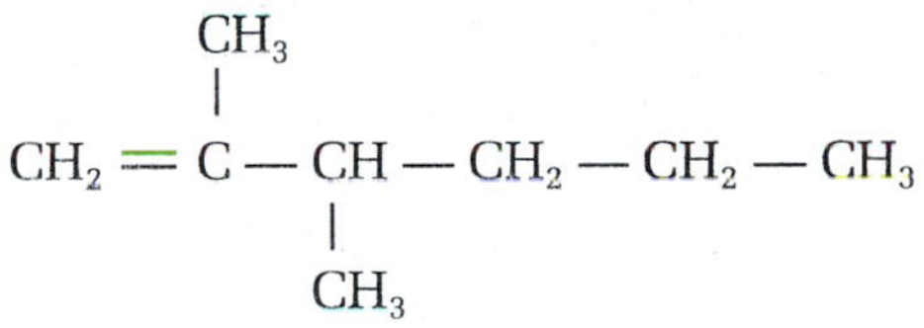

10. 2,2,6-Trimethyl-3-heptene

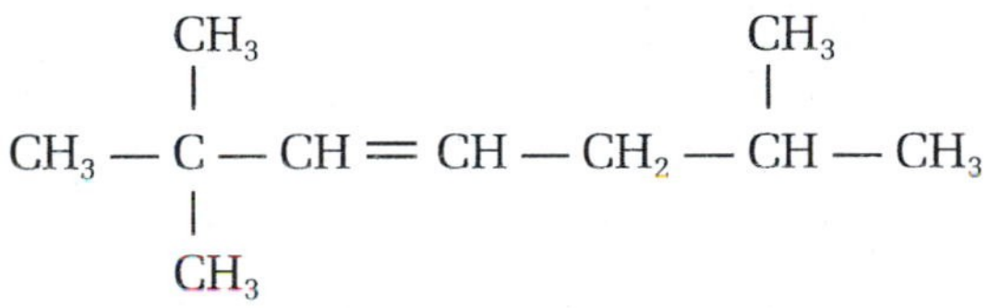

$$\overset{CH_3 \qquad\qquad CH_3}{\underset{|}{} \qquad\qquad \underset{|}{}}$$
$$CH_3-C-CH=CH-CH_2-CH-CH_3$$
$$\underset{|}{}$$
$$CH_3$$

Chapter

4

Organic Compounds II
Nonpolar Compounds

Learning Objectives

Upon completion of this chapter, you should be able to:

- Explain the benefit of understanding nomenclature in the science branch.
- Duplicate the nomenclature of hydrocarbon derivatives in general.
- Duplicate the nomenclature of alkyl halides, amines, derived ammonia compounds, ethers, epoxides, and organic peroxides.
- Identify the process of polymerization.
- Explain how branching can affect a chemical reaction.
- Describe explosive potentials.

What has been presented in the preceding chapters represents only a small portion of the chemical compounds that may be encountered by hazmat responders. This chapter and Chapter 5 cover more compounds and emphasize the need for nomenclature recognition. From the garage at home to the industrial process at the factory, these compounds are found in great numbers. Accidental releases of such chemicals pose a challenge for the emergency responder. In addition, new compounds are produced daily, and many compounds are increasingly used in military and terrorist applications. Information about the chemical helps decide on an approach to hazard mitigation and scene stabilization. Without understanding the nomenclature, research into the chemical could become laborious and time consuming. With the incident commander looking over one shoulder and the hazmat group officer looking over the other, understanding the nomenclature can provide the basic tools for rapid identification. As with any skill, repetition will give the practitioner confidence and knowledge. However, knowing your tools (reference books and databases) and how to use them will also provide you with the necessary information in the science branch.

■ Note

The top twenty chemicals produced in the United States based on volume are: (1) sulfuric acid, (2) nitrogen, (3) ethylene, (4) oxygen, (5) anhydrous ammonia, (6) lime, (7) phosphoric acid, (8) sodium hydroxide, (9) chlorine, (10) propylene, (11) soda ash, (12) urea (carbamide), (13) nitric acid, (14) Norway saltpeter, (15) ethylene dichloride, (16) benzene, (17) vinyl chloride, (18) carbon dioxide, (19) MTBE, and (20) phenylethane.

So far we have described how compounds are derived from the parent chain and the nomenclature that surrounds these compounds. This chapter and Chapter 5 cover the hydrocarbon derivatives, which, with their functional groups, can be polar or nonpolar. This polarity has to do with the tendency to have a magnetic positive and negative end. Under certain conditions, these compounds do not have this tendency. Rather, they remain somewhat neutral (in terms of polarity) in the reaction and do not readily create hydrogen bonding between molecules. Just as with covalence and ionic bonding, polar and nonpolar are terms of relative proportion. The compounds covered in this chapter are more nonpolar than they are polar, with the exception of the amines. Because amines can create hydrogen bonds in the lower molecular-weight compounds, their boiling points are elevated. Most of the other compounds in this chapter do not create hydrogen bonds, under most chemical reactions and conditions. Therefore, they are nonpolar. Nonpolar means that there is not a predominate negative or positive end of the molecule.

A hydrocarbon derivative is a compound that has a hydrocarbon backbone and a **functional group** attached. It is this functional group that gives the representative compound its chemistry and name. The backbone (as presented in Chapter 3) is a chain of carbons that has one or more of the hydrogens missing. As such, it defines the alkyl group, which now has a theoretical charge and can attach itself to the functional group. This alkyl group is referred to as the radical and is designated as "R". Our focus is on the functional group itself, so all chains of hydrocarbons are written as R to simplify the discussion.

functional group

any of the groups of atoms that identify a particular organic compound or represent a characteristic chemical entity

---

■ Note

You will find that during the referencing of hazardous compounds, many of the reference books identify compounds using the "R" designation. The use of R' and R" indicates the possibility of different radicals being attached to the hydrocarbon.

❗Safety

● When dealing with liquid spills producing a large amount of vapor, surface area plays a part in our assessment: The larger the surface area, the higher the vapor potential.

A QUICK REVIEW

In Chapter 1, we discussed volatility and vapor pressure, and in Chapter 3 we introduced the gas laws and, in particular, Henry's law. Let us take a moment to review the basic chemical terminology so that we can apply our knowledge. *Volatility* is the evaporation of a liquid, or the movement from a liquid into a gas. *Vapor pressure* is the pressure that is applied in a container due to the gas moving from the liquid to the gas state and back again. The vapor pressure of a compound is the liquid-to-gas-to-liquid equilibrium. Henry's law describes the relationship of gases in contact with a liquid and the constituents of the gases dissolving to different degrees in the liquid, which depends on the temperature and pressure of the gas and liquids involved. Simply put, Henry's law is a quantity or partial pressure of the gas in the environment such that each gas–liquid system has its own value or coefficient, which will vary with the temperature, solubility, pressure, and concentration of gases and solids that are dissolved.

Solubility depends on several variables. The first and primary variable is the polarity of the solute and solution (the polarity of the different compounds in the solution) and the temperature. Depending on polarity, temperature can change the solubility. As temperature increases, solubility will decrease between a gas and liquid environment, thus allowing the gas in solution to move easily into the vapor space or remain in solution (the result will depend on the polarity). Other material dissolved in the solution changes the solution environment and, in some cases, the polarity. Once this has occurred, the dissolved materials change the solution environment, creating gas in solution that can be released.

We can relate this back to the volatilization of a substance. We have already stated that the volatility of a compound is its ability to evaporate, which depends on vapor pressure, molecular weight, and water solubility. Water-soluble compounds have low volatilization rates as compared to non-water-soluble compounds, which have high volatilization rates. Additionally, compounds with a Henry's law coefficient of 10^{-3} or higher can be expected to evaporate readily. This concept is of vital importance when discussing environmental issues of a spilled liquid.

Consider, for example, 1,1-dichloroethylene. It has the following characteristics: MW, 96.94; BP, 89°F; VD, 3.25; VP, 591 mm; water solubility, 0.04%; and Henry's law constant, 2.61×10^{-2}. Compare acetone, which has MW, 58.09; BP, +133.7°F; VD, 2; VP, 231 mm; water solubility, miscible; and Henry's law constant, 3.97×10^{-5}. Due to the low water solubility of dichloroethylene and its high vapor pressure, Henry's law constant shows us that its volatility is higher than acetone's (greater than 10^{-3}). Thus, its movement into the vapor state.

A hydrocarbon derivative is essentially a two-part compound: the backbone and the functional groups having both polar and nonpolar qualities. Any hydrocarbon backbone may be formed with the associated functional group in order to make the hydrocarbon derivative. With each functional group, a separate set of physical and chemical properties exists. In addition, branching, low molecular-weight compounds, and the ability to form hydrogen bonding between molecules all have an effect on these properties. Just like the formulas that represent the aliphatic and aromatic hydrocarbons, the functional groups also have general formulas. It is these formulas that we use as identifying features of the hydrocarbon derivative.

A systematic naming system has to be able to accommodate many different compounds, which can become difficult, especially for compounds with double bonds. For example, we learned that $CH_2=CHCl$ has a base hydrocarbon of ethylene ("ethyl" for the two carbons and "-ene" for the double bond). However, it would be improper to name this compound chloroethylene. Instead we identify the double bonding as belonging to a vinyl group and end it with chloride: vinyl chloride. The vinyl group encompasses a procedure or naming configuration that identifies that specific molecule. It is used in the industry such that it is not confused with the normal two-carbon, double-bonded molecule. If we did not have this specific group name, confusion could arise with certain compounds and the naming configuration. Identifying the branching of hydrocarbon groups utilizes specific nomenclature, as shown in Figure 4–1.

Similarly, benzene is named as a phenyl group and is commonly referred to as such. When benzene was first discovered, it was known to have a six-carbon chain, but its properties fall somewhere between those of a cyclo-hexyane or a cyclo-hexyene. In 1865, while dozing on a train, Friedrich August Kekulé von Stradonitz theorized that if the bonds are all resonant in nature, then this resonance would accommodate what was observed in the laboratory and would suggest why benzene has specific properties. Kekulé suggested that the carbons are in a ring formation, with the electron cloud above and below the carbon plane allowing the electrons to rotate or move between each carbon and satisfying each carbon as necessary. Electronic stability is what all compounds are trying to achieve. Because of this need for stability, we see that six-carbon chains are sometimes incorporated into a circle structure, which we call a benzene ring or aromatic. In the benzene ring, a high level of stability is achieved by virtue of the ring. Because the carbon ring is more stable and stronger then the straight chain, a six-carbon structure will sometimes fall onto itself to create this structure (with a lot of help). The ring bond is resonant, or a combination of sharing, as illustrated in Figure 4–2.

$CHO-$ Formyl, a carbon with a hydrogen attached and a double-bonded oxygen (see Chapter 5, "Acyl Halides").

$$\overset{\displaystyle O}{\overset{\displaystyle \|}{CH_3C-}}$$ Acet, a methyl group plus one carbon; acetyl is a common naming configuration (see Chapter 5, "Esters").

$CH_2=CH-$ Vinyl

$CH_2=CHCH_2-$ Acryl

Figure 4–1 *Specific names for lower-weight hydrocarbon groups.*

$CH_3CH=CHCH_2-$ Croton

Figure 4–2
Resonance structures for the benzene ring.

■ Note

The prefixes that designate the benzene carbon of attachment with respect to the main attachment or point of orientation are ortho-, meta-, and para-, where ortho- is carbon number 2, or straight ahead; meta- is carbon number 3, or beyond; and para- is carbon number 4, or opposite.

■ Note

Styrene is a commonly used compound; by volume, the twenty-first highest produced in the United States. It is used in resins, protective coatings, and the like.

CH=CH₂

For the purposes of this discussion, we can call the top point of the benzene ring the number one carbon and the point of orientation. Moving to the right or left, we arrive at the second carbon. For example, if we say that an attachment occurs at the number one carbon and another occurs at the number two carbon, we have the same compound whether the number two carbon is to the left or the right of the number one carbon (see Figure 4–3).

For naming these compounds, we place names at each point of the ring. The three possible configurations for a benzene ring, other than the point of orientation, are called the ortho, meta, and para points (see Figure 4–4). These names are used to designate the carbon of attachment with respect to the main attachment or point of orientation. The base functional group is placed at the main point of attachment, and each successive carbon is called the ortho, meta, or para carbon. So, if we were to place a methyl radical on the number one carbon, we would have created a compound named toluene (methylbenzene) (see Figure 4–5). Adding a second methyl group in, say, the meta position would result in *m*-xylene. This is referred to as the BTX fraction (the commercial abbreviation for benzene, toluene, xylene). As you can see in Figure 4–6, each compound uses the components of the previous compound.

POINT OF ORIENTATION

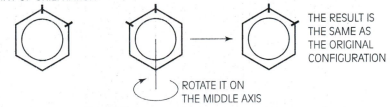

ROTATE IT ON
THE MIDDLE AXIS

THE RESULT IS
THE SAME AS
THE ORIGINAL
CONFIGURATION

Figure 4–3   *Identification of orientation.*

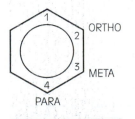

Figure 4–4   *Benzene ring orientation nomenclature.*

Benzene   Toluene   Xylene

Figure 4–5   *The BTX fraction.*

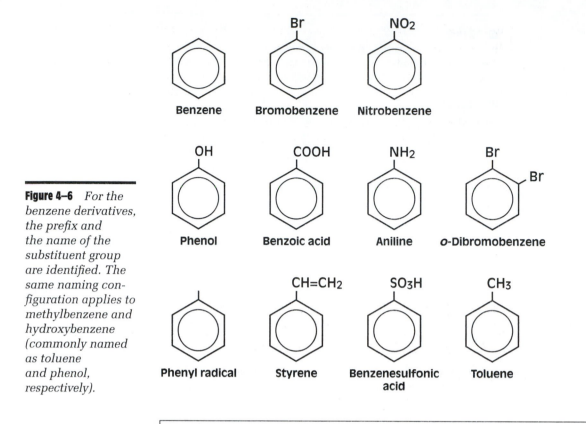

Figure 4–6 *For the benzene derivatives, the prefix and the name of the substituent group are identified. The same naming configuration applies to methylbenzene and hydroxybenzene (commonly named as toluene and phenol, respectively).*

⚠Safety

● There are significant chemical and toxological differences between the "straight ahead" ortho (*o-*), the "beyond" meta (*m-*), and the "opposite" para (*p-*) configurations. Therefore, each must be specifically identified to ensure appropriate response in a hazmat situation.

Problem Set 4.1 Draw the three isomers of xylene, naming each. Research each compound to identify its boiling point, vapor pressure, and toxicity.

Draw a benzene ring and attach a vinyl radical to it. Once you identify this compound, look up its uses and hazards.

ALKYL HALIDES

R-X

Chemistry Quick Reference Card

Family Class	Naming	Hazards
Alkyl halides	R–X (X = halogen)	Toxicity, flammability

halogenated hydrocarbon
a derivative of a hydrocarbon in which one or more of the hydrogens has been replaced by a halogen

The simplest hydrocarbon derivative involves the addition of a halogen onto the hydrocarbon backbone, resulting in the **halogenated hydrocarbons**, or the alkyl

R–X

Figure 4–7 *In the alkyl halides, R is the radical or hydrocarbon backbone and X is the halogen, F, Cl, Br, or I.*

halides (R–X). In this group, the R is the carbon–carbon chain and the X is the functional group or the halogen (see Figure 4–7). Because all the halogens react in a similar manner, the number of potential halogenated hydrocarbons is large.

The most common alkyl halides are represented by the lower-weight hydrocarbons, both saturated and unsaturated, inclusive of the isomers and cyclo compounds. Generally, we see the first four alkyl groups from the alkanes, the first three from the alkenes, and a variety of aromatic substances. Isomers are also seen in the industry, as are the halogenated cyclocompounds.

For alkyl halides, we see higher boiling points than for that of the parent chain. As a general rule, the boiling point increases with the higher-weight halogen that is attached to the carbon backbone. The alkyl halides are insoluble in water, which makes decontamination of these substances difficult if in the liquid state. Most are soluble in organic solutions with varying degrees of toxicity.

Halons are named using the number of the carbon, fluorine, chlorine, bromine, and iodine, respectively, in the compound (see Table 4–1). However, if the bromine number is zero, it is dropped from the numbered designation.

Problem Set 4.2   Draw the following compounds:

Difluorochloromethane

Trichlorofluoromethane

Dichlorodifluoromethane

Chlorofluoromethane

1,1,2-Trichloro-1,2,2-trifluoroethane

1,2-Dichloro-1,1,2,2-tetrafluoroethane

Chloropentafluoroethane

Hexafluoroethane

Phenylchloride

Vinyl chloride

Phenyldichloride

Biphenyl

Polychlorinated biphenyl

1,3,5 Tribromobenzene

2,4,6 Trichlor-toluene

Table 4–1 *Naming halons.*

Halon Compound	C	F	Cl	Br	
104	1	0	4	0	Carbon tetrachloride (freon-10)
122	1	2	2	0	Dichlorodifluoromethane (freon-12)
1011	1	0	1	1	Bromochloromethane
1202	1	2	0	2	Dibromodifluoromethane
1211	1	2	1	1	Bromochlorodifluoromethane
1301	1	3	0	1	Bromotrifluoromethane
2402	2	4	0	2	Dibromotetrafluoroethane

Methyl Chloride (Chloromethane)

In chloromethane (CH_3Cl), the functional group is named first and provides the proper name. It is used as a herbicide, topical anesthetic, low-temperature solvent, and catalyst in low-temperature polymerization. It is a colorless gas that can be easily liquefied. It is a flammable substance (LEL: 8.1%; UEL: 17.4%) with an odor threshold of 100 ppm. Its toxic parameters are a TWA of 50 ppm and a STEL of 100 ppm. The synonyms are monochloromethane and chloromethane. The common name of methyl chloride is used frequently.

> **!Safety**
> For gases under compression, such as those found in DOT Class 2, the definition for flammability is a gas at 68°F or less or a boiling point (BP) of 68°F or less, which is ignitable in a mixture of 13% or less or has a flammable range of 12% or greater.

Methylene Chloride (Dichloromethane)

A colorless, volatile liquid with an etherlike odor, methylene chloride (CH_2Cl_2) is considered nonflammable, but can ignite at 1224°F (LEL: 13%; UEL: 23%). It has an odor threshold of 300 ppm and a TWA of 50 ppm. Methylene chloride is converted to carbon monoxide in the body, making it an extremely hazardous substance. Its synonyms are dichloromethane (the common name), methylene dichloride, and methylene bichloride.

> **■ Note**
> The abbreviation CA identifies a carcinogenic compound. Although many compounds are suspected of being carcinogenic, very few compounds are actually known to be. Any compound even suspected to be carcinogenic receives the identifier of CA.

Trichloromethane (Chloroform)

Trichloromethane ($CHCl_3$) is a heavy, colorless, volatile liquid. It is one of the earliest-known anesthetics, but is no longer used because of its toxic qualities. It has a TWA of 2 ppm and a STEL of 50 ppm. Chloroform, as it is commonly known, is suspected to be carcinogenic. Its other synonym is methane trichloride.

Tetrachloromethane (Carbon Tetrachloride)

Also classified as a carcinogen, tetrachloromethane (CCl_4) was at one time used as a fire-extinguishing agent. However, during the process of extinguishment, phosgene and hydrogen chloride are produced, making it extremely dangerous. Carbon tetrachloride, as it is commonly called, in itself does not burn and is classified as a nonflammable liquid, but requires a poison label. It has an odor threshold of 10 ppm, a TWA of 5 ppm, and a STEL of 10 ppm. Perchloromethane is a synonym. It is commonly used as a refrigerant, degreasing agent, and drying agent for spark plugs.

Problem Set 4.3 Rename methyl chloride, methylene chloride, trichloromethane, and tetrachloromethane using F, Br, and I as the halogens. Draw each compound.

AMINES

RNH₂, R₂NH, or R₃N

Chemistry Quick Reference Card

Family Class	Naming	Hazards
Alkyl halides	R–X (X = halogen)	Toxicity, flammability
Amines	R$_x$NH$_x$ or ion	Toxicity, flammability

amine

an analog of ammonia that has been alkylated

The functional group in the **amine** category is the nitrogen group (RNH$_2$, R$_2$NH, or R$_3$N). This includes derivatives of the base ammonia, in which the hydrogen has been replaced. These compounds are slightly polar depending on the carbon chain attachment. This characteristic gives them high boiling points (but lower than for alcohols) due to the molecules' ability to create hydrogen–hydrogen bonding, comparable to the nonpolar compounds of the same molecular weight.

> ## ▌Safety
> Amines can be polar due to the hydrogen–nitrogen bond; however, that bond is of less strength than the hydrogen–OH bond created with alcohols. Amines have lower boiling points than alcohol, but higher than hydrocarbons of comparable molecular weights.
>
> Alcohols ↑
> Amines   Increasing BP
> Alkanes

This group of compounds has a characteristic fishy smell and is considered extremely toxic. As a class, the compounds are considered corrosives, especially if the compound is in the vapor state. Although not considered to have polar qualities, the lower molecular-weight compounds have polar tendencies. Rapid absorption can occur through all routes and through all body barriers (the skin is referenced as a barrier to some chemicals). If inhaled, the result is fatal due to the corrosive qualities. However, this fatal reaction is very chemical specific.

The reason for the toxicity in general is due to the ability to raise the pH of the affected tissue, which causes tissue necrosis similar to that of an alkali. If the amine becomes systemic, the pH change in the kidneys and liver can cause a necrotic state. If inhaled, lung tissue is severely damaged with pulmonary edema (noncardiogenic pulmonary edema). In all case studies reviewed, the exposure took weeks to months before a complete picture of symptomology appeared, especially in cases of low-dose exposures. The symptoms were most often described as dermatitis and skin discoloration. However, inhaled amines had a high mortality rate when the respiratory tree was unprotected. Some derivatives have created neurological effects such as hallucination, slurred speech, and general weakness. The greatest hazard posed by the compounds in this category is their tendency to cause intense chemical burns.

> ## ▌Safety
> The decontamination of amine compounds depends on their solubility in water. Each compound must be researched for this quality. As a general rule, the lower molecular-weight compounds will be slightly soluble in water.

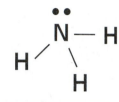

Figure 4–8

Ammonia has a tetrahedral shape, with the nonbonded electrons occupying the top portion of the tetrahedron. In an amine, one, two, or three of the hydrogens are replaced by the R group(s). It is the lone pair of electrons that gives the alkali qualities by allowing for $^+$H-N bonding.

Sometimes the classification of an amine depends on the number of R groups attached to the nitrogen. If one R group is attached, it is designated as a primary amine. Two R groups constitute a secondary amine, and three R groups, a tertiary. Like ammonia, amines are polar except for the tertiary species because of the hydrogen bonding that occurs. Amines have higher boiling points than nonpolar compounds of the same molecular weight, but lower than the alcohols and carboxylic acids. Primary, secondary, and tertiary amines are insoluble in water; however, their respective salt becomes soluble in water, thus leading to the injuries noted above.

The common name for the amines is found by naming the attached hydrocarbon chain(s), followed by amine. Notice in the following that all we are doing is replacing an H in the base ammonia compound (see Figure 4–8) with a hydrocarbon group(s):

CH_3NH_2	Methylamine
$CH_3CH_2NH_2$	Ethylamine
$(CH_3)_2NH$	Dimethylamine

The IUPAC rules of naming amines are basically the same as the common naming system. The only difference occurs when there are double bonds or isomers of the carbon chain. Then numbers are used to identify the carbon groups or the double bonds:

$CH_3(CH_3)C_2NH_2$	2-Methyl-2-propanamine (tert-Butylamine)
$CH_2=CHCH_2NH_2$	2-Propenamine (Allylamine)

Additionally, the -NH_2 group is named as the amino group and is often used when naming the more complicated amines. For example, take a six-carbon cyclo alkane (◯), attach an -NH_2 group, and it becomes cyclohexylamine. The same concept can be applied to the compounds that have a benzene ring. The ring is called the benzyl or phenyl group, depending on the chemical substitution that was used during the chemical process (see Figure 4–9).

The salts naming configurations covered in Chapter 2 can be applied with the quaternary ammonium compounds. These compounds are named with the R group, followed by adding ammonium and then the name of the anion, for example, ethylammonium sulfate, $(CH_3CH_2NH_3)_2^{+1} SO_4^{-2}$.

Derived Ammonia Compounds

Amides, nitrogen compounds derived from ammonia, are related to the amines. Because they have chemistry that is related to the carboxylic acids, a more detailed discussion is given in Chapter 5. However, keep in mind that inorganic amides

Figure 4–9   *Amine naming conventions.*

Cyclohexylamine   —NH₂

Benzylamine or Phenylmethanamine   —CH₂NH₂

Problem Set 4.4   Draw the following compounds:

Diethylamine

Isopropylamine

Diisobuylamine

Trimethylamine

Methylamine

2,4-Dimethyl-3-hexanamine

3-Chloro-2-methyl-1-propanamine

Name the following compounds:

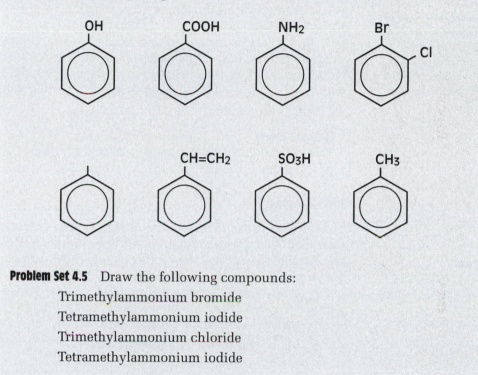

Problem Set 4.5   Draw the following compounds:

Trimethylammonium bromide

Tetramethylammonium iodide

Trimethylammonium chloride

Tetramethylammonium iodide

are produced by a reaction between an alkali metal and ammonia (see Chapter 2) Organic amides have the acyl group (-CONH$_2$) attached to the R group to yield the organic variety. In industrial chemistry, the amides of alkali metals are excellent catalysts and will produce amide by-products (see Figure 4–10).

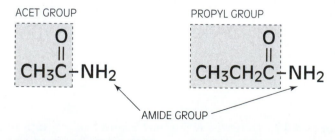

Figure 4–10   *Amide by-products.*

Acetamide or Ethanamide   **Propanamide**

Problem Set 4.6   Draw the following compounds:

Butylamide

Benzamide

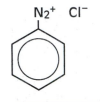

Figure 4–11

*Diazonium salt
structure, amine
derivative
(Benzenediazonium
chloride).*

Diazonium Salts

Another group of compounds that have a relationship with ammonia is the diazonium salts (see Figure 4–11). Here the nitrogen is replaced in the reaction. These products go through a combination of reactions. These salts are very useful in making intermediates in order to produce aromatic compounds. If the salts are not isolated during the intermediate reaction, thermal energy is produced, and an explosion can ensue.

Naming these compounds can become very complicated. However, the key principle to note here is the importance of identifying a diazontization process during preplanning for a response to a hazmat incident involving these compounds, or if this process is discovered during research of an actual incident.

> ❗**Safety**
> Be aware that the diazontization process can become extremely hazardous.

Mustard Agents

vesicant

an agent that causes
blistering

Because the nitrogen mustard agents fall into the general category of amines, the identification strategies employed will necessitate a knowledge of amine chemistry. Mustard agents are **vesicants** that are either nitrogen based or sulfur based (see more detailed discussion in Chapter 5). The arsenicals are also vesicants, of which lewisite, with arsenic as the central atom, is the most common family member. The following are the names of the nitrogen mustard agents, followed by their respective military designation:

Bis-(2-chloroethyl)ethylamine: HN-1

Bis-(2-chloroethyl)methylamine: HN-2

Tris-(2-chloroethyl)amine: HN-3

The nitrogen mustard agents can be detected quite effectively using colorimetric tubes. The nitrogen-based agents can be detected by organic nitrogen colorimetric tubes. For the sulfur-based vesicants, the thioether (thio- is a chemical designation for a sulfur compound) colorimetric tube must be used for detection. If lewisite (arsenic based) is suspected, then the tube of choice is the organic arsenic compound tube. The vaporized substance must be present for the tube to work. Keep in mind that the vaporization potential will increase as the temperature increases, causing further volatility of the compound.

> ■ **Note**
> The terms "bis" and "tris" are used in organic chemistry to mean taken two or three times, respectively.

Problem Set 4.7   Draw the HN-1, HN-2, and HN-3 compounds.

NITROGEN GROUPS

R-NO₂

Chemistry Quick Reference Card

Family Class	Naming	Hazards
Alkyl halides	R–X (X = halogen)	Toxicity, flammability
Amines	R_xNH_x or ion	Toxicity, flammability
Nitrogen groups	R–NO₂	Explosiveness

> **■ Note**
>
> In industry, nitro compounds are sometimes referred to as the picrates. These compounds are explosive depending on a number of factors, such as molecular weight. For example, nitro propane is a significant hazard, but mononitrobenzene is minor in comparison and trinitrobenzene is a maximum risk.

decomposition

a chemical reaction involving the breakup of a compound into a more basic compound or substance

detonation

combustion that occurs at greater than 1086 ft/s, which creates a shock wave moving at greater than 3300 ft/s

deflagration

combustion that occurs at less than 1086 ft/s, which creates a shock wave moving at less than 3300 ft/s

The nitrogen group (R–NO₂), sometimes referred to as the nitros, are used in a variety of ways, but are commonly found as explosives. The explosive principles common in this classification can also be considered when other compounds explode.

Explosive devices may contain filler material that is rated as either high-filler or low-filler explosive material. This material undergoes a sudden change in its chemical composition, or **decomposition**, resulting in a quick release of energy, heat, light, and gas. However, to achieve this end result, a sufficient amount of energy must be placed into the reaction. This energy can be provided by heat, friction, or impact. The energy entering into the reaction is used by the "unstable" chemicals, and burning results in the explosive compound. Additionally, the chemical compound itself must be under compression (or contained).

Explosive materials are classified as high order and low order. High-filler materials are further divided into high-order or low-order **detonation**. The difference between these two high-filler materials is the degree of consumption during the explosion. High-order materials detonate utilizing all of the filler material at an explosive burn rate of greater than 1086 ft/s (supersonic) or producing a shock wave greater than 3300 ft/s, destroying the target by shattering structural materials. Low-order materials undergo **deflagration** (rapid burn less than the 1086 ft/s [subsonic]), producing a shock wave less than 3300 ft/s, and may not completely consume the filler. Low-order fillers destroy targets by a push–pull–shove effect that weakens structural integrity. Low-order detonations are hazardous for the first responder because of the unconsumed explosive material, which has the potential to cause a fire hazard. The pressure waves for detonation can reach 50,000–4,000,000 psi. See Table 4–2 for various detonation velocities.

> **■ Note**
> The speed of sound is 1086 ft/s.
>
> **■ Note**
> Brisance is a measure of an explosive material's ability to shatter.

Table 4–2   *Detonation velocities.*

Material	Use	Velocity of Detonation (ft/s)
Black powder	Timed fuse	1,300
Nitroglycerin	Commercial dynamites	5,200
Ammonium nitrate	Demolition charge and composition explosives	8,900
TNT	Demolition charge and composition explosives	22,600
Teryl	Booster charge and composition explosives	23,200
C3	Demolition charge	25,000
PETN	Detonation cord, blasting caps, and demolition charges	27,200
RDX	Blasting caps and composition explosives	27,400

There is a technical difference between yield and order. Yield describes the rate or speed of the explosion, and order describes the rate of burn. The rate of burn has a relationship to the shrapnel that is propelled outward. Roughly speaking, the rate at which shrapnel is pushed away from the explosive device is approximately 75% of the rate of burn. For example, if a substance burns at 19,000 ft/s, then the shrapnel will move outward at approximately 14,000 ft/s.

❗Safety

Hurricane-force winds of 120 mph create a dynamic pressure of 0.25 psi. A shock wave of 940 mph can cause a pressure of 40 psi. Shock waves, or overpressure, are like sound waves in that they can be detected by a person behind walls, or in holes or caves. Pressure moves through material. Take cover accordingly.

■ Note

The nitro group is represented by the following structure:

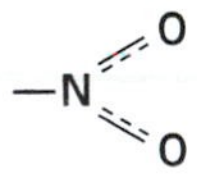

Explosiveness is due to the materials themselves and to some principles of physics. If we were to look at these materials on a microscopic level we would find them to be nonhomogenous, with different materials associated with special voids. These voids and the idea of compression are important concepts related to explosives. Explosive potential is characterized by the way the material burns in an enclosed container.

Disregarding the order of the explosive, consider the burning of a material that is occurring at an extremely fast rate in a container (see Figure 4–12). We can see that several layers are produced in the process. As the burn starts, a flame zone is recognized. This zone moves through the material; assume a direction from left to right. Behind this zone, pressure is building due to the flame decomposing the material itself. Gases, heat, and energy are produced because of the

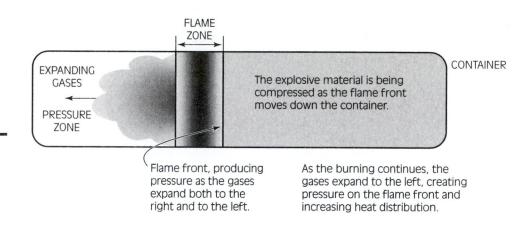

Figure 4–12
Dynamics of material burning extremely fast in a container.

flame, and cause a compression or a pressure zone. This pressure zone expands proportionally to the rate of burn and in association to the confinement.

Pressure continues to build, placing strain on the unused material and the container itself. It increases in the flame zone, forcing the flame closer and moving it faster toward the surface of the explosive material. The movement of the flame, in turn, increases the rate of heat transfer between the material that is burning and the unburned material. With increasing speed, the flame zone moves deeper into the unburned material, which causes pressure in the material. Gases are trying to get out of the container and are pushing the flame zone down the column of material. This ramming effect of the pressure wave with the flame zone increases the internal pressure of the material, causing an accelerated temperature and pressure buildup. The process continues until a point is reached at which the leading edge of the flame front pressure wave starts to crest and overtake the leading edge of the pressure wave. At this point, the movement of flame and pressure creates a shock wave, and an explosion results (see Figure 4–13).

■ Note
A blasting agent is an insensitive material that allows for little probability of an accidental explosion or burn that could create a detonation.

■ Note
Static electricity and RF are both energy sources, but each is produced by different means. Static electricity results from a difference in surface environments. A transmitting instrument is needed to produce RF.

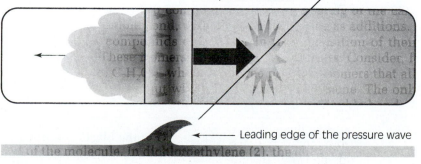

Figure 4–13   *Pressure dynamics leading to a shock wave and explosion.*

High-order explosives follow a sequence of events to detonate. This sequence of events is called an *explosive train* and has three separate segments: ignition source, initiating explosive, and main charge. Between the initiating explosive and main charge, a booster charge or delay charge may be utilized to ensure complete consumption.

Common initiating explosives include either electric or nonelectric blasting caps to start the train. An electric blasting cap consists of a shell with several explosive powder charges and an electric ignition element. The danger of electrical caps includes the possibility of accidental ignition from static electricity or energy from radio transmissions (radio frequency, RF, does have enough potential energy to start the train). Nonelectrical blasting caps utilize a fuse and initiator for ignition. A safety fuse has a combustible primer charge at the insertion end of the fuse.

❗Safety

The distance required as a safety margin between a radio transmitter and a wire blasting cap is a function of frequency, transmission power, and resistance.

● Caution

The generation of static electricity at the scene of a hazardous materials incident could provide the energy that is required to "push" a chemical reaction. As stated in Chapter 1, energy may be all that is required to initiate a reaction, and in some cases, static electricity is all the energy necessary.

In general, an electronic charge exists on all surfaces that are in different states of matter. When a surface is not grounded, the charges seek ground and will create a spark if the charge is high enough. The movement or flow of nonpolar hydrocarbons, for example, will produce a static field, which, when grounded, can result in a spark. If the gas or vapor coming off the liquid is within the flammable range, ignition will occur. Propane, as an example, has a minimum ignition energy of 0.0021 mJ; a static spark can create 1 mJ of energy. Grounding and bonding of containers and piping is imperative for safety.

Remember that temperature and humidity have an effect on the level of static conditions. In general, low temperatures and dry air create the highest potential for static electricity; humid air and high temperatures have the lowest static energy potential.

When a detonation occurs, a wave of pressure moves from the detonation point outward in all directions. The force of the initial shock wave depends on the type of explosive, confinement of the material, and the oxidizers present. This pressure continues to move out in every direction until the released energy has been equalized. Depending on the level of explosive order, a pressure wave can achieve pressures much greater than 50,000 psi. (Think of this wave as a locomotive hauling freight and striking objects while moving at 15,000 mph!) Atomization and total disintegration of material, including all biological material, will occur close to the detonation, or at ground zero. See Table 4–3 for additional shock wave effects.

■ Note

Alfred Nobel invented dynamite. Because he did not want to be remembered for his destructive invention, he created the Nobel Prize.

The initial wave caused from the blast as it moves through the air is called the *primary wave*. It consists of pressure change, heat, and fragmentation, and has two phases: the positive phase and negative phase. The positive phase includes

Table 4–3 *Shock wave dynamics.*

Potential Injury	Pressure (psi)	Structural Effects
Loss of balance, rupture of eardrums	0.5–3	Glass shatters, façade failure
Internal organ damage	5–6	Cinderblock shatters, steel structures fail, containers collapse, utility poles fail
Multisystem trauma	15	Structural failure of typical construction
Lung collapse	30	Reinforced construction failure
Fatal injuries	100	Structural failure

heat, high-pressure gases, and fragments all moving away from the center. The negative phase is created when the surrounding air around the blast rushes in to fill the vacuum created during the movement of the positive phase. Because air is rushing into an area in which a vacuum has been created, the negative phase is longer in total time duration. If a low-order explosion occurs in a confined area, such as in a building, the resulting injuries may mimic those typically seen following a high-order bomb in the open, due to the effects of implosion.

Most of the compounds in this group have common or trade names (see Table 4–4). In addition to their explosive potential, they can cause lowered blood pressure through vasodilation. Some compounds follow a naming sequence in which the nitro group is named first, then the hydrocarbon chain:

CH_3NO_2	Nitromethane
$NO_2\text{-}CH\text{-}NO_2$	Dinitromethane
$C_6H_2CH_3(NO_2)_3$	Trinitrotoluene
$C_6H_5\text{-}NO_2$	Nitrobenzene
$C_6H_5\text{-}CH_3(NO_2)_2$	Dinitrotoluene
$C_2H_5NO_2$	Nitroethane

Problem Set 4.8   Draw nitromethane, dinitromethane, trinitrotoluene, nitrobenzene, dinitrotoluene, and nitroethane.

ETHERS AND EPOXIDES

R'-O-R"

Chemistry Quick Reference Card

Family Class	Naming	Hazards
Alkyl halides	R–X (X = halogen)	Toxicity, flammability
Amines	R_xNH_x or ion	Toxicity, flammability
Nitrogen groups	$R–NO_2$	Explosiveness
Ethers	R'–O–R"	Anesthetic effects Flammability Potential to become peroxides

Table 4–4   *Representative compounds and their respective names.*

Compound Description

High Order	
Trinitrotoluene	Pale yellow solid, light cream to rust in color
Tetrytol	Alternate to TNT (2,4,6-trinitrophenylmethylnitramine), very sensitive
Composition 3	Plastic explosive with an oily texture and yellow tint
Composition 4	A white, pliable plastic explosive
Composition B	Mixture of cyclonite and TNT
Pentaerythritetranitrate	White solid, used in detonating cord, equal in force to RDX and TNT
Amatol	Mixture of ammonium nitrate and TNT or fuel oil
Cyclonite	A white solid, extremely sensitive; commonly referred to as RDX
Pentolite	Combination of pentaerythritetranitrate (PETN) and TNT
Ednatol	Mixture of halite and TNT
Semtex	Plastic explosive that can be molded; has a pinkish color
Picric acid	More sensitive than TNT; in the salt form, bright yellow crystals
Nitroglycerin	Heavy, oily liquid that resembles water; generally pale yellow and viscous
Dynamite	A mixture of nitroglycerin with sawdust and clay available in percentages of nitroglycerin
Low Order	
Black powder	Mixture of charcoal, sulfur, and either potassium nitrate or sodium nitrate
Ammonium nitrate	Fertilizer
Pyrotechnics	Variety of fireworks types and chemical types; match heads
Incendiary	
Gasoline	Mixture of petroleum products
Pyrotechnics	Variety of fireworks types and chemical types; match heads
Butane, propane	Petroleum gases
Hypergolic chemicals	Alkenes, alkynes in contact with a strong oxidizer

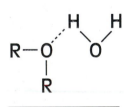

Figure 4–14
Hydrogen bonding between a water molecule and an ether. Alcohols will act in a similar manner, creating hydrogen bonding.

ether
an organic compound bonded in the middle with oxygen and two alkyl or aryl groups

In the **ether** family (R′–O–R″), there is a wide range of solvents, refrigerants, and pharmaceutical applications. Looking at this family from a structural standpoint, the ethers are related to water and the alcohols. For the most part, the lower molecular-weight compounds can form hydrogen bonds with water (see Figure 4–14), giving them a solubility factor. Although more soluble than what we have discovered in the hydrocarbon derivative thus far, their solubility is dependent on the compound and the hydrocarbon attachments.

The general formula is two alkyl groups separated by an oxygen atom. The R group can be the same or of different carbon–carbon chain lengths. The members in this family have low boiling points as compared to isometric alcohols. The weak bonding that occurs between the hydrogen and water tends to give this group extremely slight solubility factors, but only for the low carbon-weight ethers. The chemical dilemma we face with this group is the possibility of oxidation into the organic peroxides.

The naming configuration has the alkyl groups named in order of size, with "ether" at the end; for example:

$CH_3CH_2\text{-}O\text{-}CH_2CH_3$ Diethyl ether

$CH_3\text{-}O\text{-}CH_2CH_3$ Methyl ethyl ether

$CH_3CH_2\text{-}O\text{-}CH_2CH_2CH_3$ Ethyl propyl ether

In the IUPAC naming system, the R group that is the largest is the parent hydrocarbon. With the attachment of the oxygen atom, the term -oxy is placed between the parent chain and the smaller alkyl group, as in the following:

$CH_3CHCH_2CH_3$ 2-Methoxybutane
$\quad\ \ |$
$\quad\ \ OCH_3$

$CH_3\text{-}O\text{-}CH_2CH_2\text{-}O\text{-}CH_3$ 1,2-Dimethoxyethane

---

● Caution

By volume, the nineteenth most-produced chemical in the United States, MTBE (or methyl-tert-butyl-ether), is made by cracking methanol and isobutene. It is used as an antiknock agent and octane booster in gasoline. It possesses polar-like qualities that enable it to break down standard aqueous film-forming foam (AFFF). Therefore, the use of alcohol-resistant foam is suggested at any incident involving gasoline.

MTBE is classed as a possible carcinogen, resists biodegradation, and has leached into the groundwater in forty-nine states. When it burns, it produces carbon dioxide, carbon monoxide, water vapor tert-butyl formate, acetone, formic acid, and methyl radicals.

Problem Set 4.9   Draw the following ethers:

Diisopropyl ether

Methyl ethyl ether

Dimethyl ether

Di-n-butyl ether

Di-n-propyl ether

Diethyl ether

2-Methoxypentane

1,2-Dimethoxyethane

1-Methoxy-1-propene

Ethyl propyl ether

■ Note

A brief list of chemicals commonly found in clandestine laboratories:

Bases: sodium hydroxide, methylamine, piperdine

Acids: hydriodic acid, hydrochloric acid, nitric acid, sulfuric acid, hydrogen chloride gas

Flammables: diethyl ether, petroleum ether, ethanol, isopropyl alcohol, acetone, Coleman fuel

Respiratory irritants: acetic anhydride, hydriodic acid, methylamine, hydrochloric acid

Metal poisons: mercuric chloride, lead acetate

Poisons: sodium cyanide, potassium cyanide

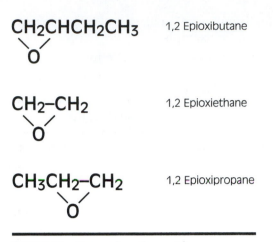

CH₂CHCH₂CH₃   1,2 Epioxibutane

CH₂–CH₂   1,2 Epioxiethane

CH₃CH₂–CH₂   1,2 Epioxipropane

Figure 4–15 *Examples of epoxides.*

epoxide
an ether in a cyclic configuration

When an ether is in the cyclic configuration we term such a compound an **epoxide** (see Figure 4–15). Specifically, they are cyclic compounds with a three-member ring. The most common is ethylene oxide, which is extremely flammable. Its uses vary, including as an intermediate to the production of ethylene glycol (antifreeze), as a fungicide, a pharmaceutical application, and a sterilizing agent of surgical utensils. Commonly called the alkene oxides, other epoxides are propylene oxide and isobutylene oxide. Common names are mostly used with the IUPAC nomenclature, for example, oxirane. Ethylene oxide (EO) is the base molecule, with all attachments named and numbered from it.

■ **Note**
THF is a cyclic ether called tetrahydrofuran, butylene oxide, or diethylene.

CH₂ — CH₂
| |
CH₂   CH₂
    \   /
     O

LEL: 2%; UEL: 11.8%

Problem Set 4.10

1. Remembering that numbering is the lowest number possible when naming, name the following using the IUPAC system:

   Propylene oxide

   Isobutylene oxide

   Cyclopentene oxide

   Cyclohexene oxide

ORGANIC PEROXIDES

ROOR and *HOOR*

Chemistry Quick Reference Card

Family Class	Naming	Hazards
Alkyl halides	R–X (X = halogen)	Toxicity, flammability
Amines	R_xNH_x or ion	Toxicity, flammability
Nitrogen groups	$R-NO_2$	Explosiveness
Ethers	R'–O–R"	Anesthetic effects Flammability Potential to become peroxides
Peroxides	R–O–O–R'	Explosiveness Oxidation

organic peroxide

a functional group in which the hydrogens in hydrogen peroxide have been replaced with alkyl or aryl groups

maximum safe storage temperature (MSST)

the highest storage temperature above which a reaction and explosion may ensue

self-accelerating decomposition temperature (SADT)

the temperature at which an organic peroxide or synthetic compound will react to heat, light, or other chemicals and release oxygen, energy, and fuel in the form of an explosion or rapid decomposition

The **organic peroxides** (ROOR and HOOR) are usually identified by recognizing the name peroxide or peroxy- without the inclusion of a metal group. The peroxide has a carbon–carbon chain origin.

The organic peroxides are sometimes referred to as the peroxo-organic compounds and have two classes, both of which are dangerous. The organic peroxides collectively have a general formula of R–O–O–R'; for example, $(CH_3)_3$–C–O–O–C–$(CH_3)_3$, called di-tert-butyl peroxide. The subgrouping of the organic peroxides called organic hydroperoxides have a general formula of H–O–O–R; for example, H–O–O–C–$(CH_3)_3$, called tert-butyl hydroperoxide.

The R–O–O–R' compounds are called peroxides and have the corresponding R groups named in order of size. The R group comes before the designation "peroxide" in the name. Compounds with the general formula HO–O–R are referred to as the hydroperoxides and are named by using the suffix hydroperoxide attached to the corresponding R group's name.

Ethers have the potential to degrade into organic peroxide if subjected to oxygen. Once the organic peroxide is made, the product becomes an extremely dangerous explosive and an exceptionally good oxidizer. Only a low level of energy is required to start the decomposition reaction: a spark, static electricity, movement, or heat. Great care needs to be afforded when dealing with organic peroxides.

The organic peroxides, like many organic compounds, have a **maximum safe storage temperature (MSST)** above which a reaction and explosion may ensue. The **self-accelerating decomposition temperature (SADT)** applies to organic peroxides or synthetic compounds that decompose at normal ambient temperatures. That is, light or moderate heat will cause the reaction to start, releasing oxygen, energy, and fuel in the form of rapid oxidation. To make sure compounds with low SADTs remain stable, they must be kept in a dark container or refrigerated to maintain their MSSTs.

Problem Set 4.11   Draw the following compounds:

   Dimethyl peroxide

   Methyl ethyl peroxide

(continues)

(Continued)

Ethyl propyl peroxide

Di-tert-butyl peroxide

Octyl peroxide

Tert-butyl hydroperoxide

polymers

compounds of high molecular weight made from repetitive linking together of monomers

POLYMERS

In today's society we cannot escape the use of plastics, which are synthetic or semisynthetic organic compounds (see Table 4–5). Compounds that we call plastic are actually **polymers**, and polymers represent a large portion of the precursor chemicals that are used in industry.

■ Note

The *Emergency Response Guidebook* identifies polymerization reactions with a "P" next to the guide number.

---

monomers

molecules considered a single unit, the combination of which results in a polymer

Polymers are long chains made up of repetitions of smaller units called **monomers**. Each monomer involves a reaction process. Special catalysts are used to make a monomer out of a double-bonded compound. The reaction starts by forming free radicals of the base product. The double bond opens up and combines with adjacent radicals. The reaction repeats to produce the longer-chained polymer in a process called *addition polymerization*.

Condensation polymerization involves a reactant with two functional groups, one an acid and the other a base (-OH or -NH₂), or one monomer containing an acid group and another containing a basic group. The addition of an equal molar mixture creates the polymer, or what is called a *polyester*.

Autopolymerization refers to the potential of a particular chemical to polymerize under heat impingement. These monomers have the capability to spontaneously polymerize. Once this reaction starts in a container, elevated pressures and the generation of heat cause catastrophic failure of the container. These substances can, under certain conditions, be controlled with the addition of carbon dioxide, nitrogen, or other inert gases. However, under fire conditions, these inhibitors do not always function in the manner in which they were intended.

Table 4–5 *Examples of plastics.*

Compound	Use
Ethylene acrylic acid (EAA)	Paperboard coating, blister packs, packing
Ethylene ethyl acrylate (EEA)	Specialty hoses, tubing
Ethylene methacrylate (EMA)	Medical packing, laminates, disposable gloves
Polyphenylene (PPO)	Appliances, pump housings, telephones
Polyphenylene sulfide (PPS)	Boiler sensors, chemical pumps, valves
Polyvinyl chloride (PVC)	Wall coverings, pipe fittings and pipe, gloves

> **■ Note**
>
> The protective suits used in the hazardous materials industry are produced similarly to the way that plastics are produced. Certain combinations of monomers give them chemical resistive qualities that must be referenced and researched during a hazardous materials event to determine the compatibility of the fabric (i.e., the polymers).

> **● Caution**
>
> Polymerization reactions are very dangerous reactions. They can start with a drop or increase in temperature or pressure, or a combination thereof. Inhibitors, which have specific life spans, are usually placed in the substance to control the reaction. Introducing an inhibitor or inerting the product is a very dangerous practice.

Inhibitors are mixed with the monomers during the transportation or storage of these chemicals. If an inhibitor is not placed in the solution or the inhibitor is allowed to escape, a reaction or autopolymerization will ensue. The monomers will start to react with themselves, creating an uncontrolled reaction termed a runaway reaction or runaway polymerization. The result is a massive display of heat and energy resembling a BLEVE. Chemical reactions double with every 18°F rise in temperature, causing the reaction to reach dangerous temperatures and pressures in the containment vessel.

> **● Caution**
>
> Some inhibited monomers, such as acrylic acid, will separate out the inhibitor if frozen. On thawing, the inhibitor must be mixed back in with the product to prevent an uncontrollable reaction.

BRANCHING AND ITS EFFECTS ON PHYSICAL PROPERTIES

The branching that occurs in a compound can affect the boiling point of the compound, thus affecting the vapor that a liquid can liberate. (See Table 4–6 for the boiling points of various chemical compounds.) In general, the lower molecular-weight compounds will either have gaseous states or will possess boiling points below or near the ambient temperature. In the liquid state, these compounds will produce a significant amount of vapors, thus causing a possible explosive atmosphere.

Problem Set 4.12   Using graph paper, graph the first ten alkyl halides listed in Table 4–6 and compare these with comparable molecular-weight amines and ethers. Let the x-axis be the molecular weight and the y-axis be the boiling point temperature.

Table 4–6   *The boiling points of various alkyl halides, amines, and ethers.*

Alkyl Halides	BP (°F)	Amines	BP (°F)	Ethers	BP (°F)
Methyl fluoride	−109.12	Methylamine	21.2	Dimethyl ether	−10.66
Methyl chloride	−10.84	Ethylamine	62.6	Ethyl methyl ether	46.22
Methyl bromide	38.48	Propylamine	120.2	Diethyl ether	94.28
Methyl iodide	108.50	Isopropylamine	91.4	Ethyl propyl ether	147.2
Ethyl fluoride	−35.86	Butylamine	172.4	Dipropyl ether	195.8
Ethyl chloride	55.58	Isobutylamine	154.4		
Ethyl bromide	101.12	sec-Butylamine	145.4		
Ethyl iodide	161.60	tert-Butylamine	113.00		
n-Propyl fluoride	27.50	Cyclohexylamine	273.20		
n-Propyl chloride	115.88				
n-Propyl bromide	159.44				
n-Propyl iodide	215.6				
Isopropyl fluoride	15.08				
Isopropyl chloride	93.2				
Isopropyl bromide	138.92				
Isopropyl iodide	192.92				
n-Butyl fluoride	89.60				
n-Butyl chloride	173.40				
n-Butyl bromide	213.80				
n-Butyl iodide	266.00				
tert-Butyl fluoride	53.60				
tert-Butyl chloride	123.80				
tert-Butyl bromide	163.94				

Summary

Chemistry Quick Reference Card

Family Class	Naming	Hazards
Alkyl halides	R–X (X = halogen)	Toxicity, flammability
Amines	R_xNH_x or ion	Toxicity, flammability
Nitrogen groups	$R-NO_2$	Explosiveness
Ethers	R'–O–R"	Anesthetic effects Flammability Potential to become peroxides
Peroxides	R–O–O–R'	Explosiveness Oxidation

To classify a nonpolar hydrocarbon derivative (alkyl halides, amines, nitrogen compounds, ethers and epoxides, and organic peroxides), analyze the compound's name by identifying the hydrocarbon backbone and functional groups in the name. Break down the areas that you have studied and produce the drawn compound. Once you identify the classification, listing hazards of that class becomes easy.

The compounds covered in this chapter have relationships to the functional groups covered in Chapter 5. Relationships are important to understand not for their ratios, but rather to ingrain in our minds the potential factors that can raise questions at the scene of a hazardous materials event. Factors such as decontamination, how weather affects the compounds, the vapor produced, and incompatibilities must all be taken into consideration. Understanding the effects that a branched compound may have is just one of the technical skills a science officer must acquire. Table 4–6 lists the boiling points of the compounds studied thus far. By figuring out the molecular weight and configuration based on the BP information, issues related to hydrogen bonding, vaporization, and flammable gas potentials can be deduced. In Chapter 5, a more complete discussion is presented, once all variables have been analyzed and discussed. Nitrogen groups and peroxides, however, will not enter into our discussion due to the extremely explosive effects these compounds normally possess.

Review Questions

1. Draw the following compounds:
   - A. Diisopropyl ether
   - B. tert-Butyl hydroperoxide
   - C. Methylethylpropyl amine
   - D. Phenyl amine
   - E. Methyl ethyl peroxide
   - F. Ethyl propyl peroxide
   - G. Propyl-sec-butyl ether
   - H. Octyl peroxide
   - I. Vinylacryl ether
   - J. Isopropylamine
   - K. Chloropentafluoroethane
   - L. Di-n-butyl ether
   - M. Trimethylammonium bromide
   - N. Vinyl amine
   - O. Isobutylene oxide

P.　Toluene

Q.　Methyl ethyl ether

R.　Butylchloride

S.　2,4-Dimethyl-3-hexanamine

T.　Propylene oxide

U.　Difluorochloromethane

V.　Trinitrotoluene

W.　Fluorobenzene

X.　1,2, Dibromobenzene

Y.　Dimethyl ether

Z.　Diisobutylamine

AA.　1,2-Dimethoxyethane

BB.　Cyclopentene oxide

CC.　Trichlorofluoromethane

DD.　Nitrobenzene

EE.　Phenylchloride

FF.　Diethylamine

GG.　Cyclohexene oxide

HH.　Dichlorodifluoromethane

II.　Di-tert-butyl peroxide

JJ.　2-Methoxypentane

KK.　Di-n-propyl ether

LL.　1,2-Dichloro-1,1,2,2-tetrafluoroethane

MM.　Diethyl ether

NN.　Methylamine

OO.　Tetramethylammonium iodide

PP.　1-Methoxy-1-propene

QQ.　Hexafluoroethane

RR.　Chlorobenzene

SS.　Trimethylammonium chloride

TT.　Trimethylamine

UU.　3-Chloro-2-methyl-1-propanamine

VV.　TNT

WW.　Chlorofluoromethane

XX.　1,1,2-Trichloro-1,2,2-trifluoroethane

2. Give the chemical name and draw the structures of the following:

A.　Halon 1011

B.　Halon 104

C.　Halon 2402

D.　Halon 1211

E.　Halon 1301

F.　Halon 122

G.　Halon 1202

3. Place the general formula and appropriate hazards in the following table:

Chemistry Quick Reference Card

Family Class	Naming	Hazards
Alkyl halides		
Amines		
Nitrogen groups		
Ethers		
Peroxides		

Problem Set Answers

Problem Set 4.1

1. Xylene

   Ortho:
   Boiling point = 292°F
   Vapor pressure = 6.9 mm Hg
   TLV-TWA = 100 ppm

   1,2-Dimethylbenzene

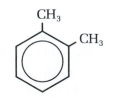

Meta:
Boiling point = 283°F
Vapor pressure = 9 mm Hg
TLV-TWA = 100 ppm

1,3-Dimethylbenzene

Para:
Boiling point = 281°F
Vapor pressure = 9 mm Hg
TLV-TWA = 100 ppm

1,4-Dimethylbenzene

2. Styrene = cinnamene, vinylbenzene, phenylethylene. Use: Polystyrene; SBR, ABS, and SAN resins; protective coatings; styrenated polyesters; rubber-modified polystyrene. Hazards: Readily undergoes polymerization when heated or exposed to light or a peroxide catalyst; explosive limits in air, 1.1–6.1%; must be inhibited during storage. Boiling point = 293.36°F; TLV-TWA = 50 ppm

Problem Set 4.2

1. Difluorochloromethane

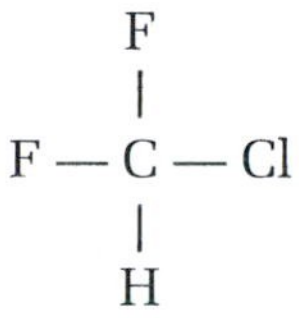

$$F - \underset{\underset{\displaystyle H}{|}}{\overset{\overset{\displaystyle F}{|}}{C}} - Cl$$

2. Trichlorofluoromethane

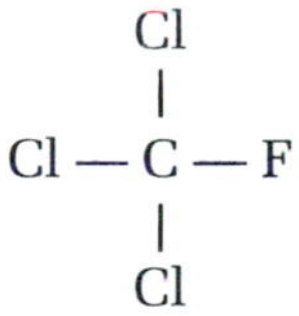

$$Cl - \underset{\underset{\displaystyle Cl}{|}}{\overset{\overset{\displaystyle Cl}{|}}{C}} - F$$

3. Dichlorodifluoromethane

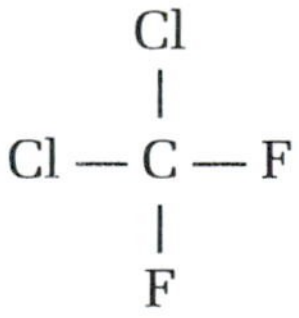

$$Cl - \underset{\underset{\displaystyle F}{|}}{\overset{\overset{\displaystyle Cl}{|}}{C}} - F$$

4. Chlorofluoromethane

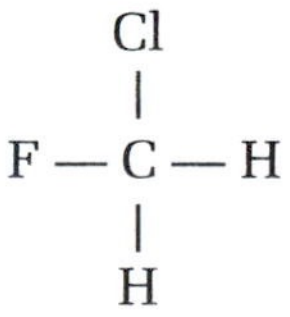

$$F - \underset{\underset{\displaystyle H}{|}}{\overset{\overset{\displaystyle Cl}{|}}{C}} - H$$

5. 1,1,2-Trichloro-1,2,2-trifluoroethane

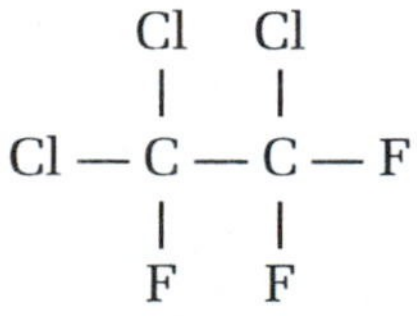

$$Cl - \underset{\underset{\displaystyle F}{|}}{\overset{\overset{\displaystyle Cl}{|}}{C}} - \underset{\underset{\displaystyle F}{|}}{\overset{\overset{\displaystyle Cl}{|}}{C}} - F$$

6. 1,2-Dichloro-1,1,2,2-tetrafluoroethane

$$\text{Cl}-\overset{\overset{\displaystyle F}{|}}{\underset{\underset{\displaystyle F}{|}}{C}}-\overset{\overset{\displaystyle F}{|}}{\underset{\underset{\displaystyle F}{|}}{C}}-\text{Cl}$$

7. Chloropentafluoroethane

$$\text{Cl}-\overset{\overset{\displaystyle F}{|}}{\underset{\underset{\displaystyle F}{|}}{C}}-\overset{\overset{\displaystyle F}{|}}{\underset{\underset{\displaystyle F}{|}}{C}}-\text{F}$$

8. Hexafluoroethane

$$\text{F}-\overset{\overset{\displaystyle F}{|}}{\underset{\underset{\displaystyle F}{|}}{C}}-\overset{\overset{\displaystyle F}{|}}{\underset{\underset{\displaystyle F}{|}}{C}}-\text{F}$$

9. Phenylchloride

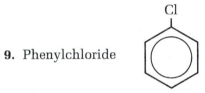

10. Vinyl chloride $CH_2=CHCl$; resonance structure: $CH_2=CH-\ddot{\underset{\cdot\cdot}{Cl}}\colon \longleftrightarrow \overset{..}{C}H_2-CH=\overset{+}{\underset{\cdot\cdot}{Cl}}\colon$

11. Phenyl dichloride

12. Biphenyl

13. Polychlorinated biphenyl   poly = many

Cl's shown do not represent the exact number

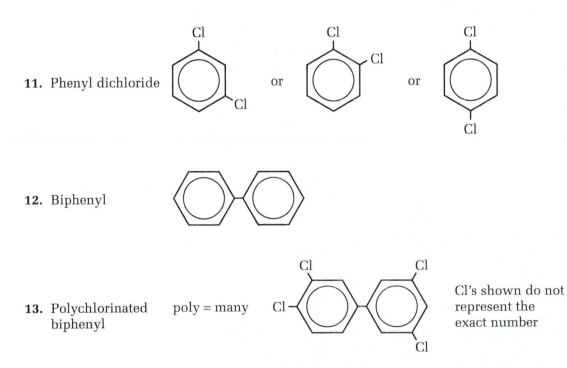

14. 1,3,5-Tribromobenzene

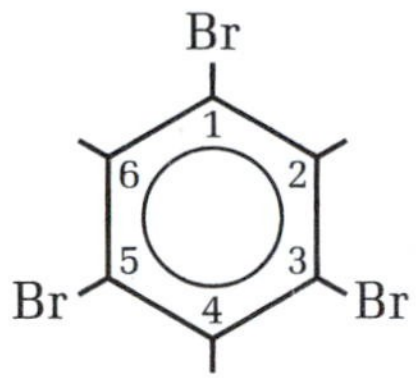

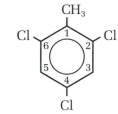

15. 2,4,6-Trichloro-toluene

Problem Set 4.3

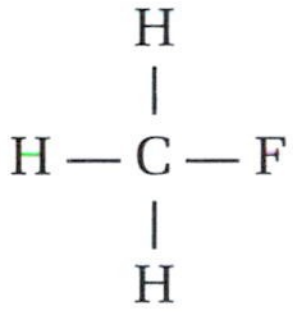

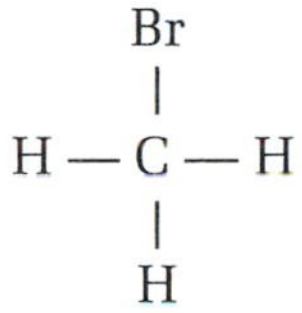

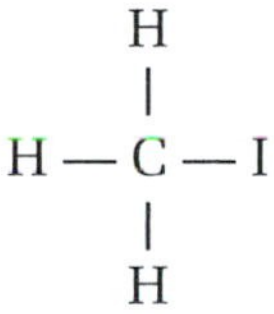

$$\begin{array}{c} H \\ | \\ H-C-F \\ | \\ H \end{array} \qquad \begin{array}{c} Br \\ | \\ H-C-H \\ | \\ H \end{array} \qquad \begin{array}{c} H \\ | \\ H-C-I \\ | \\ H \end{array}$$

1. CH_3Cl = CH_3F CH_3Br CH_3I
Fluoromethane   Bromomethane   Iodomethane
Methyl fluoride   Methyl bromide   Methyl iodide

$$\begin{array}{c} F \\ | \\ H-C-F \\ | \\ H \end{array} \qquad \begin{array}{c} Br \\ | \\ H-C-Br \\ | \\ H \end{array} \qquad \begin{array}{c} I \\ | \\ H-C-I \\ | \\ H \end{array}$$

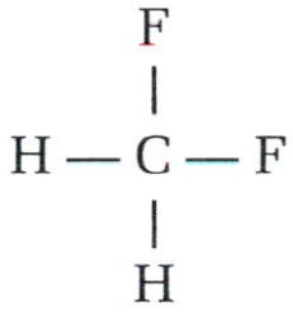

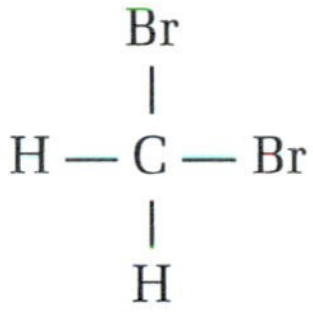

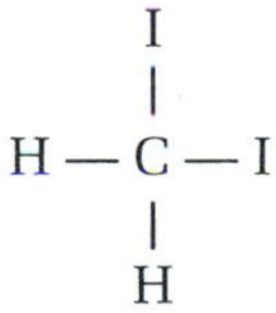

Methylene fluoride   Methylene bromide   Methylene iodide
Methylene bifluoride   Methylene bibromide   Methylene biiodide

2. CH_2Cl_2 = CH_2F_2 CH_2Br_2 CH_2I_2
Difluoromethene   Dibromomethane   Diiodomethane

$$\begin{array}{c} F \\ | \\ H-C-F \\ | \\ F \end{array} \qquad \begin{array}{c} Br \\ | \\ H-C-Br \\ | \\ Br \end{array} \qquad \begin{array}{c} H \\ | \\ I-C-I \\ | \\ I \end{array}$$

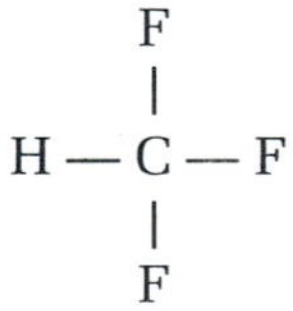

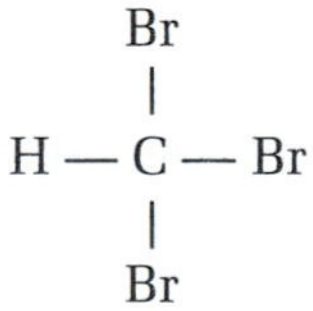

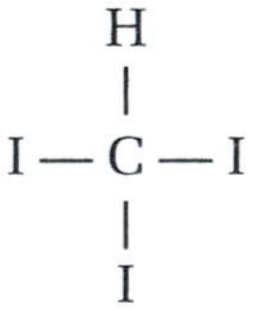

3. $CHCl_3$ = CHF_3 $CHBr_3$ CHI_3
Trifluoromethane   Tribromomethane   Triiodomethane

$$\begin{array}{c} F \\ | \\ F-C-F \\ | \\ F \end{array} \qquad \begin{array}{c} Br \\ | \\ Br-C-Br \\ | \\ Br \end{array} \qquad \begin{array}{c} I \\ | \\ I-C-I \\ | \\ I \end{array}$$

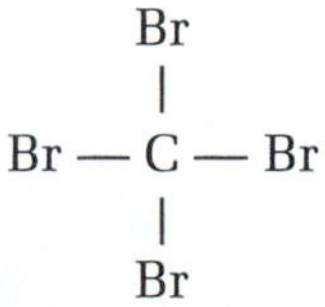

4. CCl_4 = CF_4 CBr_4 CI_4
Tetrafluoromethane   Tetrabromomethane   Tetraiodomethane

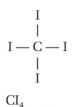

Problem Set 4.4

1. Diethylamine

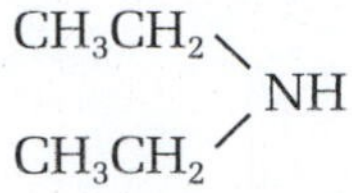

$$\begin{array}{c} CH_3CH_2 \\ CH_3CH_2 \end{array}\!\!\!\!\diagup^{\diagdown}\!\!\!\! NH$$

2. Isopropylamine

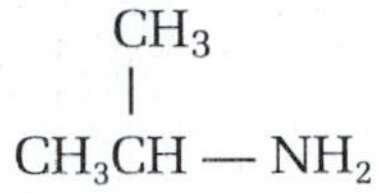

$$\begin{array}{c} CH_3 \\ | \\ CH_3CH - NH_2 \end{array}$$

3. Diisobutylamine

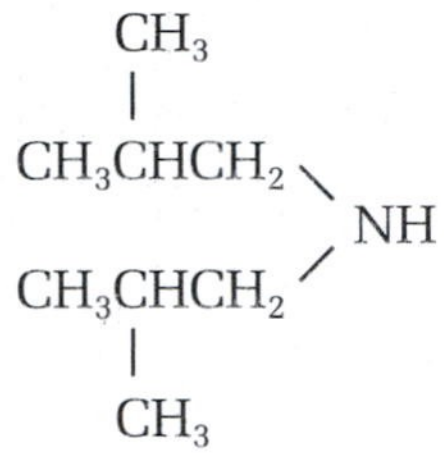

$$\begin{array}{c} CH_3 \\ | \\ CH_3CHCH_2 \\ CH_3CHCH_2 \\ | \\ CH_3 \end{array}\!\!\!\!\diagup^{\diagdown}\!\!\!\! NH$$

4. Trimethylamine

$$\begin{array}{c} CH_3 \\ | \\ N \\ CH_3 \quad CH_3 \end{array}$$

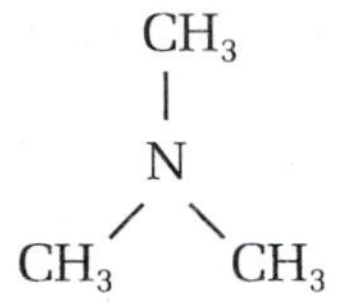

5. Methylamine

$$CH_3 - NH_2$$

6. 2,4-Dimethyl-3-hexanamine

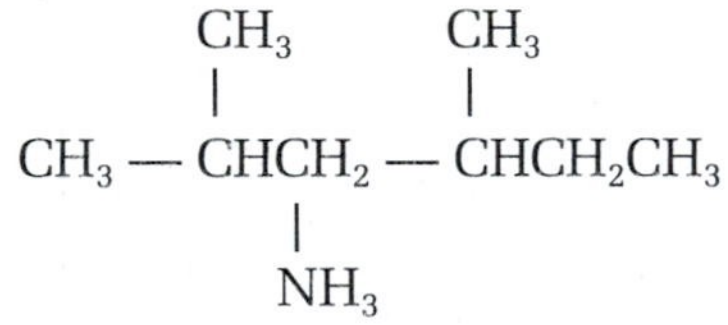

$$\begin{array}{ccc} CH_3 & & CH_3 \\ | & & | \\ CH_3 - CHCH_2 - CHCH_2CH_3 \\ & | \\ & NH_3 \end{array}$$

7. 3-Chloro-2-methyl-1-propanamine

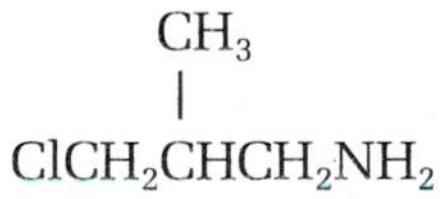

$$\begin{array}{c} CH_3 \\ | \\ ClCH_2CHCH_2NH_2 \end{array}$$

Names of structures from left to right, first line: Phenol, Benzoic acid, Aniline, *o*-bromochlorobenzene; second line: Phenyl radical, Styrene, Benzenesulfonic acid, Toluene

Problem Set 4.5

1. Trimethylammonium bromide

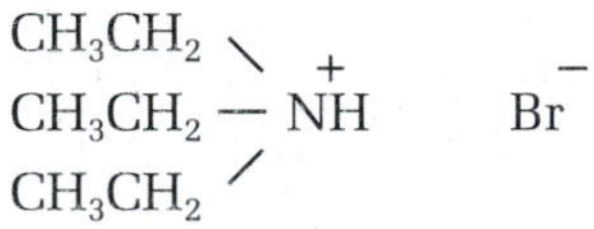

$$\begin{array}{c} CH_3CH_2 \\ CH_3CH_2 - \overset{+}{N}H \qquad Br^- \\ CH_3CH_2 \end{array}$$

2. Tetramethylammonium iodide

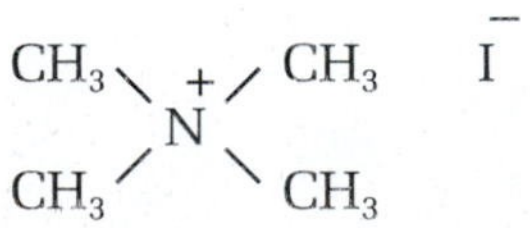

$$\begin{array}{c} CH_3 \quad CH_3 \qquad I^- \\ \diagdown \overset{+}{N} \diagup \\ CH_3 \quad CH_3 \end{array}$$

3. Trimethylammonium chloride

$$CH_3\!\!\underset{CH_3}{\overset{+}{\diagdown}}NHCH_3 \quad \overset{-}{Cl}$$

4. Tetramethylammonium iodide

$$CH_3\!\!\underset{CH_3}{\overset{+}{\diagdown}}\underset{\diagup}{N}\!\!\overset{\diagup CH_3}{\underset{CH_3}{\diagdown}} \quad \overset{-}{I}$$

Problem Set 4.6

Butylamide

$$CH_3CH_2CH_2CH_2NH_2$$

Benzamide

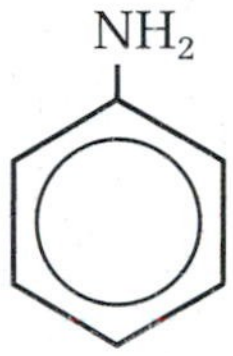

 = Aniline

Problem Set 4.7

HN-1   Bis-(2-chloroethyl)ethylamine

$$CH_3CH_2 - N\!\!\overset{\diagup CH_2CH_2Cl}{\underset{CH_2CH_2Cl}{\diagdown}}$$

HN-2   Bis-(2-chloroethyl)methylamine

$$CH_3 - N\!\!\overset{\diagup CH_2CH_2Cl}{\underset{CH_2CH_2Cl}{\diagdown}}$$

HN-3   Tris-(2-chloroethyl)amine

$$CH_2ClCH_2 - N\!\!\overset{\diagup CH_2CH_2Cl}{\underset{CH_2CH_2Cl}{\diagdown}}$$

Problem Set 4.8

Nitromethane
$$CH_3NO_2$$

Nitrobenzene
$$NO_2$$

Dinitromethane
$$NO_2 - CH_2 - NO_2$$

Dinitrotoluene
$$CH_3$$
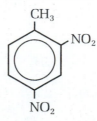
$$NO_2$$
$$NO_2$$

Trinitrotoluene

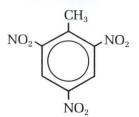

Nitroethane

$CH_3CH_2NO_2$

Problem Set 4.9

1. Diisopropyl ether

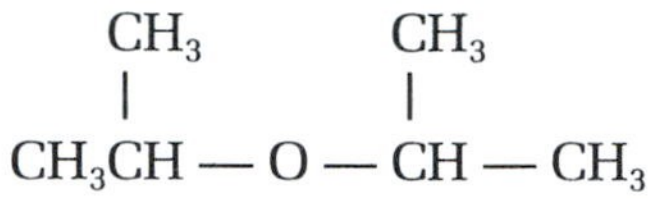

$$CH_3CH - O - CH - CH_3$$

with CH_3 groups on the first and third carbons (the $\overset{|}{CH}$ positions)

2. Methyl ethyl ether

$CH_3 - O - CH_2CH_3$

3. Dimethyl ether

$CH_3 - O - CH_3$

4. Di-n-butyl ether

$CH_3CH_2CH_2CH_2 - O - CH_2CH_2CH_2CH_3$

5. Di-n-propyl ether

$CH_3CH_2CH_2 - O - CH_2CH_2CH_3$

6. Diethyl ether

$CH_3CH_2 - O - CH_2CH_3$

7. 2-Methoxypentane

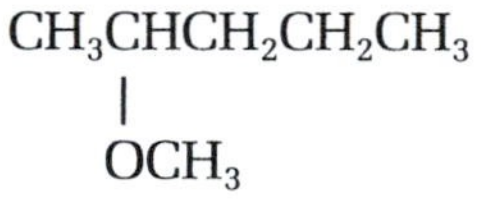

$$CH_3CHCH_2CH_2CH_3$$
$$|$$
$$OCH_3$$

8. 1,2-Dimethoxyethane

$CH_3 - O - CH_2CH_2 - O - CH_3$

9. 1-Methoxy-1-propene

$CH_3 - O - CH = CHCH_3$

10. Ethyl propyl ether

$CH_3CH_2 - O - CH_2CH_2CH_3$

Problem Set 4.10

1. Propylene oxide

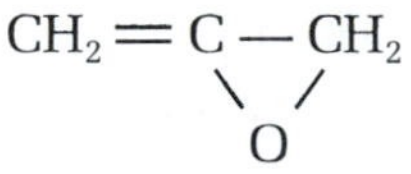

$$CH_2 = C - CH_2$$
$$\backslash \ /$$
$$O$$

2. Isobutylene oxide

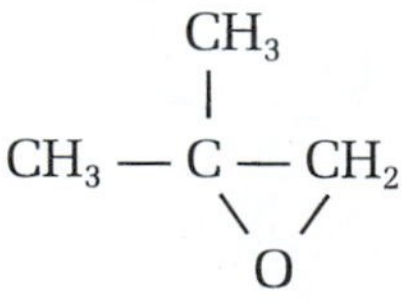

$$CH_3 - \overset{\overset{\textstyle CH_3}{|}}{C} - CH_2$$
$$\backslash \ /$$
$$O$$

3. Cyclopentene oxide

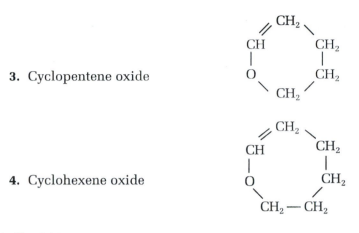

4. Cyclohexene oxide

Problem Set 4.11

1. Dimethyl peroxide

$$CH_3 - O - O - CH_3$$

2. Methyl ethyl peroxide

$$CH_3 - O - O - CH_2CH_3$$

3. Ethyl propyl peroxide

$$CH_3CH_2 - O - O - CH_2CH_2CH_3$$

4. Di-tert-butyl peroxide

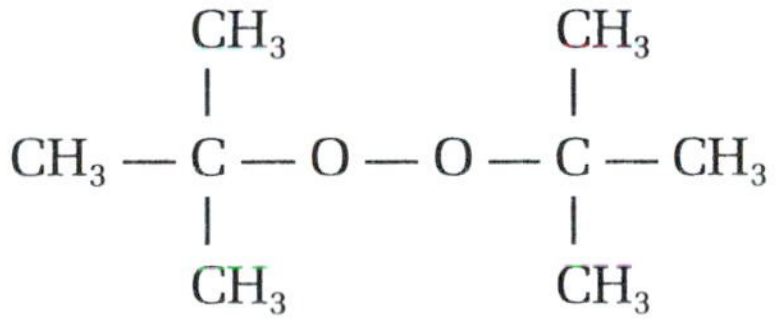

$$CH_3 - \underset{\underset{CH_3}{|}}{\overset{\overset{CH_3}{|}}{C}} - O - O - \underset{\underset{CH_3}{|}}{\overset{\overset{CH_3}{|}}{C}} - CH_3$$

5. Octyl peroxide

$$CH_3CH_2CH_2CH_2CH_2CH_2CH_2CH_2 - O - O - CH_2CH_2CH_2CH_2CH_2CH_2CH_2CH_3$$

6. Tert-butyl hydroperoxide

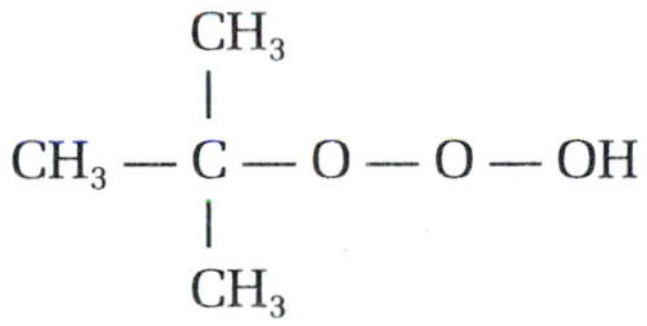

$$CH_3 - \underset{\underset{CH_3}{|}}{\overset{\overset{CH_3}{|}}{C}} - O - O - OH$$

Problem Set 4.12

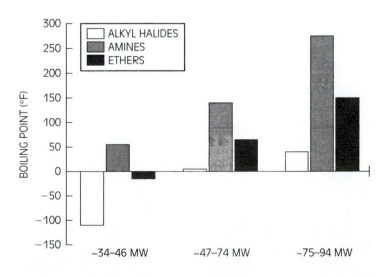

Chapter

5

Organic Compounds III
Polar Compounds

Learning Objectives

Upon completion of this chapter, you should be able to:

- Explain the benefit of understanding nomenclature in the science sector.
- Duplicate the nomenclature of alcohols, phenols, and their sulfur analogs.
- Duplicate the nomenclature of carbonyl groups, organic acids, and cyanide salts.
- Compare the branching effects on the groups covered in this chapter with the parent hydrocarbon.
- Discuss chemical warfare agents and compare them to the industrial chemicals we have studied thus far.

Drawing on the chemistry presented so far in this book, we continue our study of organic chemistry by focusing on polar compounds. These chemicals are the solvents, reactants, and by-products that we frequently encounter as industrial, household, and transported chemicals. The chemistry at this point becomes gray in terms of neatly categorizing substances. As demonstrated in Chapter 4, chemical variations can, and do, overlap.

We also cover the chemistry and nomenclature that surrounds chemical warfare agents (CWAs). Although the discussion concentrates on the chemicals themselves and their toxic qualities, true nomenclature is briefly reviewed. In practical terms, understanding the hazards of CWAs and their relationships to industrial chemicals is more important than knowing the true chemical family to which they belong.

Here, as in Chapter 4, polarity is a relative term. The weight of the compound, its branching, and its size all have an influence on polarity, as do the chemicals that these compounds may come in contact with (including those of a human being). We begin, therefore, by classifying these chemicals into relative polar compounds.

ALCOHOLS, PHENOLS, AND THEIR SULFUR ANALOGS

R-OH

Alcohols

Chemistry Quick Reference Card

Family	Naming	Hazards
Alcohols	R–OH	Toxicity Flammability

alcohol
an organic compound composed of an alkyl or aryl group and a hydroxyl group

hydroxyl group
the –OH functional group

Alcohol compounds (R–OH) are formed by the attachment of a **hydroxyl group** (OH) on the hydrocarbon. The "R" denotes an alkyl group. The OH is an ion, a hydroxide complex that collectively has lost or gained one or more electrons. This complex ion is important because the OH component can influence the outcome of a chemical reaction, especially acid–base reactions. The hydroxyl group can be thought of as a molecular fragment that comes from another chemical compound combining with the hydrocarbon.

As a group, the alcohols are flammable liquids and are polar. With a short chain, these compounds fit the flammable definition; as the chain grows longer, they meet the combustible definition. The characteristic of polarity should alert you to the problem of extinguishment, especially when considering water as the agent. Because of the polarity, alcohols are miscible in water, especially the short-chained compounds. Thus, water application only increases the volume without changing the property of flammability.

> **■ Note**
> Henry's law of solubility infers that dilution decreases vapor production and can thus reduce the vapor concentration below the LEL. Because of extreme solubility potentials, many alcohol compounds require a large quantity of water to stop vapor production and extinguish a fire.

As we attach longer chains, the alcohol group becomes a thicker liquid until a solid results. However, these longer-chained compounds are still able to burn if

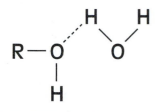

Figure 5–1   *Hydrogen bonding between a water molecule and an alcohol. Ethers act in a similar manner, creating hydrogen bonding (see Figure 4–14).*

a high temperature is applied. The short-chained alcohols are most prevalent in transported chemicals and industrial processes, as well as household uses.

For the most part, the boiling points of alcohol compounds are higher than those of the parent hydrocarbon but lower than most other molecules. Because of the hydroxyl group, lower-chained alcohols are very soluble in water, and higher-chained compounds are insoluble. The -OH group affects the boiling point, melting point, and solubility of alcohols because of the hydrogen bonding that occurs between the alcohol and the water molecule (see Figure 5–1), specifically, the placement of the -OH on the carbon chain and the length of the chain with respect to the -OH group.

Alcohols react with the same compounds that water reacts with. Reactions can produce alkenes or ethers, and, under controlled conditions, aldehydes, esters, alkyl halides, ketones, or acids. Because a wide variety of compounds can be generated from alcohols, they are frequently used in clandestine drug laboratories.

The vapor pressures in this group are relatively high, and this makes the possibility of toxic fumes a common problem. If these fumes find an ignition source, fire is yet another hazard to consider.

In nature, the occurrence of alcohol is widespread. In the fermentation of wood (or cellulose), a tremendous amount of wood alcohol is produced. Wood alcohol is sometimes referred to as wood spirits, but the scientific name is methanol. Methanol is the methyl radical with the hydroxyl group attached.

Both the IUPAC naming and the common naming configurations are used for alcohols (see Table 5–1). As for the IUPAC naming system, only two rules are applied:

1. Count the carbon chain to which the hydroxyl group is attached. In the final name of the alkane that represents the hydrocarbon, drop the "e" from the end and add -ol (for example, methane becomes methanol).

2. For cases in which the -OH group is somewhere in the middle of the chain, number the longest carbon chain. However, the hydroxyl group is the lowest carbon position, and is so identified in the name propanol, compared with 2-propanol.

Table 5–1   *Representative alcohols.*

Common Name	IUPAC Name	BP (°F)	VP (mm Hg)	TLV (ppm)
Methyl alcohol	Methanol	147	92	200
Ethyl alcohol	Ethanol	172	43	1000
Propyl alcohol	1-Propanol	207	21	200
n-Butyl alcohol	1-Butanol	244	5.5	500
Sec-butyl alcohol	2-Butanol	211	10	100

So, for example,

$$CH_3CHCH_2CH_2CH_2OH$$
$$|$$
$$CH_3$$

is a five-carbon compound with a methyl group attached. Because the hydroxyl group takes precedence in the naming sequence, the -OH is in the number-one carbon position, thus indicating that the methyl group is attached to the fourth carbon: 4-methyl-1-pentanol. We indicate the hydroxyl group by its carbon number, and all substituents inclusive of multiple bonds receive the appropriate carbon number position.

Problem Set 5.1

1. Draw propanol and 2-propanol.

2. Draw five combinations of methylpentanol by attaching the methyl group to each carbon, and name each compound.

3. Draw three combinations of methylpentanol by maintaining the position of the methyl group at the number-four carbon and moving the hydroxyl group along the chain. Name each compound. Why are there only three possibilities and not five?

4. Draw the following compounds:

    1-Propanol

    3-Chloro-1-propanol

    4-Penten-2-ol

    Ethylanol

    Vinyl alcohol

    Isopropyl alcohol

    Tertbutyl alcohol

    Sec-butyl alcohol

    Isobutyl alcohol

    Propyl alcohol

    Ethyl alcohol

    2-Hexylanol

    7-Methyl-4-octen-3-ol

Glycols and Glycerin. Two families associated with the alcohols are the glycols and glycerins. The glycols have two -OH groups, and the glycerins have three or more -OH groups. Two examples of this subgroup of the alcohols are ethylene glycol, commonly referred to as antifreeze, and nitroglycerin.

Information about the toxic values of these products is limited. Ethylene glycol is commonly used in industrial processes as well as in automobiles. A toxic event usually happens during oral siphoning of a radiator. A fairly small dose, usually a mouthful, will elicit a toxic reaction. The reaction occurs in three phases—glycol, glycolate, and nephropathy—each bringing the patient closer to death. The glycol metabolizes into oxalate crystals that enter the tissues of the brain and renal system, causing extensive damage.

Glycols are usually named according to the common nomenclature in which the carbon–carbon chain is identified and the corresponding prefix utilized. The

TREATMENT NOTE

Ethylene glycol poisoning progresses through three phases. The first phase, called the glycol phase, lasts approximately 30 minutes to 12 hours. In this phase, the glycol is metabolized in very much the same way drinking alcohol would be. However, here the principal metabolite is glycolic acid. Signs and symptoms are similar to ethanol intoxication. In 4 to 24 hours, the toxic metabolite peaks to indicate the second phase, called the glycolate phase. In this phase, noncardiogenic pulmonary edema, academia, and neurological toxicity manifest. Once renal damage is detected, the process has moved into the third phase, which occurs in 24 to 72 hours and is called the nephropathy phase. Renal edema and tubular necrosis are observed from the buildup of oxalate crystals in the form of calcium oxalate, thus causing hypocalcemia. Because the metabolic process produces an acidoic state, EKG "T" waves are pronounced, which is consistent with hyperkalemia. Renal failure soon follows. In some of the reference literature it is suggested that a second neurologic phase occurs. At this point the oxalate crystals have entered the tissues of the brain, causing further damage.

The current treatment of choice for the first-phase patient is high loading dosages of ethanol, which competes for the enzyme, alcohol dehydrogenase, that is used in the metabolism of ethylene glycol. This competition reduces and, in some cases, prevents the production of toxic metabolites. Other antidotes must be used for a patient in the later stages. For example, once in the glycolate phase, thiamine and pyridoxine are used to reduce the oxalic acid that would otherwise combine with calcium. Leucovorin, folic acid, and 4-methylpyrazole are also used.

word "glycol" is then placed to identify the complete structure (see Table 5–2). In the IUPAC naming system, the prefix is utilized to identify the carbon–carbon structure, and the suffix -diol identifies the glycol, with "glycerol" used for the glycerins.

Problem Set 5.2   Name the following compounds (Hint: draw them):

$CH_2OHCH_2CH_2OH$

CH_2OHCH_2OH

$CH_2OHCHOHCH_2OH$

$CH_3CHOHCH_2OH$

Table 5–2   *Representative glycols.*

Common Name	IUPAC Name	BP (°F)	VP (mm Hg)	TLV/TWA (ppm)
Ethylene glycol	1,2-Ethanediol	387	0.05	50
Propylene glycol	1,2-Propanediol	371	0.129	50
1,3-Butylene glycol	1,3-Butanediol	406	0.02	not listed
1,4-Butylene glycol	1,4-Butanediol	446	<1	not listed
Diethylene glycol (DEG)	3-Oxa-1,5-pentanediol	471	<0.01	50

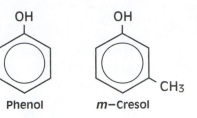

Figure 5–2 *Cresol is a phenol compound with a methyl group attached.*

> **■ Note**
> Chloral hydrate, referred to as knockout drops, has the chemical name trichloroethylidene glycol. It is used in the manufacture of liniments and sedatives in medicine.

Phenols

phenols

a group of aromatic compounds with the hydroxyl group attached directly to the benzene nucleus

Closely related to alcohols, **phenols** have the carbon chain in a ring pattern: the benzene ring (see Figure 5–2). In this classification, the boiling points are higher than the parent hydrocarbon, ranging from 358°F for phenol to 423°F for nitrophenol (see Table 5–3). Solubility in water is slight, with true solubility in alcohol or organic solvents. The acidity of this family is high in comparison to its alcohol cousin.

The simplest member of this family is referred to as a phenol, but it is also known as carbolic acid. It has a destructive action on animal tissue.

The nitrogen groups, sometimes referred to as the nitros, are commonly used as explosives, but they have other uses as well. The naming of these compounds utilizes the hydrocarbon chain beginning with the prefix "nitro," or the chemical name when utilizing IUPAC nomenclature (see Table 5–4).

> **■ Note**
> The orientation around the benzene ring is from the -OH such that each added group will follow the ortho, meta, and para designation in relation to the OH.

Table 5–3   *Boiling points of some phenols.*

	VP (mm Hg)	BP (°F)
Phenol	0.2 mm Hg @ 20°C	359
o-Cresol	1 mm Hg @ 38°C	376
m-Cresol	0.04 mm Hg @ 20°C	394
p-Cresol	0.04 mm Hg @ 20°C	394
p-Chlorophenol	0.10 mm Hg @ 20°C	428

Table 5–4   *Nitrogen-based compounds.*

Common Name	IUPAC Name	Formula
Nitroglycerin	Glyceryl trinitrate	$CH_2NO_3CHNO_3CH_2NO_3$
Nitrocellulose	Cellulose nitrate	Nitro group attached to cotton
Trinitrotoluene	TNT	$C_6H_2CH_3(NO_2)_3$
Trinitrophenol	Picric acid	$C_6H_2OH(NO_2)_3$
Cyclotrimethylenetrinitramine	Cyclonite	$N(NO_2)CH_2N(NO_2)CH_2N(NO_2)CH_2$

Problem Set 5.3   Draw and name the three designations for the following attachments to phenol:

     Chlorine

     Bromine

     Nitro compound

!Safety

● In the "Amine" section of Chapter 4, we made reference to the fact that the nitrogen mustard agents could be detected utilizing a nitrogen compound colorimetric tube. As with the nitrogen mustards, the sulfur mustards can be identified with the thioether colorimetric tube.

Sulfur Analogs

Sulfur groups are counterparts of the alcohols, phenols, and ethers. They are a combination of carbon-chained compounds and functional groups (see Figure 5–3). Sometimes referred to as organosulfur compounds, they also have a specific naming system.

thiols

compounds similar to alcohol with the oxygen replaced by a sulfur atom; generally denoted by R–SH and commonly referred to as mercaptans

Thiols. The **thiols** are also referred to as the mercaptans. These are similar to the alcohols in that the placement of the sulfur atom is the same as the oxygen that is found in alcohols. For example, if we replace the oxygen in ethyl alcohol, CH_3CH_2OH, with a sulfur atom, we have CH_3CH_2SH, or ethyl mercaptan, or more correctly, ethanethiol.

Problem Set 5.4   Describe how to change the following alcohols into thiols:

     Methyl alcohol

     Ethyl alcohol

     Propenyl alcohol

     n-Butyl alcohol

     Sec-butyl alcohol

Thiophenols. If we replace the oxygen in a phenol compound with a sulfur atom, we have a thiophenol. The SH group is sometimes referred to as a sulfhydryl group.

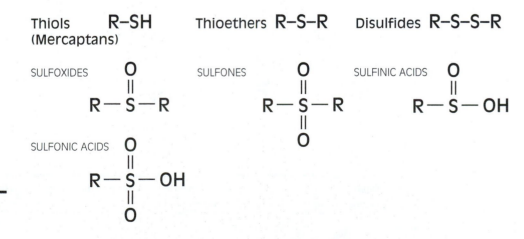

Figure 5–3 *Organic sulfur compounds.*

CARBONYL GROUPS

carbonyl group
a ketone or aldehyde functional group

A **carbonyl group** is a carbon double-bonded to oxygen, and it is due to the difference in electronegativity that this bonding configuration is very polar. This group is a very reactive cluster of atoms because of this polarity. We see a resonance-type structure (technically it is referred to as a hybrid) in that one bond of the double bond is between the carbon and oxygen atoms and the other is a "free" electron cluster ready for bonding (see Figure 5–4). This configuration creates a predominately negative side (the top of the oxygen) and a predominately positive side (the bottom of the carbon).

It is from this group that we obtain the aldehydes and ketones. In the aldehydes, an R group and hydrogen are attached to the carbon. The ketone is produced by two R groups attached to the carbon of the carbonyl group.

Aldehydes

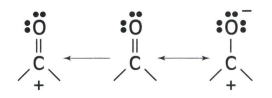

Chemistry Quick Reference Card

Family	Naming	Hazards
Alcohols	R–OH	Toxicity Flammability
Aldehydes	R–COH	Flammability Toxicity Polymerization

aldehydes
organic compounds composed of a carbonyl group, one hydrogen atom, and an R group

Aldehydes and ketones are closely related and their properties overlap. Both have the potential of forming peroxides.

Aldehydes have low boiling points: just below the boiling points of the alcohols that have comparable length, but above the boiling points of the ethers. The solubility is a mix of that of the alcohols and ethers. For the most part, the lower carbon chains give solubility in water, and the higher-chained aldehydes are insoluble in water. Aldehydes are very reactive, so much so that they can be converted to acids (carboxylic) (see Figure 5–5).

Aldehydes account for a majority of the irritant and sensitivity reactions that are observed in medicine. Chemical bronchitis, pneumonitis, pulmonary edema, and immunological effects can be caused by exposure to aldehydes.

As an example of this group, acetaldehyde is found in varnishes and photographic chemicals. It has severe respiratory injury potential, leading to bronchitis and pulmonary edema in quantities as low as 134 ppm. Narcotic-like effects are noted; however, antagonists are not suggested. There is no known antidote, and

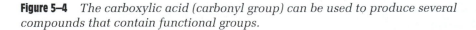

Figure 5–4 *The carboxylic acid (carbonyl group) can be used to produce several compounds that contain functional groups.*

Figure 5–5   *The carboxylic acid (carbonyl group) can be used to produce several compounds that contain functional groups.*

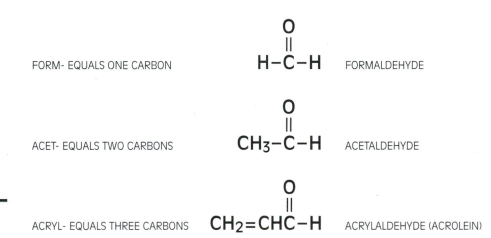

FORM- EQUALS ONE CARBON   FORMALDEHYDE

ACET- EQUALS TWO CARBONS   ACETALDEHYDE

Figure 5–6   *Naming aldehydes.*

ACRYL- EQUALS THREE CARBONS   ACRYLALDEHYDE (ACROLEIN)

treatment should be directed toward the symptoms. For example, for an inhalation injury, use high-flow oxygen, along with a nebulized bronchodilator for the bronchospasm. Endotracheal intubation may be necessary early on. Be prepared for fast-acting bronchospasm.

Naming these compounds under the common nomenclature requires a count of all carbons in the chain, including the carbon in the aldehyde functional group. After counting the carbons, the chain name is utilized to identify the number of carbons. The only exceptions to this are the first two carbons: meth(yl)- (form) and eth(yl)- (acet). For example, CH_3–CHO is acetaldehyde and CH_2O is formaldehyde. If the compound is an aldehyde and it has a double bond, as in acrolein (CH_2=CH–CHO), the prefix acryl- is used, and the name acrylaldehyde (see Figure 5–6).

In the IUPAC naming system, an aldehyde is identified by the suffix -al: formaldehyde is methanal, acetaldehyde is ethanal, and acrylaldehyde becomes 2-propenal (see Table 5–5).

Table 5–5   *Representative aldehydes.*

Common Name	IUPAC Name	BP (°F)	VP (mm Hg)	TLV (ppm)
Formaldehyde	Methanal	2.2	greater than 760	1
Acetaldehyde	Ethanal	68.4	740	100
Acrylaldehyde	2-Propenal	126.9	210	0.1

Problem Set 5.5   Name the following compounds with the common and IUPAC nomenclature:

$$CH_3CHO$$

$$CH_3CH_2CHO$$

$$CH_3CH_2CH_2CHO$$

$$CH_3CH_2CH_2CH_2CHO$$

$$CH_2CHCHO$$
$$|$$
$$CH_3$$

Ketones

Chemistry Quick Reference Card

Family	Naming	Hazards
Alcohols	R–OH	Toxicity Flammability
Aldehydes	R–COH	Flammability Toxicity Polymerization
Ketones	R–CO–R′	Flammability Toxicity

ketone

an organic compound composed of a carbonyl group bonded with two alkyl or aryl groups

With properties similar to those of aldehydes, **ketones** make up a large class of solvents. However, unlike the aldehydes, they are not affected by oxidizing agents used in synthesis. Ketones may form peroxides, and thus polymerize. They are volatile, and their narcotic effects make them extremely hazardous. These compounds are frequently encountered in clandestine operations.

The lower carbon-weight ketones are soluble in water, as well as in the usual organic solvents. Vapors are typical, resulting in toxic atmospheres and a high inhalation hazard. Avoid all skin contact. Sensory and motor neuropathy is often seen after contact with this group.

Acetone is a common solvent that is rapidly absorbed through the respiratory and gastrointestinal tracts. Toxicity manifests itself in effects similar to ethyl alcohol ingestion, but with stronger anesthetic effects. With inhalation injuries, the depression of respiratory status is most common. There is no known antidote and therapy is governed by symptomology. Ingestion of acetone usually presents no problems with the exception of vomiting; however, because of the prolonged elimination and associated metabolic effects, hospitalization may be required for observation for 30 to 48 hours.

The common nomenclature identifies the carbon–carbon chains exclusive of the carbonyl group, naming the R groups as independents. The IUPAC system identifies the ketone with the suffix -one, and the carbon of the C=O as a part of the chain (see Table 5–6).

Table 5–6   *Representative ketones.*

Common Name	IUPAC Name	BP (°F)	TLV (ppm)
Methyl ethyl ketone (MEK)	CH_3-CO-CH_2CH_3 1 2 3 4 2-butanone	175.2	200
Butyl ethyl ketone (BEK)	$CH_3CH_2CH_2CH_2$-CO-CH_2CH_3 1 2 3 4 5 6 7 3-heptanone	298.4	50
Dipropyl ketone (butyrone)	$CH_3CH_2CH_2$-CO-$CH_2CH_2CH_3$ 1 2 3 4 5 6 7 4-heptanone	290.7	50

Problem Set 5.6   Name the following compounds with the common and IUPAC nomenclature:

CH_3COCH_3

$CH_3COCH=CH_2$

i-$C_3H_7COC_3H_5$

$CH_3CH_2COCH_2CH_3$

$CH_3COC_2H_5$

$C_6H_5COC_3H_5$

$CH_3CH_2COCH_3$

$CH_3CH_2CH_2COCH_3$

$C_6H_5COCH_3$

$C_6H_5COC_6H_5$

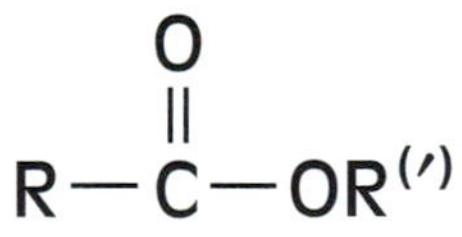

Esters

$$R-\overset{\overset{\textstyle O}{\|}}{C}-OR^{(')}$$

Chemistry Quick Reference Card

Family	Naming	Hazards
Alcohols	R–OH	Toxicity Flammability
Aldehydes	R–COH	Flammability Toxicity Polymerization
Ketones	R–CO–R'	Flammability Toxicity
Esters	R–COO–R'	Flammability Polymerization Toxicity

Table 5–7   *Representative esters.*

Common Name	IUPAC Name	BP (°F)	VP (mm Hg)	TLV (ppm)
Ethyl ethanoate	Ethyl acetate	170.6	73	400
Ethyl propenoate	Ethyl acrylate	210.9	29	5
2-Acetoxypropane	Isopropyl acetate	192.9	40	250

esters

organic compounds in which an alkyl or aryl group is attached to one side of the carbonyl group, and oxygen with an alkyl or aryl group to the other side

Esters are a bit different from the compounds covered thus far in this chapter. These compounds have some polarity characteristics; however, they do not form strong hydrogen bonds with each other. Boiling points and solubility are similar to compounds with the same molecular weight as seen in the aldehydes and ketone family. Here the R groups can be the same or different lengths.

Esters can be toxic at relatively low levels and total avoidance should be maintained. The toxicity depends on the parent chemical. For example, esters of aliphatic derivatives have low levels of toxicity; as double bonds are added, its toxic levels increase (see Table 5–7). Esters are derived from a combination of carboxylic acids and a variety of alcohols. A toxic event usually results from a reaction that occurs in the stomach once the ester is ingested. The process that occurs in the stomach separates the parent alcohol from the carboxyl group, so not only is there a toxic event from the ester, but also from the attached components. For example, if methyl carboxylate were ingested, both methyl alcohol and formic acid would be of toxic concern. Intoxication from the methyl alcohol would ensue fairly rapidly. Symptomology should guide treatment, with the addition of local protocols for cathartics. Seizure activity can be managed by the administration of diazepam or phenobarbital. Although inhalation is rare, respiratory injury is best managed by high-flow oxygen. In high concentrations, the central nervous system is depressed.

The common naming system identifies the carbon groups on either side of the compound. The carbon group that is attached to the oxygen is named first and is derived from the number of carbons in the chain. The second group is derived from counting the carbon chain that is attached to the carbon. The carbon chain and the carboxyl carbon are used in the nomenclature. The second carbon chain has the suffix -ate or -oate (see Figure 5–7).

Problem Set 5.7   Name the following compounds using the common nomenclature:

$$C_2H_5COOCH_3$$
$$CH_3COOC_2H_5$$
$$CH_3COOC_3H_7$$
$$C_6H_5COOC_3H_7$$
$$CH_3CH_2COOCH_3$$
$$CH_3COOCH=CH_2$$

Thioethers and Phosphoric Esters.   It is more important that the hazmat responder understand the detection of these compounds, rather than their naming, as well as

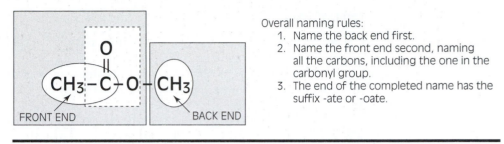

Overall naming rules:
1. Name the back end first.
2. Name the front end second, naming all the carbons, including the one in the carbonyl group.
3. The end of the completed name has the suffix -ate or -oate.

Figure 5–7   *Naming the ester.*

how these compounds can affect a hazardous materials event. The reaction in the colorimetric tube (or detection instrument) identifies the thioether or the phosphoric ester. Neurotoxins, predominately nerve agents and their derivatives (see the section "Chemical Warfare Agents"), are esters, and in particular those that are sulfur or phosphorus based. Sulfur-based agents, for example, are in the thioether category.

The bond between the phosphate is broken fairly easily, making the decontamination of these products relatively easy. The breaking is called *hydrolysis* and is accomplished by using a mild bleach solution, or water of the appropriate pH with dissolved minerals (e.g., saltwater). Additionally, these compounds undergo several different types of chemical reactions that "age" the bond between the ester and the cleaning enzyme in the nerve gap. *Ageing* is a term that is given to the attachment of a neurotoxin and the enzyme in the neural gap. After the exposure, the bond between the toxin and the enzyme becomes extremely stable (for some neurotoxins, this ageing effect occurs within a matter of minutes). Thus the antidotal therapy becomes less beneficial. As the toxin passes the blood–brain barrier in the brain and attaches to the respective enzyme there, the antidote 2-PAM (an oxime) cannot breach the barrier, making this part of the therapy routine ineffective.

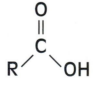

ORGANIC ACIDS

Chemistry Quick Reference Card

Family	Naming	Hazards
Alcohols	R–OH	Toxicity Flammability
Aldehydes	R–COH	Flammability Toxicity Polymerization
Ketones	R–CO–R′	Flammability Toxicity
Esters	R–COO–R′	Flammability Polymerization Toxicity
Organic acids	R–COOH	Toxicity Corrosiveness

Carboxylic Acids

If attachment occurs at the oxygen, an acid results.

If attachment occurs at the carbon, a nonpolar covalent bond results.

carboxylic acids

organic acids that contain a carboxyl group

Carboxylic acids have high boiling points, with solubility in water only in the lower molecular-weight compounds. The boiling points are characteristically higher than in the alcohols because of the molecules' ability to hold and maintain two hydrogen bonds. The lower-chained acids are very irritating and usually have toxic potentials.

These acids have K_a values of 10^{-4} to 10^{-2}. To put this into perspective, K values greater than 10^3 represent extremely strong acids such as hydrochloric or sulfuric acid. A K value between 10^3 and 10^{-2} is considered strong; sulfurous or chlorous acid, for example. A K value of 10^{-2} to 10^{-7} is weak and includes hydrofluoric or nitrous acid as examples. A K of less than 10^{-7} is considered extremely weak, for example, carbonic acid.

> ■ **Note**
> The terms "molar," "formal," and "normal" are commonly used by chemists to describe the amount of solute in solution, and should alert the emergency responder to an acid even if the name is not recognizable as such.

Incident decisions must take into consideration both strength and *concentration*. Strength, the ability to ionize, is described as weak or strong. This ionization is specified in a percentage of ionization: The closer to 100% ionization, the stronger the acid. Concentration is a ratio of the amount of a material to water. Therefore, if we place an acid of 100% concentration in a container and compare it to a solution that is 50% of the acid in water, the concentration is reduced. In contrast with the strength, ionization occurs at different rates. In the solution of acid in water, 50% of the H^+ ions are produced, compared to a concentrated solution, which has 100% of the product generating H^+ ions.

Some common carboxylic acids, sometimes referred to as organic acids, are:

Common Name	IUPAC Name
Formic acid	Methanoic acid
Acetic acid	Ethanoic acid
Propionic acid	Propanoic acid
Butyric acid	Butanoic acid
Valeric acid	Pentanoic acid
Caproic acid	Hexanoic acid
Caprylic acid	Octanoic acid
Capric acid	Decanoic acid

> ● **Caution**
> An exposure of 2.5% hydrofluoric acid on the skin surface at 100% concentration, or a 10% skin exposure at 70% concentration, is enough to cause systemic poisoning and death.

acyl halides

members of the carbonyl group, in which the hydrogen has been replaced with a halogen

Acyl Halides

Because the hydrogen is replaced with a halogen, **acyl halides** are members of the carbonyl group. The naming involves the substitution of "-yl halide" at the end for the -ic in the representative acid (see the common acid names listed in the previous table), utilizing the carbon in the carbonyl group as a counted carbon; for example:

HCO-Cl = formyl chloride or methanoyl chloride

CH_3CO-Br = acetyl bromide or ethanoyl bromide

$CH_3CH_2CH_2$CO-Cl = butyric chloride or butanoyl chloride

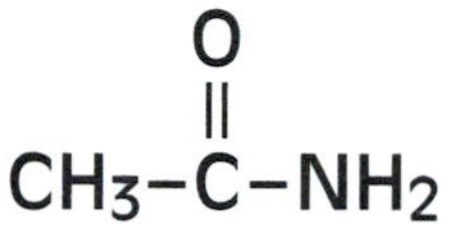

Amides

With **amides**, the hydrogen in the aldehyde is dropped and an ammonia group is attached.

For either naming convention, we drop the -oic from the corresponding acid and add the suffix -amide. For example, ethanamide is also known as acetamide. If the hydrogens in the ammonia group are named as alkyl groups, they are prefaced with an "N" in the name.

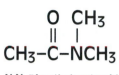

Ethanamide

$$CH_3\text{-}\overset{\overset{\displaystyle O}{\|}}{C}\text{-}\overset{\overset{\displaystyle CH_3}{|}}{N}CH_3$$

N,N–Dimethylacetamide

amides

derivatives of ammonia in which one, two, or all three hydrogens can be replaced with a hydrocarbon

Problem Set 5.8 Name the following compounds using the common and IUPAC nomenclature:

$CH_3CH_2CONH_2$

CH_3COCl

$CH_3CH_2CH_2COCl$

$CH_3CH_2CHCON(CH_3)_2$

$CH_3CH_2CH_2CH_2COBr$

$CH_3CH_2CH_2CONHCH_3$

$CH_3CH_2CH_2CH_2CH_2COCl$

$CH_2{=}CHCONH(CH_3)$

CYANIDE SALTS

Cyanide salts are placed in the organic grouping because, as with the functional groups, a carbon chain can be represented as R. This chain can be several carbons long or just one carbon. This group in general is used in the manufacturing of plastics and as solvents. The naming of these compounds can either name the carbon chain then the term cyanide, or can utilize the carbon of the cyanide ion as a basis of the name; for example:

Methyl cyanide	CH_3CN	Acetonitrile
Propenenitrile	C_2H_3CN	Acrylonitrile (vinyl cyanide)
Benzonitrile	C_6H_5CN	Phenyl cyanide

The R group may be a metal:

| Ammonium thiocyanate | NH_4SCN | Ammonium sulfocyanide |
| Barium cyanoplatinite | $BaPt(CN)_4$ | Barium platinum cyanide |

ISOCYANATES   R–N̈=C=Ö

The general formula for these compounds is R–N̈=C=Ö. When an amide follows what is called a base-promoted bromination, in other words, attacks the molecule with bromine in a basic solution, this amide becomes a brominated amide. From here chemists add more base to get the isocyanate. The examples are:

Isocyanate (crotonate), a polymer component used for particle board resin, car bumpers, shoe soles, and adhesives.

Methyl isocyanate (methyl ester), used for synthetic rubbers and adhesives, insecticide, and as a herbicide intermediate (extremely toxic—Bhopal, India).

Phenyl isocyanate (carbanil), used as an agent to identify alcohols and amines; also used as an intermediate.

Isocyanates, in the presence of alcohol, produce a urethane, which is commonly called carbamate. The term carbamate comes from the ester of an alcohol and carbamic acid, which forms the urethane. Just as in the substitution of sulfur for oxygen in the alcohols, to make the thiols, the oxygen is replaced with sulfur and the result is an isothiocyanate, which has a general formula of R–N̈=C=S̈.

BRANCHING AND ITS EFFECTS ON PHYSICAL PROPERTIES

Branching as isomers, as we have seen in all the categories, has a great effect on the boiling point and thus the ability of the compound to move from the liquid to a vapor. In general, the weight of the compound, its polarity, and the branching the R groups have will produce an effect on the boiling point and thus the production of vapors.

Polarity translates into hydrogen bonding and how well a compound can achieve this bonding. Polarity increases the boiling point of a compound and depends on hydrogen bonding and the carbonyl group. Isomers will decrease boiling points given compounds of equal weight.

Alcohols, ethers, and aldehydes all have wide flammable ranges. If an ignition source is present, the possibility of a fire or explosion increases significantly. If the boiling point and flash point are low, then the ignition temperature and vapor pressure will be high. The converse is also true.

■ Note

Chemical warfare agents, weapons of mass destruction and CBRNE (chemical, biological, radiological, nuclear, and explosive) are all synonyms.

CHEMICAL WARFARE AGENTS

Terrorist events such as the Tokyo subway and Matsumoto incidents in Japan, in which sarin was used by the Aum Supreme Truth, the use of mustard agents by the Iraqis and against the Serbs, and the use of salmonella bacteria by the Rajneesh in the northwest United States, point out the necessity of being prepared to deal with all manner of CWAs. These chemicals should not be viewed any differently than the industrial chemicals, nor as more or less hazardous than "normal" hazmat incidents encountered. In the following discussion, these agents are organized in terms of the hazardous materials categories that we usually think of. The DOT classification and the military's abbreviation accompany each hazmat category. Tables list the inherent chemical and physical properties that are applicable to the emergency responder, as well as the chemical precursors for the manufacturing of these agents. Each precursor is identified with its chemical name, the chemical abstract number, and the Department of Transportation/United Nations identification number if applicable. The highlighted chemicals in these tables are those that present a special problem: they may be gases, tend to vaporize rapidly, or are liable to produce vapor imminently. A brief discussion of the specific decontamination issues these chemicals may pose concludes each category. Not all the information about each agent is presented, but rather a quick overview highlighting the chemistry and the medical implications.

Neurotoxins

Nerve Agent	DOT Class	Chemical Name
Tabun (GA)	6.1	Ethyl N,N-dimethylphosphoroamidocyanidate
Sarin (GB)	6.1	Isopropyl methylphosphonofluoridate
Soman (GD)	6.1	Pinacolyl methyl phosphonofluoridate
GF	6.1	O-Cyclohexyl-methylfluorophosphonate
V agent (VX)	6.1	O-Ethyl-S-(2-iisopropylaminoethyl)methyl phosponothiolate

Historical Background. The earliest record of a substance that works like a nerve agent is of one that was used by tribesmen in western Africa. This tribe used an extract of the Calabar bean, which inhibits cholinesterase, both as a poison and as medicine. However, the first recorded synthesized cholinesterase inhibitor was TEPP (tetraethyl pyrophosphate), made by Wurtz in 1854. The product was tasted by his student deClermont with no ill effects, or so the story goes.

Over time, a number of scientists made modifications and investigations into organophosphorus chemistry that culminated in the 1930s with Dr. Gerhard Schrader's development of the first nerve agent, tabun, on December 23, 1936. Sarin was created in 1937. The work just before this time focused along two similar paths: one that looked at organophosphorus compounds as reversible, and one that viewed them as irreversible. These led to the discoveries in organophosphorus chemistry and carbamate chemistry. Both would bind with acetylcholinesterase, but with different progressions. **Carbamates**, as an example, were found useful in the therapy of intestinal gastric atony, glaucoma, and myasthenia gravis.

A production facility for manufacturing tabun in large quantities was established in 1942; the plant was captured by Russian forces during World War II and the facility moved to the Soviet Union, where production was restarted in 1946. Allied forces captured several thousand tons of munitions containing tabun and, to a lesser degree, sarin. In one weekend, the United States and England discovered the potential these chemicals have and the antidotes that would be needed. It has been suggested that the discovery of the antidotes resulted from an accident that occurred while investigating the chemical. Soman was synthesized in 1944, but its development was never explored until after the war. VX was developed by a British industrial concern and given to the United States in the early 1950s. In 1949, a compound by the name of GF was synthesized but was discarded as a potential military nerve agent, possibly

carbamates

organic compounds that can temporarily inhibit the enzymatic activity of acetylcholinesterase by binding to acetylcholine

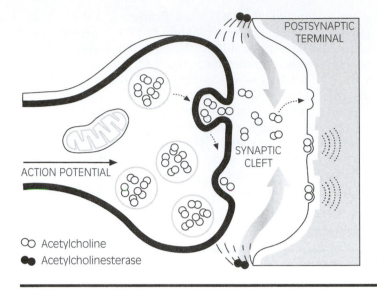

Figure 5–8   *Mechanism of action for a neurotransmitter impulse.*

because of the expense of the synthesis and the fact that it had chemical responses similar to sarin.

Activity and Chemistry.   The organophosphorus neurotoxins all act in a similar manner, but at different rates. In the human nervous system, there is a gap between each nerve cell (see Figure 5–8). As an impulse moves down a nerve, it stimulates the release of a chemical that moves through the gap, communicating with the next nerve cell. This chemical or enzyme is called **acetylcholine**. Once the next nerve is stimulated, another enzyme comes into the gap and "washes" the gap clean of acetylcholine. This chemical is called **acetylcholinesterase**. It is with this chemical that bonding occurs with the neurotoxin. Binding up the acetylcholinesterase allows the acetylcholine to continue to stimulate the receptive nerve cell, resulting in overstimulation and the signs and symptoms of salivation, lacrimation, urination, defecation, gastrointestinal, emesis, and miosis (SLUDGEM).

From a chemical standpoint, an organophosphorus compound is an ester. Organophosphorus esters have a tendency to form very strong bonds with a protein on the acetylcholinesterase enzyme. In a very short time frame, this bond increases in strength. (Some have suggested that it starts as an ionic bond and then quickly becomes a covalent bond.) Once the bond reaches its strongest point, it is said to be "aged." Once aged, the covalent bond is very difficult to break, thus reducing the ability of the antidote to remove the neurotoxin from the acetylcholinesterase enzyme.

Additionally, these compounds can have isomers that are different only in that they are mirror images of each other (called enantiomers; see Chapter 3). In some studies, this enantiomerism creates a higher toxicity for one mirror image and a lower toxicity for the other. This characteristic could explain why some individuals are more susceptible to nerve toxicity than others, although toxicity does have relationships with age, sleep deprivation, diet, gender, and previous exposure, among other factors.

The properties of nerve agents are summarized in Table 5–8.

acetylcholine

a chemical neurotransmitter that stimulates the heart, skeletal muscles, and glands

acetylcholinesterase

the enzyme that hydrolyzes acetylcholine, thereby stopping its activity

Table 5–8   *Properties of nerve agents.*

	Tabun (GA)	Sarin (GB)	Soman (GD)	GF	V Agent (VX)
Vapor Pressure	0.037 mm Hg @ 68°F	2.10 mm Hg @ 68°F	0.4 mm Hg @ 77°F	0.044 mm Hg @ 68°F	0.0007 mm Hg @ 68°F
Vapor Density	5.63	4.86	6.33	6.2	9.2
Volatility	129.65 ppm @ 86°F	5211.27 ppm @ 86°F	749.07 ppm @ 86°F	78.99 ppm @ 77°F	0.962 ppm @ 77°F
Boiling Point	428–475°F	304–316°F	388°F	462°F	568°F
Flash Point	172.4°F	N/A	N/A	201°F	318°F
Solubility in Water	9.8% @ 77°F	Miscible	2.1% @ 68°F	Immiscible	Slightly miscible
Hydrolysis	Slow with water; rapid with alkaline solutions	Rapid with alkaline solutions	Rapid with alkaline solutions	Slow with water; rapid with alkaline solutions	Rapid with alkaline solutions
Effect Threshold	0.3778 ppm Miosis	0.1749 ppm Miosis	0.0215 ppm Miosis	N/A	0.00366 ppm Miosis
TLV/TWA	0.000015 ppm	0.000017 ppm	0.000004 ppm	N/A	0.0000009 ppm
IDLH	0.03 ppm	0.03 ppm	0.008 ppm	N/A	0.0018 ppm
ED (minimum)	0.3–0.5 ppm	0.5 ppm	0.3–0.4 ppm	0.3–0.4 ppm	2.2–4.5 ppm
LCT_{50}	60 ppm	12–17 ppm	9 ppm	10–16 ppm	3–9 ppm
Odor	Fruity	Slightly fruity	Fruity with camphor	Fruity	Fruity
Persistency	1–2 days	1 day	1–2 days	1–2 days	Weeks
Physical State	Colorless to brown liquid	Colorless liquid	Colorless liquid	Liquid	Amber oily liquid

---

● **Caution**

Neurotoxins are extremely dangerous. Their manufacture is illegal and requires licensure from the appropriate authorities. The information given herein is to alert the first responders to the possibility of a clandestine operation. Law enforcement agencies should be contacted immediately and joint operations ensue if such an operation is suspected. Purchasing of these materials can also lead to police action. The technology in the production of the chemicals is beyond the realm of most chemists, as would be the expense of the level of personal protective equipment that would be required.

Treatment. Providing an open airway and supporting ventilation is the first-priority treatment to establish. Oxygenation should be accomplished before administering an antidote to avoid possibly fatal cardiac dysrhythmias. Antidotes for nerve agent poisoning include atropine IM or preferably IV, followed by pralidoxime chloride (protopam chloride, 2-PAM). Atropine is administered in high doses while under close cardiac and vital-sign monitoring. (This monitoring will not be possible if treating a large number of patients and is the reason for the IM antidote. Although somewhat slower than IVP, it can provide an adequate antidote route.)

Atropine blocks the effects of acetylcholine while the body is naturally metabolizing the organophosphate. Atropinization must be maintained until all the absorbed organophosphate has been metabolized and the body again produces sufficient quantities of acetylcholinesterase. Administering 2-PAM assists the body in this process. The treatment could last from days to weeks, necessitating the use of huge quantities of atropine.

Pralidoxime (2-PAM) is the next drug indicated. It has three desirable effects: (1) it frees and reactivates acetylcholinesterase, (2) it detoxifies the nerve agent, and (3) it has anticholinergic (atropine-like) effects. The only disadvantage of 2-PAM is that it is an oxime that cannot pass the blood–brain barrier, while the ester can. Once the organophosphate passes this barrier, the use of 2-PAM is negated. Thus, the use of quick antidotal therapy is a necessity. Additionally, the body normally will produce acetycholinesterase, but in very small quantities.

Chemical Precursors. Chemistry provides a myriad of chemical processes in order to synthesize chemicals, and CWAs are no different. Industrial chemicals that are common either in manufacturing, industrial processes, transportation, or over the counter possibly exist in all communities. Table 5–9 gives a list of chemical precursors for the neurotoxins.

Table 5–9 *Chemical precursors for CWA neurotoxins* (highlighted chemicals have a high inhalation hazard).

	Chemical Abstract Number	Department of Transportation/ United Nations Identification Number
Tabun	77-81-6	2810
Diethyl *N,N*-dimethyl phosphoramidate	2404-03-7	
Dimethylamine	124-40-3	1032
Dimethylamine	506-59-2	1160
Phosphorous oxychloride	**10025-87-3**	**1810**
Phosphorous pentachloride	**10026-13-8**	**1806**
Phosphorus trichloride	**7719-12-2**	**1809**
Potassium cyanide	**151-50-8**	**1680**
Sodium cyanide	**143-33-9**	**1689**
Sarin	107-44-8	2810
Ammonium bifluoride	1341-49-7	1727
Diethyl ethylphosphonate	78-38-6	
Diethylphosphite	762-59-2	
Dimethyl methylphosphonate	756-79-6	
Dimethylphosphite	868-85-9	
Ethylphosphonous dichloride	**1498-40-4**	**2845**
Ethylphosphonyl dichloride	1066-50-8	
Ethylphosphonyl difluoride	753-98-0	
Hydrogen fluoride	**7664-39-3**	**1052**
Methylphosphonous difluoride	753-59-3	
Methylphosphonyl dichloride	676-97-1	
Methylphosphonyl difluoride	676-99-3	
Phosphorus trichloride	**7719-12-2**	**1809**
Potassium bifluoride	7789-29-9	1811
Potassium fluoride	7789-23-3	1812
Sodium bifluoride	1333-83-1	2439
Sodium fluoride	7681-49-4	1690
Thionyl chloride	**7719-09-7**	**1836**

(continues)

Table 5–9   *(Continued)*

	Chemical Abstract Number	Department of Transportation/ United Nations Identification Number
Soman	96-64-0	2810
Ammonium bifluoride	1341-49-7	1727
Diethylphosphite	762-59-2	
Dimethyl methylphosphonate	756-79-6	
Dimethylphosphite	868-85-9	
Methylphosphonous difluoride	753-59-3	
Methylphosphonyl dichloride	676-97-1	
Methylphosphonyl difluoride	676-99-3	
Phosphorus trichloride	**7719-12-2**	**1809**
Pinacolone	75-97-8	
Pinacolyl alcohol	464-07-3	
Potassium bifluoride	7789-29-9	1811
Potassium fluoride	7789-23-3	1812
Sodium bifluoride	1333-83-1	2439
Sodium fluoride	7681-49-4	1690
Thionyl chloride	**7719-09-7**	**1836**
GF	329-99-7	2810
Ammonium bifluoride	1341-49-7	1727
Dimethyl ethylphosphonate	6163-75-3	
Dimethyl methylphosphonate	756-79-6	
Dimethylphosphite	868-85-9	
Methylphosphonous difluoride	753-59-3	
Methylphosphonyl dichloride	676-97-1	
Methylphosphonyl difluoride	676-99-3	
Phosphorus trichloride	**7719-12-2**	**1809**
Potassium bifluoride	7789-29-9	1811
Potassium fluoride	7789-23-3	1812
Sodium bifluoride	1333-83-1	2439
Sodium fluoride	7681-49-4	1690
Thionyl chloride	**7719-09-7**	**1836**
VX	50782-69-9	2810
Diethyl methylphosphonite	15715-41-0	
Diethylaminoethanol	100-37-8	
Diethylphosphite	762-59-2	
Diisopropylamine	108-18-9	
Ethylphosphonous dichloride	1498-40-4	
Methylphosphonous dichloride	676-83-5	
Methylphosphonous difluoride	753-59-3	
N,N-Diisopropyl-aminoethanethiol	5842-07-9	
N,N-Diisopropyl-(beta)-aminoethanol	96-80-0	
N,N-Diisopropyl-(beta) aminoethyl chloride	96-79-7	
O-Ethyl,2-diisopropyl aminoethyl methyl-phosphonate	57856-11-8	
Phosphorous pentasulfide	1314-80-3	1340
Triethyl phosphite	122-52-1	2323

Decontamination Considerations. When we think of decontamination, we usually think about water or a soap-and-water progressive step approach. Hydrolysis is the chemical reaction between a substance and water that can produce by-products; for example, tabun releases cyanide, sarin, and hydrogen fluoride. In addition, these agents are only slightly miscible in water. Because of the added potential hazard of the by-products, neurotoxin decontamination can become complicated. Indeed, it is a convoluted and highly controversial topic, especially when we are discussing decontamination of personnel. (See the "Decontamination" section in Chapter 6.)

■ Note

Sodium hypochlorite, an alkali solution, speeds up the hydrolysis of neurotoxins and thus completes decontamination more quickly. Note, however, that calcium hypochlorite can react with a variety of chemicals. Skin and respiratory hazard is a high possibility. Extensive flushing with water is required.

Corrosive Materials

Vesicant, Blister Agent (Mustard)	DOT Class	Chemical Name
Distilled mustard (HD)	6.1	Bis-(2-chloroethyl) sulfide
Nitrogen mustard (HN-1)	6.1	Bis-(2-chloroethyl) ethylamine
Nitrogen mustard (HN-2)	6.1	Bis-(2-chloroethyl) methylamine
Nitrogen mustard (HN-3)	6.1	Tris-(2-chloroethyl) amine
Lewisite (L)	6.1	Dichloro-(2-chlorovinyl) arsine
Phosgene oxime (CX)	6.1	Dichloroformoxime

Historical Background. A vesicant is a chemical that produces a vesicle or blister, hence the name of the military classification. These chemical agents work like corrosives in many ways, and as a group should be considered as strong corrosive chemicals. The use of this chemical has its origins on the battlefields of World War I, causing more injuries than chlorine, phosgene, and cyanide combined. Nitrogen mustards were first synthesized in the 1930s and were used as a mainstay in cancer therapy.

The name "mustard" comes from its yellow-brown color and distinct odor of onions and garlic. The Germans called it "hun stoffe," which led to its military abbreviation, HS. There are several translations of "hun stoffe," one meaning "new synthesis," and another being simply "German stuff."

Activity and Chemistry. A Sulfur mustard is a sulfide-based compound derived from the nerve agents (chlorinated thioether). It is sometimes referred to as "yellow cross" (because of the yellow cross found on the munitions), LOST (for the two scientists who suggested its use—Lommell and Steinkopf), or Yperite (for the site of its first use in France). The nitrogen mustards are amines that can penetrate skin very rapidly. However, symptoms may take 4 hours to several days before they present.

Arsenicals, in which the central atom is arsenic, are considered "blister" agents. Lewisite is the primary example in this category. The primary difference between the sulfur or nitrogen mustards and the arsenicals is that the arsenicals produce almost instant pain and eye irritation, whereas the sulfur or nitrogen mustards have a 2- to 24-hour window (in some cases, depending on concentration and type, several days) before pain is perceived by the victim.

These chemicals have no known antidote. The true mechanism by which they work is still not fully understood. For the sulfur or nitrogen mustards and Lewisite, it is thought that the agent changes the permeability of the cell membrane to allow the agent into the cell. Calcium levels are changed, affecting the homeostasis chemically. A loss of proteins and lipids that make up the membrane result in the loss of cellular integrity. DNA breaks down, thus destroying the capability of the cell to reproduce. Additionally, once the chemical enters the cell, metabolism is stopped and, to a certain degree, cellular anoxia results. (Industrial alkalis work somewhat differently, although some may have the same qualities as described.)

Urticants also fall into this basic material category. Phosgene oxime is representative of this category and is sometimes referred to as nettle gas, because of its ability to cause mild prickly sensations similar to multiple bee sting-type symptoms. Its mechanism is truly unknown, but it is suspected that the chlorine in the molecule is the cause of the direct effect. It produces skin lesions similar to those from a strong acid. Pain is immediate, in contrast to the nitrogen or sulfur mustards. When inhaled, pulmonary edema ensues. Additionally, absorption of phosgene oxime through the skin can cause pulmonary edema. Direct injury is highlighted by the corrosive injury: indirect injury through the release of peroxides causes tissue damage and activation of the immune system, thus destroying tissue.

The properties of vesicants and blister agents are summarized in Table 5–10.

Chemical Precursors. Chemical precursors for CWA corrosives are summarized in Table 5–11.

Decontamination Considerations. Decontamination of the CWA corrosives (vesicants) presents a very special problem: immiscibility in water. Depending on the temperature and the weather, some vesicants can last days to several weeks, even in humid environments. The average persistency is several days; however, they can remain in water longer because of their insoluble (immiscible) tendencies. Alkaline solutions are recommended for decontamination, and chloramine solutions are recommended for use with sulfur mustard and Lewisite. There is limited evidence that the blisters created by the vesicant hold fluid with blister-causing qualities, but only a few injuries from vesicants have reported such behavior of the chemical.

Decontamination is the key to patient care and is a high priority with both military-grade weapons and industrial-grade products. Lewisite has an antidote called BAL (British anti-Lewisite), a topical ointment for systemic placement. It is a chelator for heavy metals such as arsenic.

Chemical Asphyxiants

Blood Agent	DOT Class	Chemical Name
Hydrocyanic acid (HC)	6.1	Hydrogen cyanide
Cyanogen chloride (CK)	2.3	Chlorine cyanide
Arsine (SA)	2.1, 6.1	Arsenic trihydride

Historical Background. Cyanide has been used as a weapon for a long time. In ancient Rome, Nero used cherry laurel water to poison individuals who displeased him. Karl Wilheim Scheele, in 1786, isolated cyanide using sulfuric acid and Prussian

Table 5–10   *Properties of corrosive materials (vesicants and blister agents).*

	Distilled Mustard (HD)	Nitrogen Mustard (HN-1)	Nitrogen Mustard (HN-2)	Nitrogen Mustard (HN-3)	Lewisite (L)	Phosgene Oxime (CX)
Vapor Pressure	0.072 mm Hg @ 68°F	0.24 mm Hg @ 77°F	0.29 mm Hg @ 68°F	0.012 mm Hg @ 77°F	0.394 mm Hg @ 68°F	11.2 mm Hg @ 77°F
Vapor Density	5.4	5.9	5.4	7.1	7.1	3.9
Volatility	93.95 ppm @ 68°F	218.96 ppm @ 68°F	800.60 ppm @ 86°F	21.56 ppm @ 86°F	529.34 ppm @ 68°F	387.18 ppm @ 68°F
Boiling Point	442°F	381°F Decomposes below BP	167°F Decomposes below BP	493°F Decomposes below BP	374°F	264°F, with decomposition
Flash Point	221°F	N/A	N/A	N/A	None	N/A
Solubility in Water	Immisicible	Immisicible	Immisicible	Immisicible	Immisicible	Slowly but completely
Hydrolysis	Slight with water; increases with alkaline solutions and higher water temperature	Slight with water; increases with alkaline solutions	Slight with water; increases with alkaline solutions	Slight with water; increases with alkaline solutions	Slight with water; increases with alkaline solutions	Slow; violent reaction with alkaline solutions
Effect Threshold	15 ppm Eye pain, discomfort, injury	28 ppm Eye pain, discomfort, injury	16 ppm Eye pain, discomfort, injury	24 ppm Eye pain, discomfort, injury	35 ppm Eye pain, discomfort, injury	Irritation to eyes is extreme
TLV/TWA	0.0005 ppm	0.0004 ppm	0.0004 ppm	0.0004 ppm	0.00035 ppm	N/A
IDLH	0.0005 ppm	0.0004 ppm	0.0004 ppm	0.0004 ppm	0.0004 ppm	N/A
ED (minimum)	23 ppm	14–29 ppm	14–29 ppm	14–29 ppm	< 36 ppm	0.6 ppm
LCT_{50}	231 ppm	179–470 ppm	179–470 ppm	179–470 ppm	141–177 ppm	687 ppm
Odor	Garlic, horseradish	Faint musty, fishy	Soapy to fruity	Garlic, fishy, fruit to none	Geraniums	Irritating
Persistency	1 day to 2 weeks	1 day to 1 week	1 day to 2 weeks	1 day to 2 weeks	1 day to 1 week	Few hours
Physical State	Amber liquid	Oily, pale yellow liquid	Dark liquid	Oily liquid	Brownish liquid	Liquid

blue. This mixture, which he called prussic acid, was hydrocyanic acid, and he was killed by the fumes generated when a beaker of the substance was dropped to the floor in his laboratory.

During World War I, the French used cyanide in munitions. The problem was that the gas is lighter than air and dissipates very rapidly. The Germans rapidly learned of its use and equipped the troops with masks. Thus, given the problem of dissipation and defensive preparedness, cyanide was not an effective weapon.

Table 5–11   *Chemical precursors for CWA corrosives* (highlighted chemicals have a high inhalation hazard).

	Chemical Abstract Number	Department of Transportation/ United Nations Identification Number
Sulfur Mustards	505-60-2	2810
2-Chloroethanol	107-07-3	
Sodium sulfide	1313-82-2	1385
Sulfur dichloride	10545-99-0	
Sulfur monochloride (sulfur chloride)	**10025-67-9**	**1828**
Thiodiglycol	111-48-8	2966
Thionyl chloride	**7719-09-7**	**1836**
Nitrogen Mustards	538-07-8	2810
2-Chloroethanol	107-07-3	
Thionyl chloride	**7719-09-7**	**1836**
Triethanolamine	102-71-6	9151
Arsenicals, Lewisite	541-25-3	2810
Arsenic trichloride	**7784-34-1**	**1560**

In 1916, the more effective cyanogen chloride was introduced, which is heavier than air and has cumulative effects. By World War II, hydrogen cyanide, known as Zyklon B, found its way into the death camps. Zyklon B was originally used as a fumigant for rodents.

Activity and Chemistry.   Cyanide acts on the oxygen metabolism in the cell itself. It combines with an enzyme called cytochrome oxidase. This enzyme in its original form is necessary for cellular respiration. By binding it, cyanide causes the cells to move into anaerobic metabolism. Without the cytochrome oxidase, the cell cannot utilize oxygen. Although the half-life in the body is about an hour, death takes place before the body has the ability to detoxify or excrete the chemical.

The antidote is to use nitrites to bind the cyanide so that the body can excrete the chemical combination. The nitrites (amyl nitrite and sodium nitrite) convert hemoglobin into methemoglobin. Methemoglobin competes with cytochrome oxidase for the cyanide ion, actually attracting the cyanide away from the cytochrome oxidase and thus freeing it to again participate in aerobic cellular metabolism. The last step is to infuse sodium thiosulfate, which acts as a cleanup agent by changing the remaining cyanide into a relatively harmless substance, thiocyanate.

Arsine works differently than the cyanides. It has a delayed action on the liver and kidneys. The lethal dose varies; however, a dose of 0.5 ppm has shown symptoms. The effect is a triad of symptoms over days. The first sign is a port-wine coloration of the urine. Jaundice, an intense bronze hue over the entire body, appears on the second or third day. Severe renal failure occurs with suppression of urinary function. Destruction of the red blood cells (hemolysis) has occurred with small-dose exposure, thus causing liver and kidney failure.

The properties of chemical asphyxiants are summarized in Table 5–12.

Table 5–12 *Properties of chemical asphyxiants (blood agents).*

	Hydrocyanic Acid (HC)	Cyanogen Chloride (CK)	Arsine (SA)
Vapor Pressure	612 mm HG @ 68°F	1000 mm Hg @ 77°F	11,100 mm Hg @ 68°F
Vapor Density	1	2.1	2.69
Volatility	978,912.31 ppm @ 77°F	2,443,623.94 ppm @ 77°F	9,714,487.36 ppm @ 32°F
Boiling Point	78°F	55.04°F	−80.5°F
Flash Point	0.4°F	None	Highly flammable
Solubility in Water	Highly	Immisicible	Slightly
Hydrolysis	High	Low	High
Effect Threshold	Varies	4.7 ppm, tearing	Lethal
TLV/TWA	9.9 ppm	0.2 ppm	0.005 ppm
IDLH	45 ppm	N/A	6 ppm
ED (minimum)	Varies	2784 ppm	N/A
LCT_{50}	3600 ppm	4375 ppm	1571 ppm
Odor	Bitter almonds	Irritating	Garlic
Persistency	Short, becomes a vapor	Short, becomes a vapor	Short
Physical State	Liquid	Liquid	Gas

Treatment. The treatment for exposure to chemical asphyxiants is supportive; that is, treating the symptoms, anticipatory signs, and if appropriate, utilizing the cyanide antidote kit.

Chemical Precursors. Chemical precursors for CWA chemical asphyxiants are summarized in Table 5–13.

Table 5–13 *Chemical precursors for CWA chemical asphyxiants* (highlighted chemicals have a high inhalation hazard).

	Chemical Abstract Number	Department of Transportation/ United Nations Identification Number
Hydrogen Cyanide	74-90-8	1051
Potassium cyanide	151-50-8	1680
Sodium cyanide	143-33-9	1689
Cyanogen Chloride	506-77-4	1589
Sodium cyanide	143-33-9	1689
Arsines	7784-42-1	2188
Arsenic trichloride	7784-34-1	1560

Decontamination Considerations. The hydrolysis of hydrogen cyanide, cyanogen chloride, and arsine is limited. Under laboratory conditions, the products of decontamination have shown possible secondary contamination issues. Hydrogen cyanide, under the appropriate conditions, has degraded into formic acid and ammonia, cyanogen chloride into hydrogen chloride, and cyanic acid and arsine into arsenic acids and hydrides.

Both hydrogen cyanide and cyanogen chloride are liquids. Decontamination is thus directed toward removal of the liquid droplets. Soap and water will dilute these products adequately. Arsine as a gas dissipates very rapidly, posing limited decontamination issues.

Irritants

Irritants can be thought of as any agent that can disrupt the ability of a person to carry out a particular task. They can interfere with oxygen absorption capability (respiratory irritants) or disrupt the will of aggression due to discomfort (lacrimators). However, irritants can also be nonchemical agents such as loud noise, high-intensity light stimulation, central nervous system bombardment through high-frequency generators, and olfactory assault that changes the cognitive thought abilities. This discussion, however, focuses on the chemicals that have military applications and the potential for industrial accidents. Some of these chemicals are considered respiratory irritants; the vomiting and lacrimators are referred to as incapacitating agents.

Historical Background. There are several historical examples of incapacitating agents being used. In 200 B.C., an officer in Hannibal's army reportedly left wine laced with an atropinic plant behind during a retreat. The army returned, recapturing the camp and slaughtering its soldiers as they lay helpless. In 1672, belladonna grenades were used to defend a fortress, only to have the wind shift and blow the agent back on the population using it. Because of this and similar incidents, in 1675 the French and Germans signed an agreement to prohibit chemical warfare.

Respiratory Irritant (Choking Agent)	DOT Class	Chemical Name
Chlorine (CL)	2.3	Molecular chlorine
Phosgene (CG)	2.3	Carbonyl chloride
Diphosgene (DP)	6.1	Trichloromethyl chloroformate
Chloropicrin (PS)	6.1	Trichloronitromethane

The first category includes the true respiratory irritants, described as the choking agents. The polarity and thus the solubility of these materials provide the effects on the human subject. Although described in the literature as military-grade weapons, these chemicals are found within industry. In addition, similar industrial chemicals, such as carbon monoxide, nitrogen, liquefied petroleum gas, ammonia, and sulfur dioxide, can produce the same symptomology, but through different biological mechanisms.

Table 5–14 gives a summary of the properties of these respiratory irritants and choking agents, and Table 5–15 lists identification numbers.

Decontamination Considerations. Decontamination for these chemicals follows the same procedures as for industrial chemicals: research specific incompatibilities and deal with medical issues and substance considerations.

Table 5–14 *Properties of respiratory irritants and choking agents.*

	Chlorine (CL)	Phosgene (CG)	Diphosgene (DP)	Chloropicrin (PS)
Vapor Pressure	5 atm @ 51°F	1173 mm Hg @ 68°F	4.2 mm Hg @ 68°F	18.3 mm Hg @ 68°F
Vapor Density	2.5	3.4	3.4	5.6
Volatility	N/A	1,065,002 ppm @ 46°F	5572 ppm @ 68°F	24,800 ppm @ 68°F
Boiling Point	–30°F	46°F	260°F	234°F
Flash Point	N/A	N/A	N/A	N/A
Solubility in Water	High	Decomposes	Limited	Immiscible
Hydrolysis	High	High	Slow	None
Effect Threshold	0.01 ppm	Varies	Varies	1.3 ppm, irritation
TLV/TWA	1 ppm	0.1 ppm	0.1 ppm	0.1 ppm
IDLH	30 ppm	2 ppm	2 ppm	4 ppm
ED (minimum)	N/A	395 ppm	N/A	N/A
LCT_{50}	6561 ppm	791 ppm	N/A	300 ppm
Odor	Suffocating	Mown hay	Mown hay	Stinging irritant
Persistency	Short	Short	Minimum of hours	Minimum of hours
Physical State	Gas	Gas	Oily liquid	Oily liquid

Table 5–15 *Identification numbers of respiratory irritants and choking agents* (highlighted chemicals have a high inhalation hazard).

	Chemical Abstract Number	Department of Transportation/ United Nations Identification Number
Chlorine	7782-50-5	1017
Phosgene	75-44-5	1076
Diphosgene	503-38-8	1076
Chloropicrin	76-06-2	1580

Lacrimator	DOT Class	Chemical Name
Ortho-chlorobenzalmalonitrile (CS)	6.1	O-chlorobenzylidene malonitrile
Dibenoxazephine (CR)	6.1	Dibenz-(b,f)-1,4-oxazepine
Chloracetepehone (CN)	6.1	Phenyl chloromethyl ketone
Oleoresin capsicum (OC)	2.2 (subsequent risk 6.1)	Cayenne pepper

Generally called riot control agents, lacrimators are used to incapacitate the enemy. These chemical sprays offer a nonlethal form of protection that causes temporary extreme discomfort, and have only rarely caused severe, lasting injury. Both CN and CS are submicron (less than 1 micron) particles. They are extremely light and are carried to the target area in a carrier solution that evaporates quickly, dispersing the agent. Because of their light, fine particles, both of these chemicals are prone to cross-contamination between the victim and emergency response personnel. The submicron size may allow these irritating particles to gain access deep into the lungs, causing injury to the fine bronchioles and alveoli.

OC, or pepper spray, has become the safest and most popular of these chemical agents. It is found in police aerosol sprays and over-the-counter agents. It is a non-water-soluble agent prepared from an extract of the cayenne pepper plant. The effects from OC start almost immediately when contact with the eyes occurs. Its particles are not submicron, so access to the lower respiratory system is limited.

Vomiting Agent (Sternutator)	Chemical Name
Diaphenylchloroarsine (DA)	Chlorodiphenylarsine
Diphenylcyanoarsine (DC)	Diphenylcyanoarsine
Adamsite (DM)	10-Choloro-5,10-dihydrophenarsazine

Sternutators cause vomiting and are used to create panic. They cause bodily discomfort such that the individual wearing respiratory protection takes it off. Then agents that are more lethal or incapacitating can be introduced to ensure the desired effect.

The primary functions of these agents is to affect the population in a short time frame while at the same time producing sensory irritation that, in turn, produces inappropriate behavior.

Summary

Chemistry Quick Reference Card

Family	Naming	Hazards
Alcohols	R–OH	Toxicity Flammability
Aldehydes	R–COH	Flammability Toxicity Polymerization
Ketones	R–CO–R'	Flammability Toxicity
Esters	R–COO–R'	Flammability Polymerization Toxicity
Organic acids	R–COOH	Toxicity Corrosiveness

There are diverse chemical and physical properties in this general family of polar compounds; although, in general, we can state that the polarity (thus, the solubility) of a compound is directly related to the molecular weight and branching. These factors also are related to the boiling point and thus the vapor production. In general, the alcohols, ethers, and aldehydes have extremely wide flammable ranges, which adds another dimension to our decision-making process.

Although this discussion is limited to very specific chemical families, functional groups, and derivatives, the potential to have compounds that cross family barriers is very real. The discussion of chemical warfare agents is presented only to highlight the similarities between military chemicals and what can be found in an industrial setting. The specific chemicals mentioned are by no means the only CWAs. Understanding the basic family characteristics and applying these principles to a hazardous materials event can assist you in the planning, implementation, and evaluation process, from deciding on the appropriate respiratory and skin-protection measures to be taken, to the appropriate decontamination actions. In addition, more accurate assessments can be made of the level of medical care that can truly and reasonably be taken at the scene, during transport, and at the medical facility.

Review Questions

1. Draw the following compounds:
   A. 1-Propanol
   B. Acrylonitrile
   C. 3-Chloro-1-propanol
   D. 4-Penten-2-ol
   E. Methyl cyanide
   F. Benzonitrile
   G. Trinitrophenol
   H. Ethylanol
   I. Vinyl cyanide
   J. Vinyl alcohol
   K. Isopropyl alcohol

L.   Acetonitrile

M. Tertbutyl alcohol

N. Propenenitrile

O. Sec-butyl alcohol

P.   Isobutyl alcohol

Q. Propyl alcohol

R.   Ethyl alcohol

S.   Phenyl cyanide

T.   2-Hexylanol

U.   7-Methyl-4-octen-3-ol

2. Name the following compounds, using both the common and IUPAC names where appropriate:

A. $CH_3COOCH=CH_2$

B. $CH_2OHCH_2CH_2OH$

C. CH_3COCH_3

D. $CH_3CH_2COCH_2CH_3$

E. $CH_3CH_2COCH_3$

F. $CH_3CH_2CH_2COCH_3$

G. $C_6H_5COCH_3$

H. $C_6H_5COC_6H_5$

I. CH_2OHCH_2OH

J. CH_3CHO

K. CH_3CHCHO
$\quad\quad |$
$\quad\quad CH_3$

L. $CH_3CH_2CH_2CHO$

M. $CH_3COC_2H_5$

N. $CH_3COOC_2H_3$

O. $C_2H_3COOCH_3$

P. $CH_3CH_2CH_2CONHCH_3$

Q. $CH_2=CHCONH(CH_3)$

R. $CH_3COCH=CH_2$

S. $CH_3CH_2CH_2CH_2CHO$

T. $C_6H_5COC_3H_5$

U. $C_2H_3COOC_3H_7$

V. $CH_3CH_2CONH_2$

W. CH_3COCl

X. $CH_3CH_2CH_2COCl$

Y. $CH_3CH_2CH_2CH_2COBr$

Z. $CH_3CH_2COOCH_3$

AA. $CH_3CH_2CH_2CH_2CH_2COCl$

BB. $i\text{-}C_3H_7COC_3H_5$

CC. $CH_2OHCHOHCH_2OH$

DD. $CH_3CHOHCH_2OH$

EE. CH_3CH_2CHO

FF. $CH_3COOC_3H_5$

GG. $CH_3CH_2CHCON(CH_3)_2$

3. Fill in the Chemistry Quick Reference Card:

Chemistry Quick Reference Card

Family	Naming	Hazards
Alcohols		
Aldehydes		
Ketones		
Esters		
Organic acids		

Problem Set Answers

Problem Set 5.1

1. Propanol $\quad\quad CH_3CH_2CH_2OH \quad\quad$ 2-Propanol $\quad\quad CH_3 — CHCH_3$
$\quad\quad\quad\quad\quad\quad\quad\quad\quad\quad\quad\quad\quad\quad\quad\quad\quad\quad\quad\quad\quad\quad\quad |$
$\quad\quad\quad\quad\quad\quad\quad\quad\quad\quad\quad\quad\quad\quad\quad\quad\quad\quad\quad\quad\quad\quad\quad OH$

2. 1. Methylpentanol

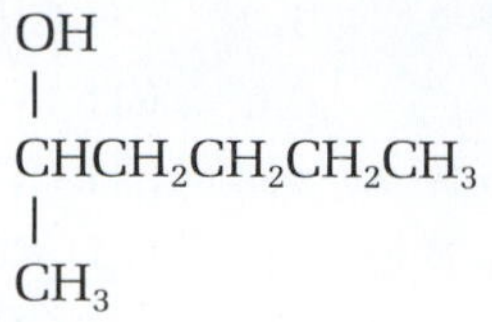

$$\underset{\underset{CH_3}{|}}{\overset{\overset{OH}{|}}{CH}}CH_2CH_2CH_2CH_3$$

2-Methylpentanol

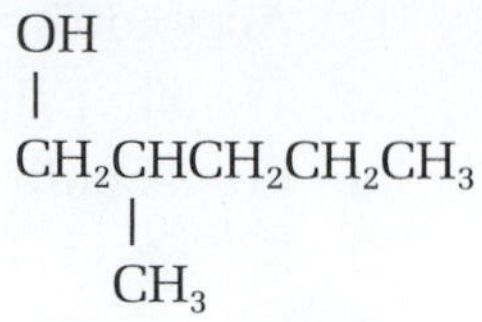

$$\underset{\underset{CH_3}{|}}{CH_2}\overset{\overset{OH}{|}}{CH}CH_2CH_2CH_3$$

2. 3-Methylpentanol

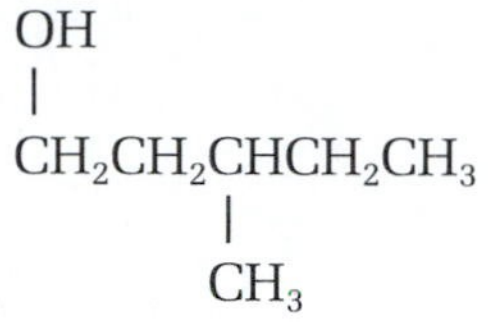

$$CH_2CH_2\underset{\underset{CH_3}{|}}{\overset{\overset{OH}{|}}{CH}}CH_2CH_3$$

4-Methylpentanol

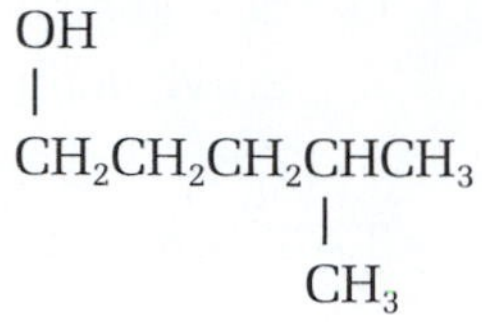

$$CH_2CH_2CH_2\underset{\underset{CH_3}{|}}{\overset{\overset{OH}{|}}{CH}}CH_3$$

3. 5-Methylpentanol

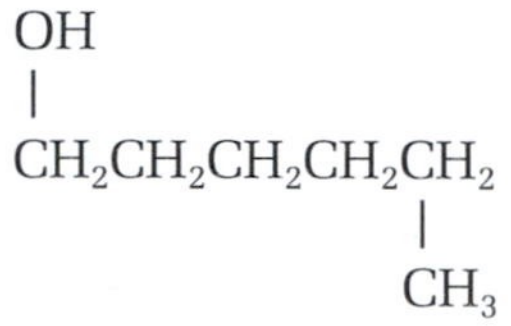

$$CH_2CH_2CH_2CH_2\underset{\underset{CH_3}{|}}{\overset{\overset{OH}{|}}{CH_2}}$$

3. 1. Methylpentanol

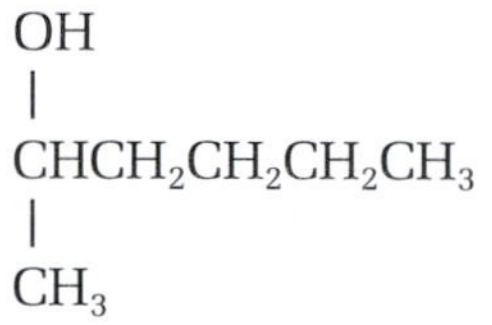

$$\underset{\underset{CH_3}{|}}{\overset{\overset{OH}{|}}{CH}}CH_2CH_2CH_2CH_3$$

Ethyl butanol

$$\underset{\underset{CH_3}{|}}{CH_2}\overset{\overset{OH}{|}}{CH}CH_2CH_2CH_3$$

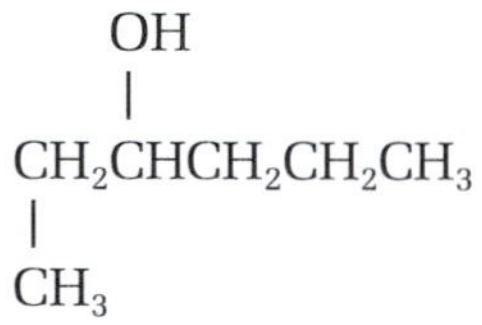

2. Equal to ethyl butanol

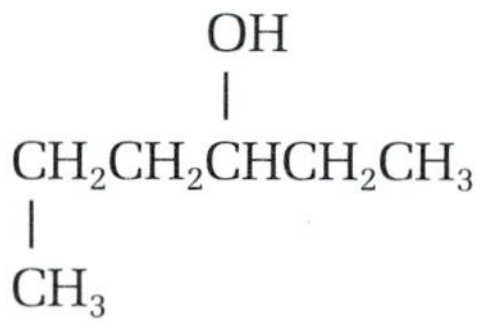

$$CH_2CH_2\underset{\underset{CH_3}{|}}{\overset{\overset{OH}{|}}{CH}}CH_2CH_3$$

Equal to methylpentanol

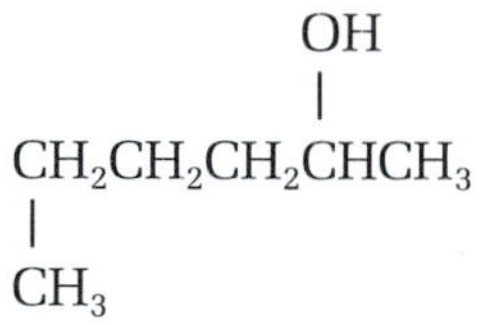

$$CH_2CH_2CH_2\underset{\underset{CH_3}{|}}{\overset{\overset{OH}{|}}{CH}}CH_3$$

3. Hexanol

$$CH_2CH_2CH_2CH_2\underset{\underset{CH_3}{|}}{\overset{\overset{OH}{|}}{CH_2}}$$

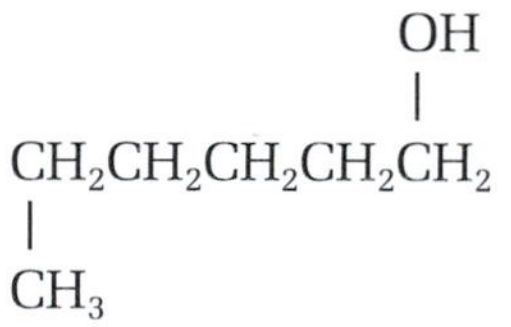

4. 1. 1-Propanol

$$CH_3CH_2CH_2OH$$

2. 3-Chloro-1-propanol

$$ClCH_2CH_2CH_2OH$$

3. 4-Penten-2-ol

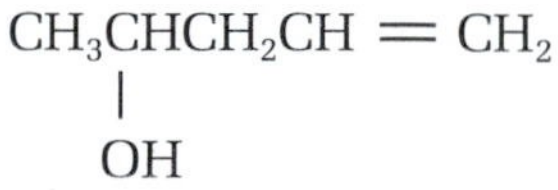

$$CH_3\underset{\underset{OH}{|}}{CH}CH_2CH=CH_2$$

4. Ethylanol

$$CH_3CH_2OH$$

5. Vinyl alcohol

$$CH_2=CHOH$$

6. Isopropyl alcohol

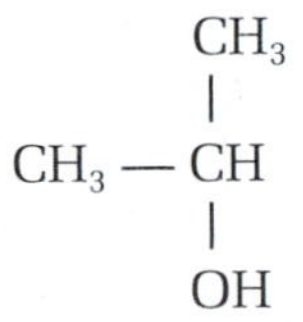

$$\begin{array}{c} CH_3 \\ | \\ CH_3 - CH \\ | \\ OH \end{array}$$

7. Tertbutyl alcohol

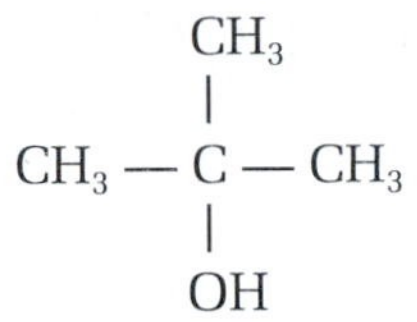

$$\begin{array}{c} CH_3 \\ | \\ CH_3 - C - CH_3 \\ | \\ OH \end{array}$$

8. Sec-butyl alcohol

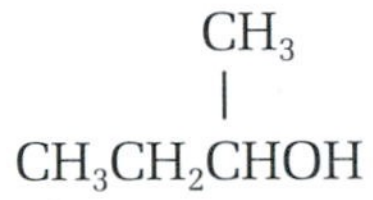

$$\begin{array}{c} CH_3 \\ | \\ CH_3CH_2CHOH \end{array}$$

9. Isobutyl alcohol

$$\begin{array}{c} CH_3 \\ | \\ CH_3CHCH_2OH \end{array}$$

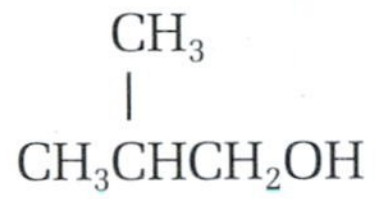

10. Propyl alcohol

$CH_3CH_2CH_2OH$

11. Ethyl alcohol

CH_3CH_2OH

12. 2-Hexylanol

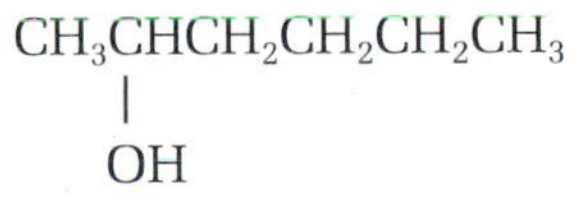

$$\begin{array}{c} CH_3CHCH_2CH_2CH_2CH_3 \\ | \\ OH \end{array}$$

13. 7-Methyl-4-octen-3-ol

$$CH_3 - CH_2 - CH - CH = CH - CH_2 - \overset{\overset{\displaystyle CH_3}{\displaystyle |}}{CH} - CH_3$$
$$\qquad\qquad\qquad\;\; |$$
$$\qquad\qquad\qquad OH$$

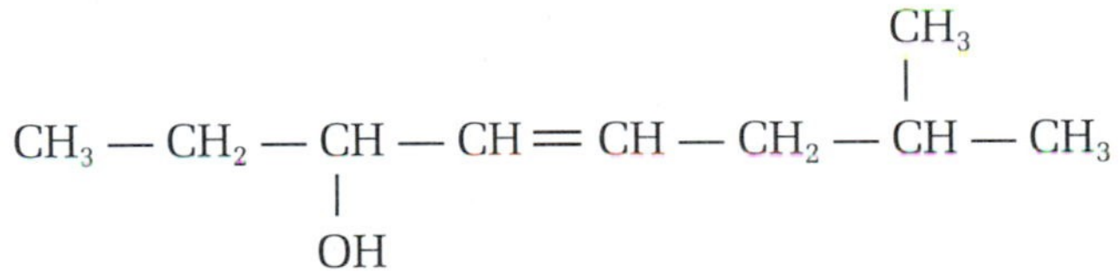

Problem Set 5.2

1. $CH_2OHCH_2CH_2OH$

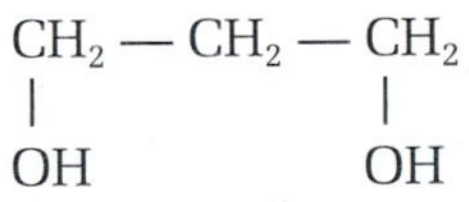

$$\begin{array}{ccc} CH_2 - CH_2 - CH_2 \\ | \qquad\qquad\quad | \\ OH \qquad\qquad OH \end{array}$$ Propylene glycol or 1,3-Propanediol

2. $OHCH_2CH_2OH$

$OH - CH_2 = CH_2 - OH$ Ethylene gl]

3. $CH_2OHCHOHCH_2OH$

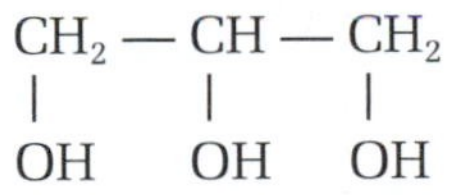

$$\begin{array}{ccc} CH_2 - CH - CH_2 \\ | \qquad\; | \qquad\; | \\ OH \quad OH \quad OH \end{array}$$ Propylene glycerin

4. $CH_3CHOHCH_2OH$

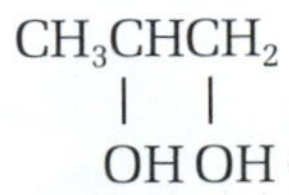

$$\begin{array}{c} CH_3CHCH_2 \\ |\;\; | \\ OH\,OH \end{array}$$ Propylene glycol or 1,2-Propanediol

Problem Set 5.3

1.

A	B	C
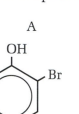	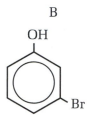	
O-Chlorophenol	M-Chlorophenol	P-Chlorophenol

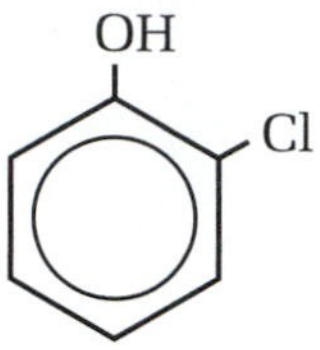

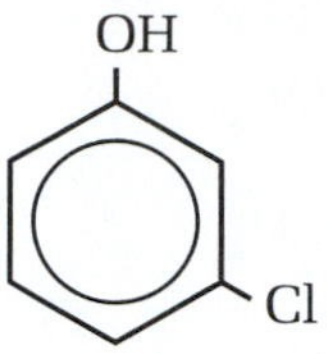

2.

A	B	C
O-Bromophenol	M-Bromophenol	P-Bromophenol

3.

A	B	C
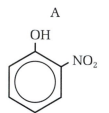	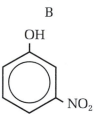	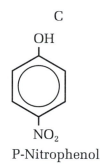
O-Nitrophenol	M-Nitrophenol	P-Nitrophenol

Problem Set 5.4

1. CH_3OH Thiol = CH_3SH

2. CH_3CH_2OH Thiol = CH_3CH_2SH

3. $CH_3CH_2CH_2OH$ Thiol = $CH_3CH_2CH_2SH$

4. $CH_3CH_2CH_2CH_2OH$ Thiol = $CH_3CH_2CH_2CH_2SH$

5. $CH_3CH_2\overset{\overset{\displaystyle CH_3}{|}}{C}HOH$ Thiol = $CH_3CH_2\overset{\overset{\displaystyle CH_3}{|}}{C}HSH$

Problem Set 5.5

1. Acetylaldehyde

2. Propanal, or methylacetaldehyde

3. Butylaldehyde

4. Pentalaldehyde

5. Isobutylaldehyde

Problem Set 5.6

1. CH_3COCH_3 $CH_3\overset{\overset{\displaystyle O}{\|}}{C}CH_3$ Propanone (acetone)

2. $CH_3COCH=CH_2$ $CH_3\overset{\overset{\displaystyle O}{\|}}{C}CH=CH_2$ 2-Butenone

3. i-$C_3H_7COC_3H_5$ $CH_3\overset{\overset{\displaystyle CH_3}{\diagdown}}{C}H\overset{\overset{\displaystyle O}{\|}}{C}CH_2CH_2CH_3$ 2-Methyl-isopropylpropylketone 3-hexamone

4. $CH_3CH_2COCH_2CH_3$ $CH_3CH_2\overset{\overset{\displaystyle O}{\|}}{C}CH_2CH_3$ Diethyl ketone 3-pentanone

5. $CH_3COC_2H_5$ $CH_3\overset{\overset{\displaystyle O}{\|}}{C}CH_2CH_3$ Methyl ethyl ketone or butanone

6. $C_6H_5COC_3H_5$ $C_6H_5\overset{\overset{\displaystyle O}{\|}}{C}CH_2CH_2CH_3$ Propyl phenyl ketone

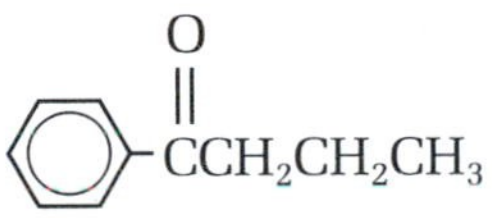

7. $CH_3CH_2COCH_3$ $CH_3CH_2\overset{\overset{\displaystyle O}{\|}}{C}CH_3$ Butanone or methyl prophyl ketone

8. $CH_3CH_2CH_2COCH_3$ $CH_3CH_2CH_2\overset{\overset{\displaystyle O}{\|}}{C}CH_3$ 2-Pentanone or methyl propyl ketone

9. $C_6H_5COCH_3$ $C_6H_5\overset{\overset{\displaystyle O}{\|}}{C}CH_3$ Acetophenone or methyl phenyl ketone

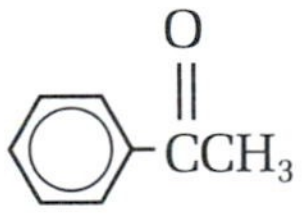

10. $C_6H_5COC_6H_5$ $C_6H_5\overset{\overset{\displaystyle O}{\|}}{C}C_6H_5$ Benzophenone or diphenyl ketone

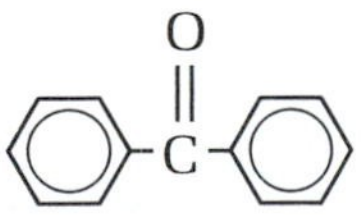

Problem Set 5.7

1. $C_2H_5COOCH_3$ $CH_3CH_2\overset{\overset{\displaystyle O}{\|}}{C}-OCH_3$ Methyl propylate

2. $CH_3COOC_2H_5$ $CH_3\overset{\overset{\displaystyle O}{\|}}{C}-OCH_2CH_3$ Ethyl acetate or ethyl ethanoate

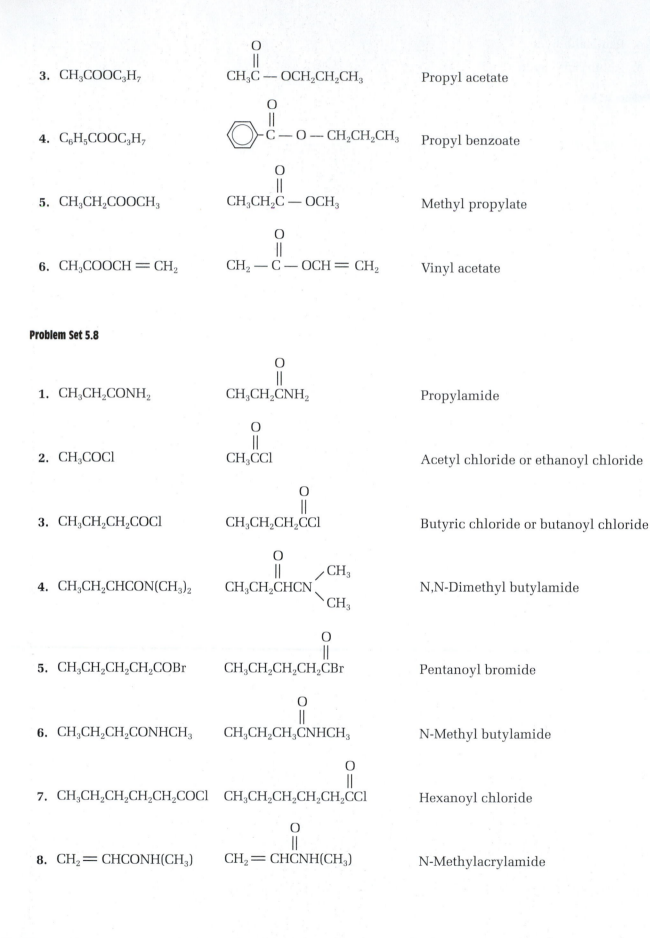

3. $CH_3COOC_3H_7$ — Propyl acetate

4. $C_6H_5COOC_3H_7$ — Propyl benzoate

5. $CH_3CH_2COOCH_3$ — Methyl propylate

6. $CH_3COOCH=CH_2$ — Vinyl acetate

Problem Set 5.8

1. $CH_3CH_2CONH_2$ — Propylamide

2. CH_3COCl — Acetyl chloride or ethanoyl chloride

3. $CH_3CH_2CH_2COCl$ — Butyric chloride or butanoyl chloride

4. $CH_3CH_2CHCON(CH_3)_2$ — N,N-Dimethyl butylamide

5. $CH_3CH_2CH_2CH_2COBr$ — Pentanoyl bromide

6. $CH_3CH_2CH_2CONHCH_3$ — N-Methyl butylamide

7. $CH_3CH_2CH_2CH_2CH_2COCl$ — Hexanoyl chloride

8. $CH_2=CHCONH(CH_3)$ — N-Methylacrylamide

Chapter 6

Science Officer's Reference Guide

Learning Objectives

Upon completion of this chapter, you should be able to:

- Analyze a hazardous materials scene based on the chemicals involved.
- Evaluate the safety perimeters at the hazmat incident.
- Formulate an action plan for the incident utilizing information gained through chemical research.
- Determine the appropriateness of the hazmat team's capability based on the referenced chemical(s).

After working through the preceding chapters, we arrive at this chapter with an understanding equivalent to second-year college chemistry. Beyond this level of chemistry, biochemical relationships are introduced. The important concept to understand is that chemistry starts at the basic level of the atom and continues on a spectrum to and inclusive of the biologies. Additionally, much of the information about viruses, bacteria, bacterial toxins, and the sciences of pharmacology and toxicology actually start here with this level of understanding. Looking back at the atom and where we began this adventure into chemistry, we can see that physics also plays a role (see Chapter 7). In physics, we find the developments of detection and monitoring capabilities and their limitations, and with the physics comes the mathematics.

Thus, there is a current trend in the fire service today to understand the concepts and theory of how our instruments work, to clarify why we would choose a strategy over a more common one, and to be able to research and identify the problems at hazardous materials incidents. Understanding our individual response capabilities and limitations leads us toward the strategies and tactics that we employ as a team, agency, and local response group.

THE DECISION-MAKING PROCESS

Successfully managing a hazardous materials incident depends heavily on the information that can be gathered at the scene, or better yet, through the preplanning efforts of the agency. It is through preplanning that the overall approach to and handling of the incident is developed and control is maintained. Part of this process has to do with the information that the science officer* coordinates and the strategic priorities that the incident commander identifies. It is a collaborative effort between individuals. As the scene becomes more involved, the list of individuals who are required to deal with it increases. Under a command structure, each individual must coordinate efforts for the safe and effective mitigation of the incident.

How often have you wondered why some officers come to a decision so easily while others seem to have a problem with the decision-making process? Or, why is it that you arrive at one emergency scene and the orders seem to flow with the incident, while at other incidents everything seems to move in a direction opposite to what should have occurred? The decision-making process is sometimes thought of as a complicated operation, but if a few principles are maintained, the process can become second nature.

Some would argue that the ability to make effective decisions depends on the historical experience of the decision maker. Others state that education alone will provide the skills that are required to proceed in the process. In actuality, a little of both is needed. However, you can have the most experienced officer with the highest level of education and still witness inappropriate decision making if the knowledge gained though experience and education are not used as a resource. Let us say, for example, that you arrive on the scene of a well-involved house fire. You give your report and order the appropriate action utilizing a $1\frac{3}{4}$-inch attack line. The firefighters engage in the action, but the action is not putting out the fire. So you order another $1\frac{3}{4}$-inch attack line. Still the fire is not going out. Why?

The answer lies in the approach. It is not uncommon to have been at fires where the same diameter attack line is pulled a second and even a third time, and still the fire dances around the firefighters. The key to the problem is the

*The science officer may be an interested chemical engineer or industrial chemist or industrial hygienist who provides experience and advices.

process and how the process is utilized. Let us look at the scenario again. You arrive on scene and analyze the incident. The officer has a plan and a set of actions. The plan is placed into action or implemented. At this point, most fire officers fail. They analyze their approach. Instead of analyzing the plan and implementation, an evaluation should be done. If the initial plan worked and the fire is going out or is placed at bay, then the first analysis was correct, the plan worked, and the implementation of that plan was appropriate. If not, the decision-making process begins again.

An effective decision-making process utilizes the APIE system:

- Analysis
- Planning
- Implementation
- Evaluation

Make an analysis, plan the method of attack, implement this plan, and evaluate the progress of the analysis, plan, and implementation. You are probably already familiar with this system without realizing it. Think of standard operating guidelines (SOG), which are devised based on past experiences and refinements of approaches used. These guidelines evolve through the process of analysis, planning, implementation, and evaluation.

Taking this common process a step further, we see that once the SOG, training, and refinement have met our expectations, we then apply them in the real world. If this application works (or seems to work), then the operation is critiqued and refined using the input of the responders with the observations of the command staff. If issues are brought forward, then a revision is placed into the plan. Again, the APIE system of decision making has been used.

However, utilizing this system does not involve a linear thought process, but rather a continuum of thought with a constant movement in time. Let us look at this idea using the analogy of a Slinky to illustrate the relationship between the concept of APIE and the movement in time (see Figure 6–1). Imagine the beginning of the Slinky at the 12:00 position to represent the analysis stage at an incident. As we move toward the 3:00 position, we are moving from the analysis position of the problem into the planning stage. Moving toward the 6:00 point, we go from the planning stage into implementation. At 9:00 we have moved into the evaluation component. As we approach the 12:00 position once again, we see that the spiral configuration of the Slinky does not land us at the same point at which we started.

All decision making follows this movement through the process, but because of the continuous movement of time, we never truly end up at the same place. In each emergency scene, the process of analysis, planning, implementation, and evaluation is a continuous one that moves with the incident. Thus we as responders engage in a dance of decision making.

Continuing with the analogy, we can see that as the scene progresses and becomes more complicated, the decision-making process seems to accelerate. The analysis, planning, and implementation become, or seem to become, one single moment in time. It is at this point that we see the breakdown of the incident: at the evaluation or reevaluation of the process. Here, as in most decision making, the evaluation (reevaluation) is the pulse of the operation. Going back to the Slinky analogy, we can see that we have moved through the process in time, analyzing, planning, implementing, and evaluating. However, if we take a more global perspective and place the Slinky end to end, there seems not to be a beginning or an end. We can say that at our imaginary beginning is the start of the incident and at the end is our termination. However, in a global sense, our beginning is the

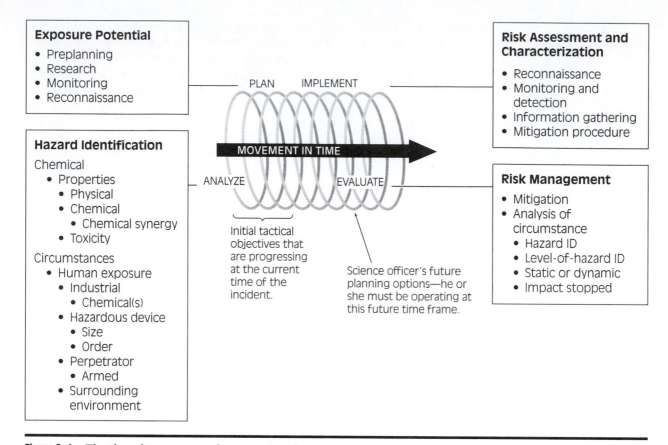

Figure 6–1   *The thought process at the scene is a dynamic one. Keeping ahead of the situation in terms of analyzing, planning, implementation, and evaluation is most important.*

training we provide to our personnel, which is inclusive of company inspection, preplanning, and fire safety checks. In essence, we are creating this Slinky in order to utilize the knowledge we gain by performing the company inspection, preplan, or safety check. When a situation is encountered, we have done our homework in order to provide the expected appropriate decision-making process.

Therefore, the APIE decision-making process is like a Slinky in a circle. As we process the information in order to provide the decision, we have moved though time, learning and acquiring knowledge as we go. As the process moves from the preincident stage into the mitigation and termination phases, it has gone full circle. What is important to note is the frequent lack of an evaluation phase at a hazardous materials incident, which disrupts the decision-making process. The thought process should be well developed moving into the future, continuously analyzing future options and evaluating each as time moves forward, while the scene is progressing.

Utilizing the APIE system for decision making with the direct application of your knowledge of chemical properties will guide you through the incident. In addition, a general catalogue of reference considerations may improve the mission objectives. The rest of the information in this chapter is intended to help you analyze situations quickly and effectively.

Effective communication at the hazardous materials event is imperative. Clear, concise text is always preferred over signals, codes, or radio jargon. Specifically, a clear and precise understanding of the chemical name is required. Because the difference of one letter can mean a completely different chemical (alkane versus alkyne, for example), it is important to spell the chemical using

Table 6–1   *The universal phonetic alphabet.*

A	Alpha	G	Golf	M	Mike	T	Tango
B	Bravo	H	Hotel	N	November	U	Uniform
C	Charlie	I	India	O	Oscar	V	Victor
D	Delta	J	Juliette	P	Papa	W	Whiskey
E	Echo	K	Kilo	Q	Quebec	X	X-ray
F	Foxtrot	L	Lima	R	Romeo	Y	Yankee
				S	Sierra	Z	Zulu

universal phonics (see Table 6–1). Spelling is the safest method to transmit the chemical name because, for example, "alkane" and "alkene" can sound the same when talking over the radio or on a phone.

MEASUREMENTS AND CONVERSIONS

Effective communication includes communicating the correct information. Table 6–2 is provided to help you make conversions of various physical measurements.

Radiation

Radiation is a complicated concept because it is entirely based in the theories of physics and higher math. Table 6–3 lists measurements and levels of acute radiation syndrome for gamma radiation exposure. Table 6–4 lists radiation measures and their conversion factors.

PPM/PPB

What does one part per million mean? What does that piece of information actually tell us? Parts per million (ppm) and parts per billion (ppb) are measurements of the relationship between the amount of a substance and the air, liquid, or solid that the substance is in. For example, 1 ppm would be equivalent to approximately an ounce of cream in 10,000 gallons of coffee, and 1 ppb would be equivalent to approximately a drop of cream in 22,000 gallons of coffee. As you can see, the quantities are very small.

Percentage is based on 100 total parts, or one in a hundred. It is a mathematical ratio. For example, 50 parts in 100 is 50/100 = 0.50, or 50%. Both ppm and ppb have a direct relationship with percentage:

ppm		%		ppb
1,000,000	=	100	=	1,000,000,000
500,000	=	50	=	500,000,000
250,000	=	25	=	250,000,000
150,000	=	15	=	150,000,000
75,000	=	7.5	=	75,000,000
10,000	=	1	=	10,000,000
5,000	=	0.5	=	5,000,000
500	=	0.05	=	500,000

Table 6–2 *Conversions of miscellaneous measurements.*

SURFACE MEASUREMENTS

	Statute Miles	Meters	Yards	Feet	Inches	Centimeters
Statute Mile	1	1609.3	1760	5280	63,360	
Kilometer	0.6214	1000	1094	3280.8	39,372	
Nautical Mile	1.1516	1853	2027	6080.2	72,960	
Meter		1	1.094	3.281	39.37	182.9
Yard		0.9144	1	3	36	91.44
Foot		0.3048	0.3333	1	12	30.48
Inch		0.0254	0.0277	0.08333	1	2.54
Centimeter				0.0328	0.3937	1

	Square Miles	Acres	Square Meters	Square Yards	Square Feet
Square Mile	1	640	2,589,998	3,097,600	
Square Kilometer	0.3861	1	1,000,000	1,195,985	
Acre	0.00156	1	4,046.9	4,840	43,560
Square Yard		0.00021	0.8361	1	9
Square Foot			0.0929	0.1111	1

VOLUME

	Cubic Inches	Cubic Feet	Cubic Yards
Cubic Inch	1	0.00058	
Cubic Foot	1,728	1	0.037
Cubic Yard		27	1
Cubic Meter		35.314	1.3079

LIQUID VOLUME

	Cubic Inches	Liters	U.S. Pints	U.S. Quarts	U.S. Gallons	U.K. Pints	U.K. Quarts	U.K. Gallons
Liter	61.025	1	2.1134	1.0567	0.2642	1.76	0.88	0.22
U.S. Pint	28.875	0.473	1	0.5	0.125	0.8327	0.4164	0.1042
U.S. Quart	57.75	0.9643	2	1	0.25	1.665	0.8327	0.208
U.S. Gallon	231	3.785	8	4	1	6.66	3.33	0.8327
U.K. Pint	34.668	0.5688	1.201	0.6	0.15	1	0.5	0.125
U.K. Quart	69.335	1.1365	2.402	1.201	0.3	2	1	0.25
U.K. Gallon	277.34	4.546	9.616	4.808	1.201	8	4	1

(continues)

Table 6–2   (Continued)

	DRY MEASURE							
	Cubic Inches	**Liters**	**U.S. Pints**	**U.S. Quarts**	**U.S. Bushels**	**U.K. Pints**	**U.K. Quarts**	**U.K. Bushels**
Liter	61.025	1	1.8162	0.908	0.0284	1.759	0.8795	0.0275
U.S. Pint	33.6	0.55	1	0.5	0.156	0.969	0.4845	0.015
U.S. Quart	67.2	1.101	2	1	0.0313	1.938	0.969	0.03
U.S. Bushel	2150.42	35.238	64	32	1	62.016	31.01	0.969
U.K. Pint	34.68	0.5679	1.032	0.516	0.0164	1	0.5	0.0156
U.K. Quart	69.35	1.1359	2.064	1.032	0.0323	2	1	0.0313
U.K. Bushel	2219.34	36.367	66.052	33.026	1.032	64	32	

	WEIGHT						
	Grams	**Kilograms**	**Ounces**	**Pounds**	**Metric Tons**	**Short Tons**	**Long Tons**
Gram	1	0.001	0.0353	0.0022			
Kilogram	1000	1	35.274	2.2046			
Ounce	28.349	0.0284	1	0.0625			
Pound	453.59	0.4536	16	1			
Metric Ton		1000		2204.6	1	1.1023	0.9842
Short Ton		907.2		2000	0.9072	1	0.8929
Long Ton		1016		2240	1.016	1.12	1

Table 6–3   *Measurements of gamma radiation exposure.*

Dose in RAD	Symptoms	Observations
0–25	Usually none, but nausea and fatigue if any	Usually no detectable effects, supportive medical care
25–100	A small percentage of the population may exhibit nausea and vomiting, anorexia, and fatigue	Damaged bone marrow; decrease in red and white blood cells, platelet counts, and lymphocytes; lymph system may be affected
100–300	Mild to severe nausea and vomiting, anorexia, and fatigue; infection is a concern	Severe damage to the hematological system
300–600	Massive bleeding, probable infection, diarrhea, in addition to previously listed symptoms	50% fatality rate
600+	Central nervous system collapse	Death most probable

Table 6–4   *Radiation measures and conversions.*

Conversion Factor	Conventional Name	International Name	Conversion Factor
$1\ Bq = 2.7 \times 10^{-11}\ Ci$	Curie (Ci)	Becquerel (Bq)	$1\ Ci = 3.7 \times 10^{10}\ Bq$
$1\ C/kg = 3876\ R$	Roentgen (R)	Coulombs/Kilogram (C/kg)	$1\ R = 2.58 \times 10^{-4}\ C/kg$
$1\ Gy = 100\ RAD$	Radiation absorbed dose (RAD)	Gray (Gy)	$1\ RAD = 0.01\ Gy$
$1\ Sv = 100\ REM$	Roentgen equivalent man (REM)	Sievert (Sv)	$1\ REM = 0.01\ Sv$

Toxic values are often identified as milligrams per cubic meter (mg/m^3). Although not as accurate as the conversion numbers presented in Table 6–2, the following equation gives a close estimate of the mg/m^3 conversion into ppm. When conversion numbers are not identified, this calculation can assist the science officer with important decisions.

$$\text{mg/m}^3 = \frac{\text{ppm} \times \text{molecular weight}}{24.5} \rightarrow \text{ppm} = \frac{\text{mg/m}^3 \times 24.5}{\text{molecular weight}}$$

SCENE CONSIDERATIONS

There are five clues that hazardous materials responders use to help identify the product in question at the time of an accidental release. Some of these elements can be used in an intentional release; however, other conditions and circumstances must be considered. These clues are collectively called recognition and identification.

Occupancy and Location

Depending on the occupancy or location of a retail store, storage facility, tanker, or rail car, one can start to identify the possible chemicals that may be involved. Knowing that communities are in close proximity to an industrial area or a main corridor of transportation indicates the types of commodity that may travel in, through, or around said community. Knowing the commodity flow to and from these locations can assist the responder in determining chemical probabilities.

Container Shapes and Sizes

Hazardous materials are transported by a variety of methods, and along with this are the containers that these chemicals are stored in during the transportation, distribution, and collection phases. Each container has specific requirements for the type of material that can be carried. Additionally, each container is built with a certain class of chemicals in mind. By knowing these classification and structural requirements of containers, the chemical possibilities are reduced from all potential chemicals to a handful of possibilities.

Placards and Labels

Each container, depending on the amount that the container carries and the potential danger of the chemical, has a placard or label. Not all chemical commodities have these warning labels or chemical descriptions. Only if the chemical falls within certain parameters found in the Code of Federal Regulation are they placarded or labeled. These placards fall into the following nine classes; some examples are found in Figures 6–3 through 6–7.

Class 1—Explosives

Division 1.1—Articles and substances having a mass explosion hazard

Division 1.2—Articles and substances having a projection hazard, but not a mass explosion hazard

Division 1.3—Articles and substances having a fire hazard, a minor blast hazard, or a minor projection hazard, but not a mass explosion hazard

Division 1.4—Articles and substances presenting no significant hazard (explosion limited to package)

Division 1.5—Very insensitive substances having a mass explosion hazard

Division 1.6—Extremely insensitive articles that do not have a mass explosion hazard

Class 2—Gases

Division 2.1—Flammable gases

Division 2.2—Nonflammable, nontoxic gases (under pressure, inert, etc.)

Division 2.3—Toxic gases

Division 2.4—Corrosive gases (Canada)

Class 3—Flammable and combustible liquids

Class 4—Flammable solids; spontaneously combustible materials; dangerous when wet

Division 4.1—Flammable solids

Division 4.2—Substances liable to spontaneous combustion

Division 4.3—Substances that, in contact with water, emit flammable gases

Class 5—Oxidizers and organic compounds

Division 5.1—Oxidizers

Division 5.2—Organic peroxides

Class 6—Toxic (poisonous) infectious substances

Division 6.1—Toxic substances

Division 6.2—Infectious substances

Class 7—Radioactive materials

Class 8—Corrosive materials

Class 9—Miscellaneous dangerous goods

Division 9.1—Miscellaneous dangerous goods (Canada)

Division 9.2—Environmentally hazardous substances (Canada)

Division 9.3—Dangerous wastes (Canada)

Shipping Papers and Facility Documents

When the commodity is transported under the above classifications, shipping papers accompany these chemicals. The shipping papers can give vital information to the responder, such as the chemical name, the quantity transported, the manufacturer and the receiver of the commodity, and emergency contact numbers. Facilities that have large amounts of chemicals not considered to be consumer quantities are required to have facility documentation called material safety data sheets. Under Title III of the Superfund Amendments and Reauthorization Act (SARA), facilities are required to submit a tier II report to the local responders: a listing of the chemicals that they have onsite along with a site plan. Responders can use these documents to plan the response prior to the accidental release, plan for potential targets of opportunity (intentional releases), and prepare community emergency management plans.

Monitoring and Detection Equipment

In order to identify, detect the presence of, or understand the parameters of chemicals, responders must have an assortment of detection and monitoring equipment.

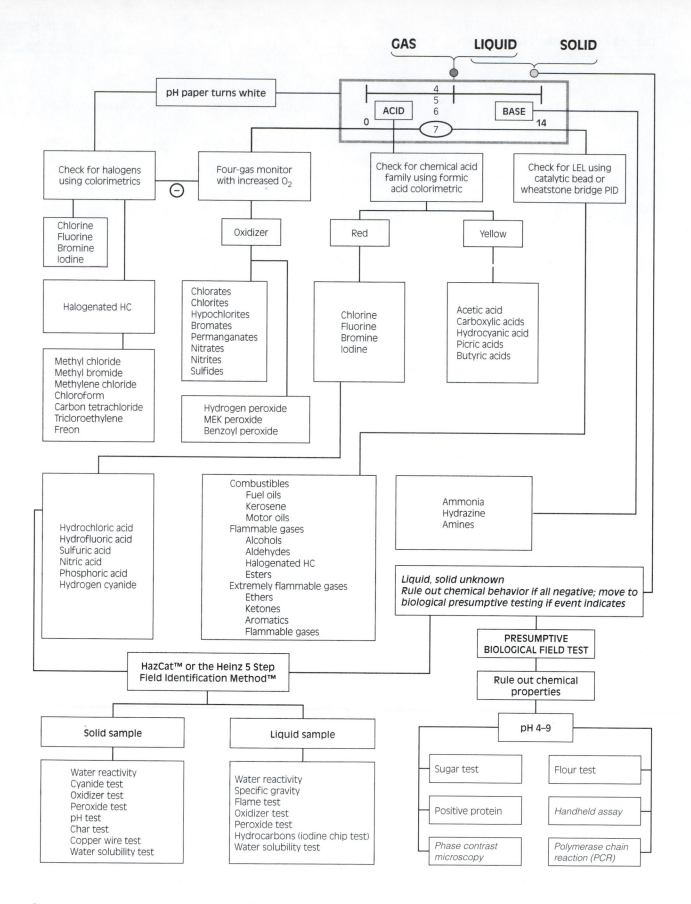

Chart adapted from hazardous material air monitoring and detection devices by Chris Hawley.

Figure 6–2 *Monitor/detection decision tree matrix.*

The sophistication of these devices has increased in recent years. Not only are hazmat teams concerned with chemicals, but also with the infectious biological agents. Because of biological agents and the threat of terrorism, first responders must have and be proficient in the use of detection and monitoring devices. In Chapter 7, infrared spectroscopy, the newest technology to hit the hazmat market, is discussed at length. Many different technologies exist, such as gas chromatography—mass spectroscopy (GC–MS). Although a proven technology, GC–MS requires a tremendous learning curve. For those teams that have dedicated team members, this technology is fairly easy to maintain proficiency in. However most hazmat units are a combination team, using individuals from other units in the response agency or external to the response agency. In this case, the GC–MS then becomes a technological hurdle that personnel must overcome at the scene of an incident.

Detection and monitoring devices should be fairly easy to operate and have only moderate learning curves, with a directed approach of strategies for detection. The understanding of how chemicals behave is the basis for detection and monitoring strategy. Figure 6–2 provides an example of such a detection strategy.

ENVIRONMENTAL CONSIDERATIONS

Temperature

Temperature has an impact on a substance's chemical and physical properties. Therefore, accurate knowledge of ambient temperatures related to an emergency scene is critical to the correct response. Much of the reference literature describes properties of chemicals in the Celsius or Kelvin scale depending on the chemical type and its inherent temperature under certain conditions. For example, in some industrial processes that involve cryogenics, the Kelvin scale may be utilized. However, the Celsius scale is common in the scientific community. The Fahrenheit scale is used only when the temperature information is being relayed to the general public, and even then, temperature is sometimes left in the Celsius scale.

The conversion factors are:

$$F = 1.8C + 32$$
$$C = F - 32/1.8$$
$$K = C + 273$$
$$C = K - 273$$

Weather-Related Factors

Dew Point. The dew point is the temperature at which the moisture in the air will become visible, for example, as fog. (As the temperature drops below freezing, the term *frost point* is used instead of dew point.) The larger the difference between the dew point temperature and the ambient temperature, the less saturated the air is with water; if the difference is small, the saturation is high.

Relative Humidity. The ratio of the amount of water vapor actually present in the air to the greatest amount possible at the same temperature is the measure of relative humidity. Table 6–5 shows the relationship between temperature and humidity as the heat index.

Additionally, some chemicals *may* create a saturation in air. This occurrence depends on the humidity, of course, but it has to do with hydroscopic

Table 6–5   *The heat index.*

Temperature (°F)	Relative Humidity								
	10%	20%	30%	40%	50%	60%	70%	80%	90%
104	98	104	110	120	132				
102	97	101	108	117	125				
100	95	99	105	110	120	132			
98	93	97	101	106	110	125			
96	91	95	98	104	108	120	128		
94	89	93	95	100	105	111	122		
92	87	90	92	96	100	106	115	122	
90	85	88	90	92	96	100	106	114	122
88	80	84	85	87	90	92	96	106	109
86	80	84	85	87	90	92	96	100	109
84	78	81	83	85	86	89	91	95	99
82	77	79	80	81	84	86	89	91	95
80	75	77	78	79	81	83	85	86	89
78	72	75	77	78	79	80	81	83	85
76	70	72	75	76	77	77	77	78	79
74	68	70	73	74	75	75	75	76	77

Add 10° when protective clothing is worn and add 10° when in direct sunlight.

Humature	Danger Category	Injury Threat
Below 60	None	Little or no danger under normal circumstances.
80–90	Caution	Fatigue possible if exposure is prolonged and there is physical activity.
90–105	Extreme caution	Heat cramps and heat exhaustion possible if exposure is prolonged and there is physical activity.
105–130	Danger	Heat cramps or exhaustion likely, heat stroke possible if exposure is prolonged and there is physical activity.
Above 130	Extreme danger	Heat stroke imminent!

droplet formation. Aerosols are a good example of hydroscopic droplet formation. When the water saturation is high, the hydroscopic effect increases and results in unusual behavior of the aerosol; that is, it behaves similarly to a gas. What happens is that when the air has a high concentration of water and is close to the dew point, the chemical and the water can attach to dust particles, causing a high respiratory hazard. Saturation of the chemical within the volume of air can also occur. This condition depends on the chemical itself. There are, unfortunately, no hard-and-fast rules concerning chemical saturation. It is, at the least, something to consider in a confined area (building or closed space) that has little or no air exchange. The decision to ventilate an area should be made carefully and after appropriate research.

● Caution

Due to the minerals it contains, a droplet of water has a vapor pressure less than that of pure water when the air is in a saturated condition. At the point of air saturation, the droplet will grow in size. This growth is called hydroscopic formation. When this same drop is exposed to an environment that is less saturated or dry, the droplet will evaporate, adding to the water saturation of the air. Thus, whether a water droplet will grow depends on the water vapor in the environment.

This growth is a result of the water condensing on the surface of an aerosolized chemical. The overall concentration of the chemical in the drop will decrease, and the vapor pressure of the solution that constitutes the droplet will increase. When the vapor pressure equals that of the surrounding environment, the droplet becomes stable. This equilibrium is important and has to do with the relative humidity; if equilibrium is not reached, vapor is liberated out of the droplet (solution) and further increases the saturation of the chemical in the environment.

The temperature of a liquid that is released under pressure will affect the size of the droplet. If a liquid forced into droplets is cooler than the ambient temperature, the droplet will stay in the air for great distances, evaporating as it goes. If the droplet is warmer than the surrounding air, it will fall fairly rapidly and evaporate on the ground. This principle has its basis in the saturation of vapor in air, and condensation and evaporation potentials.

The situation involving droplet formation and the fate of aerosols is additionally complicated when considering inhalation possibilities. The respiratory tract has a high degree of moisture, inherent to the system. If the chemical has not gone through hydroscopic development, or is in the beginning stages, the size and polarity of the chemical will determine its fate in the respiratory tree.

Wind Speed. Table 6–6 shows how the combination of wind speed and still-air temperature produces a chill factor. It is important to keep this factor in mind because cold conditions affect the chemistry at the incident site.

Barometric Pressure. As the atmospheric pressure decreases, the evaporation of a chemical will increase. Likewise, as pressures increase, the potential for evaporation will decrease. Realistically, the fluctuation of barometric pressures is so slight that on-scene practitioners need not be aware of a change. However, if a weather front is about to move through the area, or has already passed through, the consideration of atmospheric pressures should be part of the research sector's responsibility. The following questions need to be answered: How will this slight

---

Table 6–6   *The chill factor chart.*

EQUIVALENT TEMPERATURE ON EXPOSED FLESH (°F)

WIND SPEED (mph)													
40	1	−4	−15	−22	−29	−36	−45	−54	−62	−69	−76	−87	−94
35	3	−4	−13	−20	−27	−35	−43	−52	−60	−67	−72	−83	−90
30	5	−2	−11	−18	−26	−33	−41	−49	−56	−63	−70	−78	−87
25	7	0	−7	−15	−22	−29	−37	−45	−52	−58	−67	−75	−83
20	12	3	−4	−9	−17	−24	−32	−40	−46	−52	−60	−68	−76
15	16	11	1	−6	−11	−18	−25	−33	−40	−45	−51	−60	−65
10	21	16	9	2	−2	−9	−15	−22	−27	−31	−38	−45	−52
	35	30	25	20	15	10	5	0	−5	−10	−15	−20	−25

AIR TEMPERATURE (°F)

increase or decrease affect the chemical properties? What type of hazards are going to be perpetuated from this change? Will my instruments give accurate readings under present, and future, conditions?

Atmospheric Stability

The CAMEO (computer-aided management of emergency operations) program identifies six classes of stability that represent a degree of turbulence in the air. With solar radiation, the incoming sunlight heats the air near the ground and causes it to rise and generate movement in the atmosphere. The air is considered to be unstable or turbulent, and such atmospheres are identified as Classes A and B.

When the solar radiation is weak and the air near the surface has less tendency to rise, less turbulence develops. The atmosphere is stable, or inversion is possible, and is considered as Class E or F stability. The middle categories, Classes C and D, represent the neutral atmosphere or moderate turbulence, and are associated with either strong wind speeds and moderate solar radiation (thus heating the air space), or a neutral air gradient.

■ Note

Because first responders are familiar with the conditions that CAMEO presents, these can be related to temperature, humidity, and weather. Cloud cover, wind speed, and incoming solar radiation all have a bearing on the weather determinates of the scene.

! Safety

● Inversion can occur in a tall structure. For example, under certain atmospheric conditions, clouds have formed in the upper areas of the vehicle assembly building at Cape Canaveral. Buildings with tall atriums can have temperature gradients. A skyscraper involved in a fire can also create weather conditions and inversions within the structure. As with all emergency actions, ventilation and movement of air in a structure should be backed up with research and sound logic.

Ventilating a structure in which there has been a hazardous chemical release depends on the life safety in and outside of the structure. For example, if a chemical has spilled in a structure and is evolving fumes, HVAC needs to be shut down. Additionally, if the structure has an isolated room within, protection in place could be a strategic option. However, if positive pressure ventilation is used, the chance of "pushing" the contaminated air into the protected area is high. In this case, a negative vent would be indicated. If the vent can be controlled without the life safety exposure, then a positive vent would be appropriate. High concentration areas will flow into areas of low concentration; flammables such as natural gas and propane should be pushed from a structure, whereas toxins should be pulled from a structure.

● Caution

According to the Gaussian mathematical air model, wind and turbulence move a released gas through the environment. It thus provides for the prediction of downwind turbulent mixing of the gas. However, the hydroscopic displacement of heavy gases, vapors, mists, and fogs can affect calculations based on the model. The CAMEO model, therefore, has severe limitations in situations involving high-humidity sources, street canyon displacement, and shifts around buildings and up hills.

Determination of the category of atmospheric stability is important for predicting how gases that are buoyant in the atmosphere disperse.

Air stability will become a critical consideration when a large chemical release occurs. Think about a box of air in a column. As the box of air rises through the imaginary column, the gases in the box (atmospheric air) will expand. Because of the expansion, the temperature of the gas will decrease. As we move up the column, the gas in the box is cooler than the surrounding air. As it cools, it becomes denser than the surrounding air and tends to sink back down. Whatever caused the box of air to move up the column will immediately be opposed to the condition of falling, and the atmosphere is called stable.

Now consider the same box of air heated up, or reinforced to move upwards, and the air condition is referred to as unstable. There are four terms used for the degree of atmospheric stability: neutral, lapse, inversion, and elevated inversion.

Neutral conditions are when the difference in air temperature between low altitudes and higher altitudes is small. Overcast days or nights and rainy conditions are examples. *Lapse* is when cooler conditions exist at higher altitudes. These conditions exist on clear days with little or no cloud cover that have winds less then 5–7 miles per hour. A calm moonlit night is an example of a lapse condition. *Inversion* occurs when the upper atmosphere is relatively warmer than the temperatures at lower altitudes. Air convection is limited. A low cloud cover or a blanket of air pollution with little air movement is an example of inversion. *Elevated inversion* occurs when a body of cooler air has settled under a warmer body of air, generally in relation to warm and cold weather fronts. The inversion gradients are considered to be stable, whereas the lapse is considered unstable.

The type of weather conditions in terms of cloud cover should also be considered. For example, inversion layers tend to limit the dispersion of a chemical if released. Add to this factor the poor wind conditions that are normal in an inversion weather condition and high humidity, and you have a chemical cloud that remains fairly close to the surface. You need to think about whether the weather conditions increase exposure to a population or decrease it.

Topography

The lay of the land, its topography, can bring a product toward us or push it away from us. Depending on the size of a mountain or hill, wind can be redirected and velocity increased by these topographic changes. Obviously, an increase in the wind and a change in its direction can increase or decrease problem potentials. We must remain uphill, upwind, and upstream at a hazardous materials incident. Because topography can change the weather in the immediate area, continual weather updates are an important consideration. Fluctuations in temperature, humidity, dew point, and barometric pressure all may affect the scene.

■ Note

Urban areas can change wind direction and wind speeds. In larger cities with heavy urban development and tall buildings, additional weather conditions must be considered. Heavy concrete construction can increase the temperature throughout the day because of heat absorption.

Time of Day

In certain areas in a city, town, or neighborhood, population densities change throughout the day. With the increase or decrease of population density, the number of potential victims can increase or decrease. The number of vehicles in the immediate area, evacuation problems, available shelters in place, utilities demand, and the like are but a few of the problem areas that time of day can present. Depending on the incident, evacuation may be the first consideration. Be aware that not many communities have a transportation system capable of handling all of the people leaving their homes and businesses at the same time.

FLAMMABLE LIMITS

Weather conditions and the vapor that is produced from a spill will influence the potential to have a toxic and flammable environment. Table 6–7 lists chemical groups that can produce flammable limits. Depending on molecular size and polarity, branching influences the chemical environment. Petroleum products will also produce a flammable limit.

Figures 6–3 through 6–7 show additional groupings of hazardous substances.

The molecular weight of a compound can be derived from the chemical formula (or chemical name, if you have studied the chemical nomenclature). By adding the molecular weight of each element in the formula, the weight of the compound can be derived; for example, o-dichlorobenzene has six carbons ($\times 12$), four hydrogens ($\times 1$), and two chlorines ($\times 35$). So:

$$(6 \times 12) + (4 \times 1) + (2 \times 35) = 146$$

which is close to the referenced figure of 147.

If the exposure limit is 300 mg/m^3, the ppm can be calculated by utilizing the following formula and rearranging algebraically. We can see that 300 mg/m^3 equals 50 ppm:

$$\text{mg/m}^3 = \text{ppm} \times \text{molecular weight}/24.5$$

So, for our example of 300 mg/m^3:

$$300 \text{ mg/m}^3 = \text{ppm?} \times \text{mw (147)}/24.5$$
$$300 \text{ mg/m}^3 \times 24.5 = \text{ppm?} \times \text{mw (147)}$$
$$\frac{300 \text{ mg/m}^3 \times 24.5}{147} = \text{ppm?}$$
$$50 = \text{ppm}$$

Table 6–7   *General flammable limit conditions.*

Functional Group	Flammable Limit Average	
	LEL (%)	UEL (%)
Esters	1	10
Alcohols	1	30
Aromatics	2	10
Ketones	2	14
Amines	2	14
Ethers	2	40
Aldehydes	3	50

Petroleum products general LELs:

   Flammable gases, 3–6%

   Flammable liquids, 1–3%

Flammable and combustible liquids are further subdivided based on their relative hazard. The low-flash-point group consists of substances having a flash point below 0°F (−18°C). These materials present a high degree of danger because the material almost always gives off flammable vapors. Some of these substances include:

Acetaldehyde	Diethylamine	Heptene
Acetone	Diethyl ether	Hexane
Acrolein	Diisopropyl ether	Iso-pentanes
Allyl chloride	Dimethyl ether	Methyl pentane
Amyl nitrate	Ethyl amine	Octane
Carbon disulfide	Ethyl ether	Petroleum spirit
Cyclohexane	Ethyl mercaptans	Propylene oxide
Cyclopentane	Furans	Tetrahydrofurans
Diethoxymethane		

The intermediate-flash-point group consists of materials having a flash point from 0°F (−18°C) to 73°F (23°C). These materials are also considered hazardous, but are not as high a hazard as those in the first group. Some substances found in this intermediate flash point group include:

Acrylonitrile	Diethyl ketone
Alcohols (methyl, ethyl, normal + isopropyl)	Dioxane
Allyl ethyl ether	Ethanol
Amyl acetates	Ethyl methyl ketone
Bromopropanes	Heptane
Butyl acetates	Methyl ethyl ketone (MEK)
Butyl methyl ether	Naptha, petroleum
Chloromethyl ethyl ether	Propanol
Cycloheptane	Toluene
Cyclohexene	Xylenes

The high-flash-point group consists of substances having a flash point from 73°F (23°C) to 141°F (61°C), which are regarded as presenting minor danger. Some high-flash-point liquids include:

Anisole	Diethylbenzene	Hydrazine (concentrated)
Bromobenzene	Ethyl butyrate	Nitroethane
Butanol	Formalin	Turpentine
Chlorobenzene	Furfural	

Substances having a flash point above 141°F (61°C), such as the various oils, are not considered to present a fire hazard according to this classification scheme.

Figure 6–3

Categories of flammable and combustible liquids based on their relative hazard according to flash points. (From R. Schnepp and P. W. Gantt, Hazardous Materials: Regulations, Response, and Site Operations *[Albany, NY: Delmar, 1999, p. 89].)*

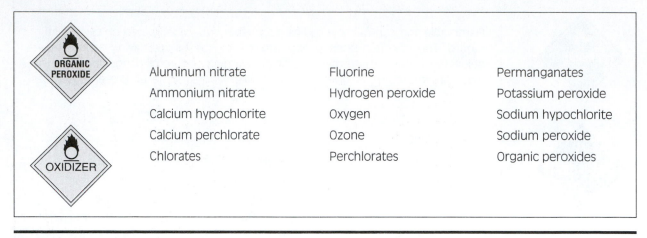

Figure 6–4 *Common industrial oxidizers. Some oxidizers are stronger than others, and concentration has a lot to do with how reactive an oxidizer may be. (From R. Schnepp and P. W. Gantt,* Hazardous Materials: Regulations, Response, and Site Operations *[Albany, NY: Delmar, 1999, p. 101].)*

Aluminum nitrate	Fluorine	Permanganates
Ammonium nitrate	Hydrogen peroxide	Potassium peroxide
Calcium hypochlorite	Oxygen	Sodium hypochlorite
Calcium perchlorate	Ozone	Sodium peroxide
Chlorates	Perchlorates	Organic peroxides

Acetone cyanohydrin	Epibromohydrin
Acrylamide	Ethyl bromide
Aldol	Hexachlorobenzene
Ammonium fluoride	Hydrocyanic acid
Aniline	Lead arsenates
Barium cyanide	Nickel carbonyl
Benzonitrile	Nitroanilines
Bromoform	Nitrotoluenes
Calcium cyanide	Phenol
Chloroform	Phenyl mercaptan
Cyanogen bromide	Potassium fluoroacetate
Dichlorobenzenes	Sodium cyanide
Dichlorodimethyl ether	Tetraethyl lead
Dichloromethane	Toluidines
Dinitrotoluenes	

Figure 6–5 *Representative chemicals in the poison classification. (From R. Schnepp and P. W. Gantt,* Hazardous Materials: Regulations, Response, and Site Operations *[Albany, NY: Delmar, 1999, p. 111].)*

Another way of looking at the problem is to take the LEL percentage. Percentage has a relationship to ppm. Using *o*-dichlorobenzene as an example, the LEL is 2%. Utilizing our knowledge of the relationship between percentage and ppm, 2% equals 20,000 ppm. As you can see in this example, the toxic levels for *o*-dichlorobenzene are well below the LEL. In this case (as with most), we must transition through the toxic range well before the LEL is reached (or even 10% of the LEL, which would be 0.2%, or 2000 ppm!).

Vapor density can be calculated using the following formula when the molecular weight of the compound is small:

$$\text{Vapor density} = MW/29$$

	Ammonia, anhydrous	
POISON 6	Arsine	Hydrogen chloride, anhydrous
	Boron trifluoride	Hydrogen fluoride, anhydrous
	Bromine chloride	Hydrogen sulfide
	Carbon monoxide	Methyl bromide
	Carbonyl sulfide	Methyl chloride
	Chlorine	Nitric oxide
INHALATION HAZARD 2	Diborane	Phosgene
	Dichlorosilane	Phosphine
	Ethylene oxide	Selenium hexafluoride
	Fluorine	Sulfur dioxide
	Germane	

Figure 6–6 *Some of the toxic gases that pose an inhalation hazard. (From R. Schnepp and P. W. Gantt,* Hazardous Materials: Regulations, Response, and Site Operations *[Albany, NY: Delmar, 1999, p. 112].)*

As the examples listed in Table 6–8 show, the heavier compounds do not indicate an accurate calculation. As an additional example of the divergence, *o*-dichlorobenzene = 146, as previously calculated. It has an estimated VD calculated at 146/29 = 5.03, whereas the actual value is 1.30.

> ■ **Note**
> Molecular weights above 29 indicate that the vapor will drop; conversely, molecular weights of less than 29 indicate that the vapor will rise.

In some of the textbooks the vapor density value is referred to as specific density for the gas (RgasD, or relative gas density). For most purposes, specific density and vapor density can be considered the same. Sometimes a conversion factor is given for calculating the ppm. This figure can be used by dividing it into the referenced number and thus calculating the ppm. For example, the TWA for divinylbenzene is referenced as 50 mg/m³. The conversion is given as 1 ppm = 5.33 mg/m³. Therefore, the TWA can be calculated as follows:

$$50/5.33 = 9.38 \text{ ppm}$$

Table 6–8 *Vapor density calculations.*

	MW	Calculated VD	Actual VD+
Methane	16	0.5517	0.554
Ammonia	17	0.5862	0.597
Hydrogen	2	0.0689	0.069
Butane	58	2.0000	2.0461
Sulfur dioxide	64	2.2068	1.4503
Hydrogen sulfide	34	1.1724	1.190

Some particularly hazardous corrosives are found in the following list:

Allyl chlorocarbonate	Selenium oxychloride
Bromine	Sulfur chlorides
Chromium oxychloride	Sulfur trioxide
Chromosulfuric acid	Sulfuric acid
Fluorosulfonic acid	Sulfuryl chloride
Hydrofluoric acid	Thionyl chloride
Nitric acid	Trifluoroacetic acid
Nitrohydrochloric acid	Vanadium tetrachloride
Selenic acid	

Substances are regarded as presenting a medium danger if they can cause skin destruction from 3 minutes to 60 minutes under test conditions but have no particular additional hazard. Examples of some of these chemicals include the following:

Acetic acid (50% to 80%)	Ethyl chlorthioformate
Acetic acid, glacial	Ethylphenyldichlorosilane
Acetic anhydride	Ethylsulfuric acid
Acetyl bromide	Fluoroboric acid
Acetyl iodide	Fluorosilicic acid
Acrylic acid	Formic acid
Alkyl, aryl, or toluene sulfonic acid	Hexafluorophosphoric acid
N,N-Diethyethylene diamine	Hexyl trichlorosilane
Diethyldichlorosilane	Thiophosphoryl chloride
Diphenyldichlorosilane	Trimethylacetyl chloride
Dodecyl trichlorosilane	Vanadium oxytrichloride

Substances presenting a minor degree of danger (but still meeting the definition of skin destruction in a 4-hour test) include the following:

N-Aminoethylpiperazine	Methacrylic acid
Amyl acid phosphate	Phosphorous trioxide
Butyric acid	Phosphorous trichloride
Cyanuric chloride	Phosphorous oxychloride
3-(Diethylamino)-propylamine	Tetraethylenepentamine
Ethanolamine	Triethylenediamine
2-Ethylhexylamine	Vanadium trichloride
Hydroxylamine sulfate	Zirconium tetrachloride

Figure 6–7 *A sampling of corrosives. (From R. Schnepp and P. W. Gantt,* Hazardous Materials: Regulations, Response, and Site Operations *[Albany, NY: Delmar, 1999, p. 131].)*

CRITICAL TEMPERATURE AND PRESSURE

Kinetic molecular theory states that all molecules are in constant motion. This movement ranges from the slight vibration in a solid to the rapid constant motion in a vapor or gas. The theory states that if a material—either solid, liquid, or gas—is brought down to a particular temperature, all motion in the atom, and thus the molecule, will stop. This temperature is referred to as the absolute temperature. Absolute temperature, or Kelvin (K), is −459.67°F. This number is referred to as the temperature of absolute zero (0 K). Recall the definition of standard temperature and pressure (STP): a temperature of 32°F (0°C or 273 K) at a pressure of 1 atm (760 torr or 760 mm Hg) at which a reaction is tested or occurs. The critical temperature and critical pressure are the temperature and pressure needed to change states (remember the phase diagram). The following are all equivalent:

$$1013.2 \text{ millibars} = 29.92 \text{ in Hg} = 14.7 \text{ psi} = 760 \text{ mm Hg}$$
$$= 1 \text{ atm} = 760 \text{ torr} = 1.013 \times 10^5 \text{ pascal}$$

We have studied the gas laws and have discussed how the inversion of temperature and pressure relates to the volume of a gas. These laws can also be used to approximate the volatility of a given compound. This approximation is important when we reference products that have extremely high toxic potential with relatively low vapor pressure. Previously we have discussed that the vapor pressure has a correlation to the amount of vapors that are given off from a liquid product. So why does a product with low vapor pressure pose a life safety hazard?

First, we must give some general assumptions for our discussion. When we think about volume of air, we usually discuss it in gallons or cubic meters. Here we have to identify the relationship between the volume of air and its components such that 1 mole of any gas at STP will occupy 22.4 liters. At a temperature of 104°F, we have a barometric pressure of 760 torr or mm Hg. Correcting our STP for the given temperature of 104°F (40°C), we have to use Kelvin for the adjustment:

$$(273 + 40)/273 = 1.1465 \times 22.4 \text{ liters} = 25.68 \text{ liters of air at } 104°F \text{ for 1 mole}$$

■ Note

The volume of air that is required for 1 mg/m³ is 1 mole or 22.4 liters.

Let us say that chemical XYZ has a gram weight derived from the periodic table, and for our example occupies 25.68 liters with a pressure of 760 torr. However, XYZ has a vapor pressure of 0.5 mm Hg. The volume of our vapor is then calculated as:

$$\frac{25.68 \times 760}{0.5} = 39{,}034 \text{ liters}$$

This calculation indicates that if the chemical XYZ is fully dispersed into the environment (e.g., as an aerosol), then its saturation is such that 39,034 liters of saturated air contains the molecular weight of XYZ. Alternatively, 1 liter of air saturated with XYZ contains $x/39{,}034$ = grams of XYZ (if x equals 140 grams [molecular weight of 140], then we would have 0.00358 grams of chemical in the air).

VOLATILITY

Volatility is the degree of evaporation under certain atmospheric conditions. Temperature and humidity (saturation of water vapor in the air) have a direct bearing on the volatility a chemical may posses. We can calculate this chemical quality by using the ideal gas law. Remember, this is an average and not a specific measurement:

$$\text{Volatility (mg/m}^3) = \frac{16{,}020 \times \text{molecular weight} \times \text{vapor pressure}}{\text{Temperature (K)}}$$

Using the chemical XYZ, which has a molecular weight of 140 grams and a vapor pressure of 0.5, we get 3583 mg/m³. Plugging this figure into our ppm conversion formula, we can see that we have a chemical that has the potential to produce 626 ppm in the working environment.

TOXICOLOGICAL CONSIDERATIONS

Reactions in the Body

Substances can be inherently toxic or converted into toxic compounds in the body during the metabolic process (protoxic). As previously discussed, shape, polarity, and orientation in space all have an effect on a chemical and its physical properties. These same factors also affect the metabolism of that chemical, which may manipulate the chemical into a toxic or nontoxic compound. Chemicals also have the ability to enhance, cancel, support, and give an additive response to toxicity. The primary reason for these reactions are due to what are called *Phase I* and *Phase II* reactions.

In a Phase I reaction, we have to remember that the chemistry of the functional groups depends on the surrounding available compounds. An assortment of products can be created from a single compound. Now multiply the possibilities if we have several chemicals invading the body (which possesses its own variety of chemicals).

● **Caution**

Inhalation hazards are identified in the *Emergency Response Guide Book* as A, B, C, and D. These are arbitrary representations of toxicity, as follows:

A $LC_{50} \leq 200$ ppm
B $LC_{50} > 200 \leq 1000$ ppm
C $LC_{50} > 1000 \leq 3000$ ppm
D $LC_{50} > 3000 \leq 5000$ ppm

In some reactions, the Phase I reaction takes a highly active polar chemical and converts it into a lipophilic toxic chemical. Once this chemical becomes modified, the product is a water-soluble compound. However, the solubility is relative to the original compound; if the compound is slightly soluble, then the product will more than likely have high water-soluble qualities. Conversely, if the compound is not soluble in water at all, then, although the product compound may have water-soluble qualities, it may not be as soluble as is necessary for the excretory process to handle. The following are examples of Phase I reduction reactions:

Aldehyde	R–CHO	RCH_2OH
Ketone	R–CO–R′	RCH(OH)R′
Alkene	C=C	–CHCH(OH)
Nitro groups	$R–NO_2$	$R–NO$; RNH_2; RNH–OH

If the substance does not become as soluble as it needs to be for the kidney to take over, then an "active site" is placed on the compound so that the substance can be eliminated through the kidneys. This attachment of an active site is termed the Phase II reaction.

We must remember that these reactions are the body's attempt to detoxify an intruding chemical. In some cases, the chemical intermediate is a highly toxic compound that will start to destroy cells and tissue. In other cases, the cells are hindered in accepting oxygen and nutrients, or in releasing carbon dioxide along with bound receptor sites. The intermediates may have synergistic, additive, potentiative, or antagonistic qualities. At what level does the chemical in question harm us? Does the chemical become enhanced, or does it cancel out the effects of toxicity?

We can describe these chain reactions or the exposure of numerous chemicals in the following terms:

- *Synergistic.* The combined effects are more severe than those of the individual chemicals. In this case we may have a chemical that by itself is moderately toxic, but in combination with another has enhanced toxic qualities ($1 + 1 = 3$).

- *Antagonistic.* The combined effects cancel each other, decreasing the toxic event. One of the chemicals is acting to decrease the effects of the other chemical ($1 + 1 = 0.5$).

- *Additive.* Some chemicals that are different in chemical structure, shape, and polarity may have the same physiological response in the organism. Thus the effect is a twofold (or more) enhancement ($1 + 1 = 2$).

- *Potentiative.* A chemically inactive species acts on another chemical, which enhances the chemically active substance ($1 + 0 = 2$).

Most antidotal treatment is based on the organism's ability to excrete the end product. There are a limited number of antidotes as compared to the vast variety of chemicals that pose exposure hazards. Antidotes that are not typically discussed but that need to be mentioned because of the dangers associated with their use are the chelating agents. Chelating agents are utilized to form a chemical complex between the insulting chemical and the agent. Either another chemical is introduced to eliminate the newly formed complex or the chelating agent and the insulting chemical combination is eliminated. With these particular agents, the attachment of the antidote to the toxic chemical is how the chemical is removed from the body, thus making it water soluble (in the Phase I and Phase II reactions).

Human beings are made mostly of water. All substances excreted from our body are water-soluble. In order for an introduced substance to be excreted from our body via the kidneys, it must be water soluble. Remember that polarity limits the solubility to like substances. With some chemicals, the body looks for a way to convert the nonpolar (thus nonsoluble) substance into a polar molecule so that the excretory process can handle the metabolite. Unfortunately, sometimes a nonpolar, nontoxic substance is converted into a polar and possibly toxic substance. In order for the body to manipulate the nonpolar substance into a polar substance, several reactions must take place. During this reaction cycle, a toxic intermediate compound may be produced that starts to destroy the required chemical processes in the body.

For the most part, these reactions occur at the enzyme, or protein, level. The alteration of the metabolism of these chemicals (enzymes) interrupts the body's ability to recognize particular unwanted chemicals. If this cycle of reactions continues (repeated exposure), the body is unable to return to the preexposure state.

If the exposure to the chemical is minimal, specific enzymes in the body that are needed to recognize the unwanted chemical act on the toxic metabolite. The chemical reactions continue to take place until the unwanted substance is excreted from the body.

Testing for Exposure Effects

The numbers we use for toxic levels are not absolute numbers. There is a misconception in the emergency field that these numbers will tell you the level that is safe and the level that may be dangerous. This is a false assumption. Predictable outcomes when dealing with the toxic levels of chemicals are based in the testing procedure. Although these tests show a level of possible insult, they do not truly identify at what level this insult will take place. Some of this inaccuracy is due to the limited knowledge we have on biological processes and some lies in the testing procedures.

Acute Exposure. The term *acute* is used to describe a sudden onset or exclusive episode. It relates to a hazardous materials exposure as a single event that causes an injury. This single exposure is usually of short duration and is sometimes classified as *nonpredictable*. It can occur within a 24 hour period or can be a constant exposure for 24 hours or less. In certain cases, it could also be multiple exposures within the 24-hour time frame.

Standard testing routines exist for determining exposure effects. The first step in the procedure usually is to reference all of the information that is currently known about a chemical. The testing technique is then based on the historical reference of the material.

Oral toxicity tests are relatively easy to conduct. Groups of rats or mice are used to study the chemical or drug in question (after the reference literature has been searched). These tests are done initially to *rate* the chemical as to the toxic levels. In other words, the test animals are given increasing amounts of the chemical until a lethal dose is attained.

The test animals are divided into four, five, or six groups. In group one, the rats or mice are given a specific amount of the chemical that may or may not be at the threshold limit value. It is, in general, a no-death dose; all the animals are expected to live. However, there may be an observable toxicity event. The remaining groups (depending on the chemical and the testing technique) are given increasing dosages. The last group is given a dose that is known to be completely lethal. From this observation a further subranking of the chemical is done, and for the next 14 days, the death rate, sickness, fur loss, and so on, are observed. The dose that produces a 50% death toll over 14 days is referred to as the LD_{50}.

Each test group will produce a statistical picture of the chemical dose versus the death rate, fur loss, sickness, and so on. This data is compared to effective doses that are known for the chemical or like chemicals. Both graphs are statistically smoothed into a straight line called the *dose response curve*.

Through statistical modeling, the effective dose and the lethal dose are calculated and the published LD_{50} is established. Both oral toxicity and inhalation toxicity are measured in this way. The oral toxicity levels are usually for a one-time dose. For cases in which the inhalation exposure is an exposure that occurs for 1 hour to each test group, each test group receives the appropriate dose. In each case the groups of animals are observed for a period of 14 days. At the end of the 14-day cycle the 50% death rate must be observed. If the percentage is higher or lower, the doses are reevaluated until the end result is 50%.

Because its skin closely resembles that of a human being, a pig is usually used for dermal testing, but rats and mice are also used. The animal is exposed for a period of 24 hours on the bare skin (if hair is present, it is shaven off). Only 10% of the total body surface area is used. Observed reactions are matched with the known level of the tested chemical or chemical family.

If a chemical needs to be studied further or the results are such that the human element needs to be established, then the procedure is repeated utilizing different species. In general during the testing process, if all the test animal species respond in a similar manner and the statistical slope of the dose response curve is steep, then the results are considered to be accurate for LD_{50} or LC_{50}. Conversely, if the testing battery showed that there was a diversity of slopes across the animal species spectra and shallow slopes were observed statistically, then the accuracy of the outcome is questioned. The problem is that the numbers that we as emergency responders use are some point in time in which a toxic dose was received. Not only does it give us a small window into the dose response, but it does not tell us the angle of the testing slope. Was the slope steep and thus a good correlation to the human event, or was it a shallow slope, which does not tell us the true toxicity of the chemical in man?

Lethal concentration or dose tests are not performed on humans. If the reference literature states a level that caused death in man, this information was obtained through an autopsy performed on the victim of a suicide, homicide, or accidental release. It is hypothesized that the variation of exposure will give the true toxic picture. In simpler terms, without the documentation of a human experience, if the chemical in question has been demonstrated to be toxic to most all plant and animal life, then it is considered to be a health hazard to humans.

● Caution

The LD subscripts are maintained as close to 50 as possible. These numbers represent the population that showed a response, and, in most cases, the response is the death rate of the population. The actual number as it moves toward zero represents a more lethal toxic value. In other words, low numbers as toxic values are extremely dangerous chemicals. However, consider a low toxic value and a high number subscript, and the lethality of this particular chemical is very dangerous.

In general, the smaller the number that the lethal concentration or dose projects, the more toxic the chemical (see Table 6–9). In addition, most chemicals are weight dependent. In other words, the greater the mass of the animal or person in relation to the chemical exposure, the higher the likelihood of successfully combating the exposure. Looking at, say, the LD_{50} of a chemical in order to estimate the lethal dose, we could take the weight of the person and multiply it by the quantity of the material. For example, for a small child weighing 40 lb (18.18 kg),

Table 6–9   *Hodge and Sterner table of toxicity.*

Reference	Nomenclature	ppm	LC_{50} Reference (mg/m^3)
10 drops of vapor in 1000 ft^3	**Extremely toxic***	1–10	3–30
	Highly toxic	10–100	
¼ cup of vapor in 1000 ft^3	**Moderately toxic***	100–1,000	300–3,000
	Slightly toxic	1,000–10,000	
1 quart of vapor in 1000 ft^3	**Practically nontoxic***	10,000–100,000	30,000–300,000
	Relatively harmless	Greater than 100,000	

Note: These descriptions of toxic levels are relative terms used to convey a scale of toxicity. The specific language does not imply that there are no harmful effects. The values are based on a 4-hour exposure.

*These terms are frequently used on an MSDS.

the LD_{50} would be multiplied by 18.18 kg. So if this child ingested sevin (a moderately toxic pesticide with an LD_{50} of 500 mg/kg), we would multiply 18.18 kg by 500 mg/kg and obtain 9090 mg (9.090 grams or 0.02 lb). If our patient were a 220-lb man, for example, it would take a little over a tenth of a pound to reach the LD_{50} (220/2.2 = 100 kg; 100 × 500 = 50,000 mg or 50 grams, which is 0.11 lb). Metabolism, sensitivity, humidity, temperature, and vapor pressure are but a few of the influencing factors not considered here. This calculation is based only on body weight as it relates to the LD_{50} of the chemical.

Because most chemicals are in the solid or liquid state and do not change states of matter, there are more dose problems than there are concentration problems. Very few become a vapor problem, unless we are experiencing flame impingement on a product or it has been aerosolized (as with chemical warfare agents). There are relatively few chemicals that are airborne contaminants. However, this brings up yet another problem: the testing procedures that these chemicals are experimented with. Emergency responders have to deal with the health, flammability, and reactivity issues that chemicals *may* possess. Under fire conditions, what will a chemical represent? What are the synergistic, additive, antagonistic, or potentiative reactions? The problems of how the chemical may react, combust, or jeopardize our health integrity are all significant management problems.

Lethal concentrations and doses (including TLVs, PELs, and RELs) are only *estimates* of the potential health problems that may exist. They are by no means all-or-none limits. That is, they should not be understood as set limits above which health hazards exist and below which no health effects will occur.

■ Note

Dalton's law of partial pressures gives us a valuable tool for establishing the total partical pressure when we have a mixture of gases in an enclosed space. OSHA and EPA have used this basic concept to calculate the toxicity concentrations of mixtures:

$$E_c = C_n/L_n$$

where E_c is the exposure concentration, n represents the number of gases, C_n is the on-scene actual concentration, and L_n is the reported exposure limit value. Thus, for three gases we would have:

$$E_c = C_1/L_1 + C_2/L_2 + C_3/L_3$$

or, in general:

$$E_c = C_1/L_1 + C_2/L_2 + \cdots + C_n/L_n$$

Subchronic Exposure. Subchronic and subacute are two terms that are sometimes used interchangeably to describe the same type of exposure. However, the appropriate term is *subchronic*. This type of exposure is an acute exposure that is repetitive. It is a recurring event. It is an exposure that happens during approximately 10% of the organism's total life span.

The testing procedures for subchronic exposures are based on the LD_{50} or LC_{50}, which established death in 50% of the test group. The testing procedure is carried out in a minimum of two species, with each having a control group. Three to four groups are tested in a similar manner. At the top end, a dose under the LD or LC and high enough to show signs of injury, but not death, is given. At the low end of the dose range, the treatment is such that there should not be any noticeable effects. Depending on the test chemical, a middle point is chosen. If the curve from the acute testing is shallow, then the two or more middle points are picked. If the curve is steep, one point is selected.

The exposure is given a period of 90 days. At the end of 90 days, all the animals are autopsied for any histological evidence of effect, which is the chemical's effect as compared with the control group(s).

Chronic Exposure. When an exposure occurs during 80% of the total life span of an organism, it is called *chronic* exposure. Chronic, or long-term, effects are much harder to establish than acute effects because so many more factors come into play. Acute exposure and the organism's sensitivity are limits to the testing procedure. Our understanding of toxicological responses and knowledge of biochemistry is limited. Chronic exposure can also influence some, any, or all offspring. Accumulation, or acclamation, may lead to hyposensitivity or hypersensitivity.

As in the acute testing process, a certain degree of inaccuracy is also seen in the chronic exposure figures. Most of what we see and read about is an after-the-fact chronic event. In other words, someone or a group of individuals has become ill or has died. An immediate and obvious cause and effect will not be obvious with all chemical exposures.

● Caution

Exposure level numbers should not be considered as safe levels, but rather as a statistical analysis of risk that may result in injury from an exposure. They provide an expedient method for estimating a chemical's potential toxic effects.

As a starting point, the chronic toxicity of most chemicals is determined through oral acute studies. There is good reason for this particular starting point. The first studies of chronic toxicity were undertaken to isolate those chemicals that could potentially become a problem if introduced into food, such as preservatives and production additives. This method of exposure study is easy to do. Chronic exposure studies have been expanded in recent years to include chronic inhalation and topical exposures.

Similar to the evaluation of subchronic (subacute) exposures, the chronic-exposure experiment starts out by finding the toxicity range of the chemical in question. Most of the information is gathered from the acute studies. The chemical is introduced into the animal by placing it in food. This process lasts for 90 days, during which time a variety of dose ranges are used to establish the high, medium, and low toxic ranges. In the high toxic range, the effect is limited. In the low toxic range, no undesirable effect is noted. In the moderate range, the effect is mild. In addition to these three groups under study, there is a control group.

Typically a large number of animals and two or more species are used. The oral ingestion is started from birth and continues through approximately 80% of the animal's life expectancy. Every day the animals are observed for negative effects, and a battery of tests is performed weekly. At the end of the experiment, all animals are autopsied and histological tests are analyzed. From the data, a statistical model is used to give the dose effect response.

This type of testing can become extremely expensive. It can take years to observe some effect. If, for example, new concepts were introduced in the experiment during the testing period, the results would be skewed.

ACID–BASE REACTIONS

We earlier identified the process of acid–base reactions as a relationship between hydronium ions, H_3O^+, and the production of a hydroxide ion, OH^-. We described the acid as a compound capable of transferring or donating a hydrogen ion in solution, and the base as a group of atoms containing one or more hydroxyl groups. In essence, the hydrogen is replaced by the acid, or the base is referred to as the proton acceptor. It is this acceptance and donation that produces the acid–base reaction.

These reactions are classified according to their particular ionization in water. The scale used to measure acidity or alkalinity (ionization) is the pH scale, where 0–6.9 is an acid and 7.1–14 is a base. Seven indicates a neutral substance, or one with an equal number of hydronium and hydroxide ions (see Chapter 1 for a more detailed discussion).

The idea of pH is a mathematical representation of how hydronium and hydroxyl ions move in a solution, which is represented by the following formula:

$$H_2O \leftrightarrow H^+ + OH^- = 1 \times 10^{-14} \quad \text{or} \quad pK_w = pH + pOH = 14$$

For our discussion we can say:

$$pH = -\log[H^+] \quad \text{and} \quad pOH = -\log[OH^-]$$

So, for example, if we have a solution that we know has an H^+ concentration of 2.3×10^{-4}, we can place it in the $-\log[H^+]$ equation and obtain the pH:

$$\begin{aligned}
-\log[H^+] &= -\log(2.3 \times 10^{-4}) \\
&= -\log 2.3 - \log 10^{-4} \\
&= -\log 2.3 - (-4) \\
&= 4 - \log 2.3 \\
&= 4 - 0.3617 \\
&= 3.638
\end{aligned}$$

Scientists use a method of concentration when they speak about acid–base solutions. For example, a bottle of HCl (hydrochloric acid) that you might find on the shelf in the science lab has a 0.00023 N solution. The N means normal solution and is the mole/liter concentration in the solution. You can see from the preceding mathematics that the HCl 0.00023 N solution has a pH of 3.6

Now let us consider the application of this knowledge in a real-life scenario. You respond to a spill of an acid or base and wonder if diluting with water or a neutralization process is the appropriate mitigation objective. It may not be feasible to move a chemical with a pH of 2 to that of 7, especially when the movement from 2 to 7 may create a larger problem. A pH of 6 may be achievable, and in this scenario would be considered environmentally safe according to the EPA. Therefore, the prime objective may not be to bring the product into a completely neutral state, but rather to arrive at a state having minimal environmental impact or one that can be absorbed.

Once the primary objective has been determined, you next need to decide whether water dilution can be used or if the pH can be changed. Consider first the spill of an alkali solution. In order to move the pH, you have only two choices: to add an acid to the spill (a dangerous procedure) or to dilute it with water. In either case, obtaining input from a chemist or, preferably, an environmental chemist would be the most appropriate first step.

For an acid spill (transporting of chemicals more often involves acids than bases), water dilution or neutralization toward an achievable pH are again mitigation options. Suppose you are faced with a spill of 1000 gallons of acid with an "out-of-the-can" pH of 2. From a dilution standpoint, you can calculate that if you increase the pH by 1, you in effect reduce the concentration to 1/10 of the original value. Let us say the spill is in an environmentally sensitive area where a pH of 7 would be the level of choice, but you have to settle for a pH of 6. In this case, you would have to use 10,000,000 gallons of water! With pH, the movement from one whole number to another is in powers of ten. A pH of 7 is described as 0.0000001 or 10^{-7}. For 1 gallon of an acid with a pH of 0, 10 gallons of water would be needed to move the pH to 1; 100 gallons to move the pH to 2; 1000 gallons to 3; 10,000 to 4; 100,000 to 5; and 1,000,000 to 6. (The opposite is true pertaining to bases.) Therefore, a pH of 14 would require 10 gallons to move the pH to 13; 100 gallons to 12; 1000 to 11; 10,000 to 10; 100,000 to 9; and 1,000,000 to 8. To identify the dilution, subtract the desired pH from the existing pH and add that number of zeros to obtain the gallons of water per 1 gallon of product. For the example of an acid at a pH of 2 needing to be moved to a pH of 6: $6 - 2 = 4$; add four zeroes to get 10,000 of water per gallon of acid; calculate $10,000 \times 1,000$ gallons of acid and end up with 10,000,000 gallons—not a viable option.

Suppose the spill is 1000 gallons of sulfuric acid at 50%. You have already figured that dilution is too impractical so you need to figure out how to neutralize the spill. Start by calculating the density by utilizing your knowledge of specific gravity and the density of water. The specific gravity (SG) of sulfuric acid is 1.84. Knowing that the density (or weight) of water is 8.33, multiply the two numbers to give pounds per gallon of sulfuric acid:

$$1.84 \times 8.33 = 15.3 \text{ lb/gal}$$

Because the sulfuric acid in the example is at 50%, you have 50% acid and 50% water, and you need to divide the densities of each component to obtain the volume of each:

$$50 \times \left(\frac{1 \text{ gal}}{15.3 \text{ lb}}\right) = 3.27 \text{ gal}$$

$$50 \times \left(\frac{1 \text{ gal}}{8.33 \text{ lb}}\right) = 6.00 \text{ gal}$$

The total volume is 9.27 gallons.

Percentage is expressed by the gallons of acid over the total volume multiplied by 100:

$$\left(\frac{3.27 \text{ gal of sulfuric acid}}{9.27 \text{ gal of total volume}}\right) \times 100 = 35.275\%$$

This percentage is then multiplied by the total amount spilled to get the volume of the acid that has been discharged:

$$1000 \text{ gal} \times .35275 = 352.75 \text{ gal}$$

Convert back into density:

$$352.75 \text{ gal} \times \left(\frac{15.3 \text{ lb}}{1 \text{ gal}}\right) = 5{,}397 \text{ lb of sulfuric acid}$$

Dividing the pounds of acid by 2.2 and multiplying by 1000 gives the weight of the acid in grams:

$$\left(\frac{5{,}397 \text{ lb}}{2.2}\right) \times 1000 = 2{,}453{,}181 \text{ g}$$

Then divide the gram molecular weight of sulfuric acid (98) into the grams of acid:

$$2{,}453{,}181 \div 98 = 25{,}032 \text{ moles}$$

■ **Note**

The number of protons released by sulfuric acid is derived from the chemical formula.

Table 6–10   *Field-method conversions for the calculation of neutralizing material.*

	Acid			
Base	**Sulfuric** **(SG = 1.84)**	**Hydrochloric** **(SG = 1.19)**	**Nitric** **(SG = 1.50)**	**Phosphoric** **(SG = 1.87 Pure,** **1.33 at 50%)**
Sodium hydroxide	0.8163	1.096	0.6349	1.224
Sodium carbonate	1.082	1.452	0.841	1.622
Sodium bicarbonate	1.673	2.247	1.302	2.541
Potassium hydroxide	1.145	1.537	0.8904	1.717
Calcium hydroxide	0.7551	1.0137	0.587	1.133

Sulfuric acid releases 2 moles of protons or H^+. In this example, 50,064 moles of hydronium ($2 \times 25{,}032$) will be needed to neutralize the sulfuric acid. If you use potash (potassium hydroxide), which has a molecular weight of 56 grams per mole, you will need 2,803,584 grams of it ($56 \times 50{,}064$), or 6167.89 pounds, to neutralize the spill.

Obviously, these types of calculations can become very involved, and under the pressures of emergency response, errors can occur. To simplify things, an estimate of the neutralization agent required can be obtained by using the logic behind the calculations. First, multiply the quantity of the acid that has been discharged by the specific gravity, then multiply by the weight of water, then multiply by the percentage, to get the corrected weight in pounds. Then multiply that number by a conversion factor for each acid versus the base utilized (see Table 6–10) to end up with a fairly close approximation of the amount of neutralization agent that would be required. Utilizing the preceding example, the calculation would look like:

Quantity of Product Discharged		Specific Gravity of Product Discharged		Weight of Water (8.3334)		Percentage of Concentration		Weight of Product Discharged 7663.6 lb
1000 gal	$\times$	1.84	$\times$	8.334	$\times$	50%	=	of sulfuric acid

and then:

Weight of Product Discharged 7664 lb		Acid–Base Conversion Factor 1.145 for potash		Estimated Weight of Neutralizing Agent 8775 lb of potash
	$\times$		=	

Thus we arrive at an estimate of 8775 pounds of potash to neutralize the 1000-gallon spill of 50% sulfuric acid. Obviously, this calculation (see Figure 6–8) is not as accurate as the technical calculation, but it is much easier and quicker and will provide for an adequate amount of neutralizing agent required for an incident.

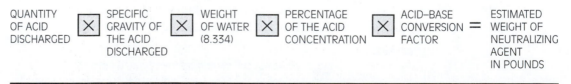

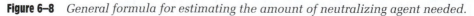

Figure 6–8   *General formula for estimating the amount of neutralizing agent needed.*

As you add a neutralizing agent to an acid, a reaction is taking place. Heat and gases will be liberated depending on the atmospheric conditions and the particular acid and base involved. Movement toward a pH of 7 starts out slowly but occurs rapidly after a certain point. Every application of a base to an acid must be done with great care and with a pH assessment as the procedure is being carried out.

BASIC CHEMISTRY APPLICATION FOR DEFENSIVE OPERATIONS

Atomic theory allows us to understand the principles of bonding, which is the basis of molecules, and thus allows the first responder to determine what type of chemical he or she is dealing with. We have looked at the chemical, physical, and toxicological properties of materials, which gives us some basic clues at the scene of an incident. Other clues exist in our identification process, such as occupancy and location, placards and labels, container shapes and sizes.

Defensive operations are those functions that require the responder to either confine or contain the material in question. Containment is the process by which we prevent the further expansion of the spill or leak, and confinement is the restriction of the material(s) in a protective system. We can use either or both of these concepts to limit the amount of material(s) released into the environment.

The specific activities in this area of hazardous materials response are absorption, diking, damming, diverting, retention, dilution, vapor dispersion, and vapor suppression. We can use our knowledge of the chemical and physical properties to decide which of these activities is appropriate in a given situation.

Absorption

Spilled material and its resulting cleanup is a common response for hazardous materials teams. Two general categories of materials fall into this typical response: petroleum products and corrosives. Ground-up newspaper, clay litter, synthetic fibers, and cornhusks are all examples of absorption materials. Understanding the material in question is essential for the application of the proper absorption materials. Petroleum products have a specific gravity of less than one, and thus float on water. We would like to use a material that absorbs the petroleum product and not the water (synthetic booms and pads), and that floats on the water rather than sinks. Understanding the chemical and physical properties of the absorption material and the spilled product can lead us to this defensive action conclusion.

Damming, Diking, and Diverting

The same principles of specific gravity apply here as with absorption. Depending on the weight, corrosivity (pH), and polarity (solubility) of the leaking material(s), we can apply a dam (overflow or underflow) or dike, or divert the material into a body of water. In each of these cases we need to know if the spilled material(s) is miscible in water (such as acids) or insoluble (nonpolar). Depending on its properties, we may be able to apply this defensive action. With the overflow dam, the contaminant sinks and the flow of water continues moving, thus reducing the amount of uncontaminated material flowing past the problem area. With an underflow dam, the contaminant is on top of the water, and the water flows underneath the contaminant. Absorption materials may also be used for the overall cleanup. Diverting is when we use a system, such as sand, dirt, tarp, hoses, and so on, to keep a spilled product from entering a body of water, such as a storm drain. It stops the product from entering an area that will increase our problem, but it does not stop the overall flow of the material into the environment.

Retention

In diverting a spill, we still need to maintain some control over the spilled product, so retention is the application of choice. This may be as simple as digging a hole downhill from the spill for capture, or as sophisticated as having a portable retention tank brought to the scene or digging a large hole with heavy equipment.

Vapor Dispersion and Suppression

In most of the situations described so far, vapor may be a consideration. When a vapor is produced, vapor dispersion and vapor suppression may be employed. Vapor dispersion may include the use of a fog nozzle to move the vapor to the ground, or a ventilation fan. Other considerations come into play if a fog nozzle is used, for example; a larger spill area will be created, thus possibly creating a higher degree of public concern. With vapor suppression, polarity (solubility) is important; if the product is polar, then foam that is resistant to mixing with the product will be needed.

At the scene of a hazardous materials incident, both the chemical and the containment system (container, pipe, or delivery system) influence the tactical employment. For example, petroleum products can create a static charge when moving through pipes. Grounding the pipes and containment system for offloading should be considered when dealing with these products.

Understanding flammable limits and how we can qualify these limits can give us clues on the potential areas of toxic concern. Remember that the toxic values hover around the lower explosive limits. By analyzing the flammable range, potential toxic environments can be determined. The vapor that a chemical releases during a spill is also hard to control. The chemical's boiling point has a direct relationship with the amount of vapor a chemical may release. Understanding the boiling point will provide clues towards the type of vapor production you can anticipate. Looking at the ambient temperature in relation to the boiling point with the identification of a flammable range can indicate a potential hazard zone. The lower the boiling point of a compound in relation to the ambient temperature, the greater the vapor production.

Corrosives

Corrosives (pH) have their own set of considerations. For example, acids are acids because of the disassociating hydrogen. In some cases where acids are released, this hydrogen can produce a hydrogen cloud (additionally, the attached halogen in binary acids can become part of the vapor cloud), thus creating a potential flammable (and toxic) environment. The first thing that a responder may look at in this situation is this flammable vapor cloud. Although we have an acid, allowing water into the acid (as with vapor dispersion techniques) will create a larger event; when the water mixes with the acid, a violent reaction could occur (depending on the solution and concentrations). However, if the acid has been reduced to a salt (solid), then dilution or neutralization may be employed.

Pressure and Containment

An ammonium nitrate explosion decimated Texas City, Texas in 1947. The understanding of chemicals in the fire service was not widely understood at this time. They applied water to the ammonium nitrate, and closed the ship's hatches, thus created the world's biggest pipe bomb. Considerations of reactivity and the pressure buildup with a containment system were not analyzed due to the lack of chemical education.

Previously we looked at hazards contained in each chemical family. The identification of chemicals and their inherent hazards can lead to detection strategies and safe mitigation efforts, and hopefully reduce the number of tragic instances.

It is up to the emergency responder to consider every choice before deciding on a single mitigation strategy. It is the identification of the chemical(s) that will assist you in this endeavor. Without a clear understanding of the chemical(s) you are dealing with, the strategy you choose may not be the most appropriate one for the event.

GENERAL MONITORING AND DETECTION

The concepts of monitoring and detection could fill a book by themselves. The discussion here is limited to emphasizing that detection and monitors utilize concepts from chemistry and physics to provide the information needed by emergency responders (see Chapter 1). Each device has its own inherent mechanism for detection. These technologies *must* be understood before the responder utilizes such equipment. As with referencing, three technologies should be used to confirm a hazardous environment. Table 6–11 lists equipment and the associated technology currently in use.

Table 6–11   *Monitoring and detection devices.*

Device	Technology	Remarks
Combustible gas indicator	Wheatstone bridge or catalytic bead	Detects through a difference in filament or bead heat
Colorimetric tubes	Reactants in a sealed tube	Chemical in the air interacts with the reactants in the tube
Photoionization detector	Ionization by light	Ionization potential created and compared to a calibrated gas
Flame ionization detector	Ionization by heat	Ionization potential created and compared to a calibrated gas
Metal oxide device	Ionization that occurs with the metal oxide	Usually supported by a microprocessor; read against a calibrated gas
Electrochemical sensor	Degree of ionization in an alkali bath	Chemical-specific to the sensor; e.g., an alkali bath
Sound acoustical wave device	Harmonic frequency	Amplitude vs. time during absorption on a chemical-specific polymer
Radiation device	Ionization in a gas tube	Degree of ionization change in gas tubes
Mass spectrometer-gas chromatograph	Ionization, vibration in the molecule	Identifies molecule through pattern recognition of ionization and molecule vibration
Organic vapor analysis	Ionization	Ionization potentials
Chemical analysis	Utilizes chemical reactions for analysis	Decision tree for possible chemical
Chemical agent detector	Ionization from a radiological source (ion mobility)	Ionization of particles vs. placement on a charged screen (time of flight)
Optical remote sensing	Infrared spectral analysis	Either laser or infrared spectrometer
Detector paper	Specific litmus-type paper	Chemical dependent
pH paper	Litmus paper reacting to positive hydroniums	Strength and concentration dependent

PERSONAL PROTECTIVE EQUIPMENT

Hazardous materials response is all about chemistry, right down to the personnel protective equipment (PPE) worn in a hazardous environment. PPE is made from chemicals; it is the resistance qualities that we are researching when we reference the compatibilities of an entry garment with the chemical(s) in question. The resistance qualities of a suit are affected by penetration, degradation, and permeation (see Figure 6–9).

Penetration refers to the physical movement of a chemical through the natural openings of the suit. Zippers, exhalation ports, seals around the face shield, and connections can provide a route for the chemical to enter the suit. Most commonly, liquids, finely divided particles (aerosols), and gases under pressure can penetrate suits. Abrasions, punctures, tears, and degradation can contribute to penetration. *Degradation,* the physical destruction of the suit, can be caused by temperature, concurrent chemical exposure, an inappropriate storage environment, and incompatibility with a chemical environment. Signs of degradation include discoloration, bubbling, chaffing, shrinking, and any visible evidence of destruction.

Permeation is the movement of a chemical through the protective materials of the suit at the molecular level. An atom has space between its center (nucleus) and its electrons (outer region); there is also space between the bonded atoms in a molecule. Chemicals can move through or permeate the protective material between these spaces. When concentration on a suit builds—either through prolonged exposure, repetitive use, or incompatibilities with a chemical—the chemical diffuses through the outside of the fabric. Eventually the chemical breaks through the fabric (a process known as *desorption*), leading to exposure within the suit (see Figure 6–10).

Suits are tested by the manufacturers for compatibility with specific chemicals utilizing established testing processes. The maximum limit on suit breakthrough is identified as 480 minutes, or 8 hours, of use. To assist the responder in determining the compatibility of a suit with the chemical(s) to be mitigated at a hazmat incident, the manufacturers provide compatibility charts listing breakthrough times. Obviously, knowing the exact chemicals involved at an incident is a critical safety factor.

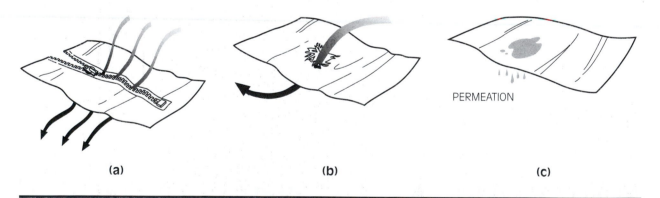

(a)   (b)   (c)

Figure 6–9 *Permeation, the movement of liquid through a fabric.* (a) *Penetration of a chemical through a suit zipper;* (b) *degradation, the quickest and most complete type of suit failure;* (c) *permeation, the movement of a chemical through the protective material.* (From R. Schnepp and P. W. Gantt, Hazardous Materials: Regulations, Response, and Site Operations [Albany, NY: Delmar, 1999, pp. 257 and 259].)

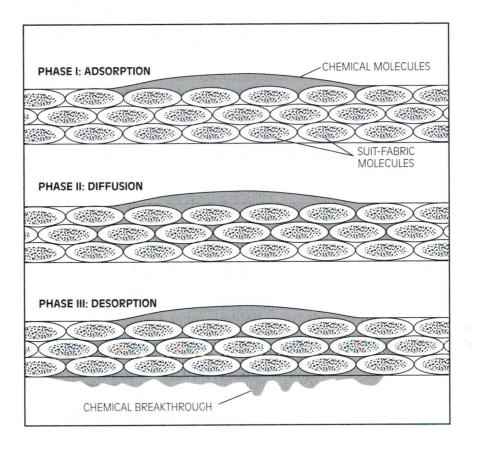

Figure 6–10

Permeation theory. (From R. Schnepp and P. W. Gantt, Hazardous Materials: Regulations, Response, and Site Operations *[Albany, NY: Delmar, 1999, p. 256].)*

TRIAGE CONSIDERATIONS

Because a chemical incident may produce victims needing treatment, response systems are set up in communities to engage incidents as they happen with the available resources. Response profiles are commonly arranged to provide each patient and each emergency the same or a similar level of service. Response, treatment, transports, and hospitals are all geared toward limited patient load. In the case of a multiple-casualty incident, plans are utilized for response patterns that can accommodate the increase in patient care. As part of these plans, a system of sorting has developed over the years, which we call *triage*, derived from the French word *trier,* meaning to select, cull, or sort.

This method of sorting patients categorizes patients into priorities of treatment. It prioritizes patient load in terms of criticality and the resources that are immediately available in order to maximize effectiveness of response. It is a way to provide the best delivery of service for the largest group that will benefit from that service.

Although triage is traditionally thought of in connection with a multiple-vehicle accident, high-rise fire, or structural collapse, it is rarely considered during a hazardous materials event. But whether it is an accidental release or intentional act of terrorism, a chemical incident can overwhelm the local medical capabilities.

A hazardous substance response must adjust the response profile to meet the additional challenge of caring for the chemically injured patient. Thus, it will become necessary for the most seasoned emergency medical system (EMS) officer to organize, implement, and deliver a structured medical echelon such that medical care can be delivered to the greatest number of injured given the resources that are provided. To accomplish this goal, the medical officer must maintain four priorities:

1. An understanding and direct knowledge of the available physical and human resources and the ability to deploy these resources and initiate variances in medical protocol.

2. Knowledge of the level of patient load (including potential rescuer injuries) that may be generated from the incident, inclusive of critical, moderately stable, stable, and walking wounded patients.

3. An awareness of the difference between positive and negative outcomes in treatment modalities as related to the natural course of the injury or disease process.

4. An understanding of the medical resources and their capabilities (basic life support versus advanced life support) at the scene, during transport, and at the hospital(s).

The need to make choices, to help those who will gain the greatest benefit and set aside those who have negative outcome possibilities, stems from the simple fact that each system, agency, and community has, at best, limited resources. It is not only the level of these resources that often influences the scene outcome, but also the mismanagement of these resources. Triage is basically a match between the resources that a system can deploy and the situation at hand. It is a dynamic process requiring that every level of command set priorities consistent with the chemical injury to ensure the success of the incident response.

Figure 6–11 is presented as an aid to triage decision making and effective management of available resources. It is adapted from the START (simple triage and rapid transport) model. Although it is based on dealing with chemical warfare agents and thus utilizes the general categories of neurotoxins, chemical asphyxiants, corrosives, and respiratory irritants, it also can be used in general for chemicals in those categories.

The first two priorities are vital during a chemical catastrophe. Resources are the limiting factor in a chemical release. At each level, medical resources, such as antidotal therapy, should be considered. Many chemicals will require the use of supportive advanced treatments. Without these treatments, patients may not fit the triage priority but must be considered as such.

DECONTAMINATION

In preparing for the decontamination of equipment and personnel, several facts must be kept in mind. The chemical found in the decontamination corridor is at a lower concentration level than that found in the hot zone. The warm zone must be treated with the same respect as the hot zone. By utilizing water in the decontamination process, the chemical characteristics may not change. The concentration of the material will decrease but the strength of the material may remain the same. Therefore, all runoff fluids should be confined and tested prior to disposal. They must be tested and compared to the EPA's recommendations of toxicity. If the runoff is below this standard, then the solutions may be disposed of in a normal manner. If the levels are above the recognized level, or if it is felt that it may endanger the public or environment, the material must be disposed of as a hazardous commodity.

1. Look at:
  a. Safety considerations—These include, but are not limited to, the use of personal protective equipment during triage and decontamination, and ensuring that decontamination has produced a clean patient.
  b. Needed resources—Management of medical resources and the magnitude of these resources escalate as patient load increases. Consider that antidotal therapy is a labor-intensive task.
  c. Space requirements—Have all ambulatory patients move to a predesignated area for decontamination. Consider that separate ambulatory male and female decontamination should be provided for. Additionally, nonambulatory corridors and a contamination reduction zone will have to be provided for rescue personnel.

DELAYED EVAC

2. No respiratory effort—nonsalvageable—including agonal respiration

RESPIRATORY STATUS — Respiration: > 30 or < 10 per minute

- Neurotoxins—Severe dyspnea, fasciculations, nausea, and vomiting.
- Chemical asphyxiants—Severe distress, tachypnea, hyperpnea, dyspnea, convulsing, bradypnea.
- Corrosives—Any airway involvement (inhalation injury); nose, sinus, pharynx burning; pulmonary edema; pseudomembrane formation; extreme immediate irritation; airway obstruction due to edema.
- Respiratory irritants—Any airway involvement (inhalation injury).

Each chemical will present with different symptomology. Decontamination should be done prior to treatment. Treatment modalities should be emphasized for basic and advanced life support.

IMMEDIATE EVAC

Respiration normal (between 10 and 30 per minute)—go to next step

3. No circulation—nonsalvageable—including agonal pulse

PERFUSION STATUS — Capillary refill > 2 seconds

- Neurotoxins—Not talking, full DUMBELS symptomology (diarrhea, urination, miosis, bronchospasm, emesis, lacrimation, salivation).
- Chemical asphyxiants—Flushing (circulatory system fully oxygenated), hypertension, reflex bradycardia, AV (arterio-ventricular) nodal or intraventricular rhythms.
- Corrosives—5–50% Body Surface Area, erythemia, pruritis, burning, blisters, blanching, erythematous ring, weals, greyish epithelium.
- Respiratory irritants—Allergic-type reaction compromising circulation or having the potential of circulatory failure.

Chemicals that affect circulatory status act within minutes of exposure. Advanced Life Support must be on scene to ensure rescue and not recovery.

IMMEDIATE EVAC

Circulation normal (capillary refill < 2 seconds)—go to next step

4. Mental status is dependent on chemical involvement

NEUROLOGICAL STATUS — Cannot follow instructions

- Neurotoxins—Fasciculations, nausea, and vomiting.
- Chemical asphyxiants—Post-ictal.
- Corrosives—N/A.
- Respiratory irritants—Post-ictal.

Many chemicals can affect mental status and stimulate or depress the central nervous system. Through metabolism, intermediates can cause a variety of neurological manifestations.

IMMEDIATE EVAC

Mental status normal (alert/orient x 3)—go to next victim, ensure decon

Figure 6–11 *A triage decision-making model.*

> ■ **Note**
> Casualties sent to hospitals or self-evacuated without the appropriate level of decontamination have the potential to limit access to a local medical facility, thus placing a strain on the community's medical infrastructure. If resources permit, gross decontamination should take place at the scene, and a detailed, patient-specific decontamination should occur at the medical facilities.

Methods

There are several categories of decontamination. *Gross decontamination*, sometimes referred to as physical decontamination, is the removal of contaminants by either brushing off the product or removing clothing. By removing clothing, more than 80% of the contaminants are removed.

Dilution decontamination refers to the washing (usually with soap and water) and rinsing off of the contaminants. This method is used for a chemical that is liquid, soluble or semisoluble in water, or that can be removed utilizing copious amounts of water and soap.

Technical decontamination involves the use of a chemical (see Table 6–12) to neutralize or remove the contaminants. This is a dangerous procedure and can cause other reactions if not done properly. This form of decontamination should *never* be utilized on a patient. It is meant for equipment only. Technical decontamination is done using the process of chemical degradation, which alters the chemical makeup of the hazardous chemical. The chemicals used in this process can be harmful to the decontamination crew. Training and complete research must be completed before their use at the hazardous materials incident. For further information about solutions, contact the EPA or U.S. Chemical Corps.

Technical decontamination solutions are quite useful in area decontamination. This is particularly true for organophosphate and carbamate insecticides and nerve agents, as well as the class of synthetic pyrethroids. All of these are high-molecular-weight compounds and have relatively low volatility; in other words, they are contact toxins that require someone to be sprayed with or fall into them to be at risk. All of them are sensitive to alkaline hydrolysis and most of them become thermally unstable at temperatures between 100°C and 150°C. The decomposition products may still be toxic, but much less so than the original compound. If these products are suspected or identified, an application of a

Table 6–12   *Equipment decontamination solutions.*

Solution	Usage	Preparation
A (used for unknowns or knowns)	Organic solvents and compounds, inorganic acids, polybrominated biphenols (PBBs), and polychlorinated biphenols (PCBs)	Solution of 5% sodium carbonate and 5% trisodium phosphate
B (used for unknowns or knowns)	Cyanides, chlorinated phenols, dioxins, heavy metals, ammonia, pesticides, and nonacid-producing inorganic waste	Solution of 10% calcium hypochlorite
C (used for knowns)	Solvents, oils, organic solvent compounds, PBBs, PCBs, and unspecified wastes that are not contaminated with phosphoric esters	Solution of 5% trisodium phosphate, which can be used as a rinse
D (used for knowns)	Caustic waste of an inorganic nature	Solution of diluted hydrochloric acid

mixture of 4 gal of 10% bleach, 1 gal of ethanol (80 proof vodka will do), and 8–10 oz of dishwashing liquid (e.g., Palmolive liquid, Ivory liquid, Joy, or even shampoos) is a very effective area decontamination mix if applied via a foam eductor set at 3% or 6% (depending on the spill volume) at 100 gallons per minute through a firefighting fog nozzle from directly above the contaminated area. If only 5.2% bleach is available, halve the quantities of the other components and spray at 6%. Technical decontamination solutions are used on equipment only and are *not* used for decontamination of people.

Specialized decontamination (see Table 6–13) uses diluted technical solutions to remove the contaminant from a patient. This method is primarily geared toward the chemical warfare agents. The solution is applied to remove the toxin, and then the patient is rinsed using a soap-and-water solution. The decontaminated individual is established as "clean" by detection devices, or decontaminated again if contaminants still exist.

■ Note

Decontamination Considerations: One must look at the physical state of the product in question. A well-established risk-benefit analysis must be done prior to deciding which type of decontamination procedure should be performed. Gases, although the most difficult to contain, are the easiest state of matter to manage from a decontamination point of view. With gases, the level of chemical contaminant is limited to the amount that has contacted the body and become impregnated into the clothing. Liquids and particulate solids tend to be more difficult to contend with. Issues of polarity and the size of the particulate compared to the solution (e.g., soap and water) direct the decontamination procedure. When life safety is involved, containment of decontamination solutions and runoff become issues after patients have been taken care of. The amount of chemical that can be found in the decontamination solution is negligible at best; however, the environment should be considered after life safety has been addressed when dealing with an extremely toxic chemical. If decontamination of equipment is a concern, the information in Table 6–12 should guide you toward the appropriate solution, remembering that these solutions are for equipment only and should not be used on human tissue.

Chemical Warfare Agents (CBRNE)

Decontaminating victims of chemical warfare agents is difficult to address. The military argument is that technical solutions should be used. The hazardous

Table 6–13 *Solutions used for specialized decontamination.*

Solution	Usage	Preparation	Remarks
Soap and water	Flush and rinse	Nonfragrant soap in water	Must follow the use of the chemical solutions
Sodium hypochlorite	Chemical warfare agents: nerve, blister, and biologicals	Bleach or diluted sodium hypochlorite solution of 0.5%	Skin and respiratory hazard and corrosive qualities; contact time depends on chemical
Calcium hypochlorite	Chemical warfare agents: nerve, blister, and biologicals	Diluted calcium hypochlorite solution of 0.5%	Can react with a variety of chemicals; skin and respiratory hazard; extensive flush rinse required; toxic vapor production and contact time required
Chloramine	Chemical warfare agents: blister and V nerve agent	Diluted solution of chlorine and ammonia for a 10% solution	Skin and respiratory hazard; extensive flush rinse required

materials community states that technical solutions have no advantage. The military has been looking at this problem since the end of World War II, and the hazardous materials community has been in the business for only about thirty years. So the argument comes down to philosophy, tactics, and physiology and medical ethics.

Philosophy. The idea of utilizing technical solutions on civilian populations came from the historical extrapolation of military conflict and warfare doctrine. These solutions are intended to break the covalent bond in the warfare agent and render the phosphoric ester, amine, and thioester harmless. This method evolved to destroy the compound as fast as possible so that the integration within the protective ensemble could (1) be used again and the soldier placed back into battle or (2) render the PPE as safe as possible such that the soldier could take it off and put it back on again, placing him or her back into the engagement. In cases that present the possibility of skin contact, antidotal treatment and aggressive technical decontamination is used, again with the objective of returning the soldier to battle. Although limiting factors reduce the affected population, there is no way that responders in the field can determine which individuals represent that fraction, thus *all* potential victims must be triaged, decontaminated, and treated for symptoms. This must be weighed in light of delayed symptoms from organophosphate, biological, blister, or radiological responses. There is still the potential for contaminated individuals to present themselves at an emergency room. It is then incumbent on response command to ensure dispatch of decontamination teams to all local hospitals in order to interdict these walk-ins before they contaminate the emergency room, keeping in mind that contamination resulting in a hospital closure will upset the medical infrastructure of the community. In the civilian population, "returning" the individual is also an issue, but the health and safety of both the responders and the community take precedence.

Tactics. In the military environment, a large volume of a warfare agent is required to achieve the desired effect. For example, if a hill or tactical advantage needs to be gained, several hundred pounds of nerve agent would have to be disseminated in order to achieve the desired effect. In addition, the military uses 99.99% pure substances in its munitions. In the civilian or terrorist environments, this requirement to maintain a toxic cloud is neither feasible nor probable due to the expense of the homemade munitions and the amount that is required. The sarin used in both of the incidents in Japan was a 25–30% concentration. In both cases, the plume could not be maintained, thus limiting the potential harm within the population that was affected. Another limiting factor is that nerve, mustard, and combinations of warfare agents are (very low vapor pressures) liquids and *must* have a dissemination device. Based on this fact alone, the expected ratio of affected to unaffected is 1:5. Disrobing the individuals as much as they will allow, combined with the preceding statistic, results in only a 10–20% level of concern on the high side, and as little as 5% on the low side. In a civilian population, aggressive CBRNE treatment may allow a return to normal; thus, compound destruction is a priority. Civilians tend to lack PPE and full turnout gear, and because a self-contained breathing apparatus offers 10,000X protection that provides approximately 30 minutes of responder protection, public safety and rescue should be the first priority.

Physiology and Medical Ethics. Considering the chemical nature of CWAs versus the chemical qualities of the technical solutions is like choosing between the lesser of two evils. Most of the technical solutions are alkali in nature and can cause significant injury, particularly to the structures in the mucosa and skin through liquefaction of the fatty substances in the tissues. This liquefaction process allows the alkali to penetrate deeper into the underlying structures, causing a much more devastating injury (and possibly introducing military agents deeper into the individual) than acids, for example. The skin is a protective covering, and when severe trauma is present, the exposed tissue has a higher degree of devastation than the intact skin tissue once these solutions are used.

When decontamination of victims occurs, one must look at the issue as a spectrum. One side of the argument suggests that under no circumstances should a technical solution be used on patients. The other side suggests that technical solutions are required. In general, the following considerations will guide the emergency responder when chemical warfare agents are employed:

1. Decontaminating the patients as soon as possible includes disrobing. An immediate water (soap and water, if readily available) wash must be provided, as fast as one would pull a $1\frac{3}{4}$ for fire attack.

2. Expect that the affected-to-unaffected ratio is 1:5, leaving 10–20% truly contaminated patients. Add to this quantity the fact that by disrobing you have, in effect, achieved upwards of 80% decontamination. Thus you are left with approximately 5–10% of the population.

3. The level of severity of the chemical agent must be considered in association with trauma. Open chest wounds or abdominal wounds change the method of decontamination to water irrigation (normal saline) only. Mucosa and the mucous membranes should never have contact with technical solutions. Tissue destruction is high in these areas with the use of such solutions.

The primary objective of decontamination is to remove the agent from the patient's skin, thereby reducing further effects of the chemical and protecting the emergency responder and medical personnel from secondary contamination. Taking action will also provide the patients with psychological support and comfort. The primary decontamination solution should be flushing with water, then soap and water only, because applying soap and water takes time. Time in this scenario is against you. Utilize technical solutions only when a support mechanism is in place (such as air monitoring and detection), when the resources to produce the solution are available, and when extravagant decontamination corridors have been set up (see Figure 6–12). Because technical solutions must never reach the mucosa and trauma-related injuries, the practicality of their use is significantly reduced. Therefore, in a nonmilitary setting, technical solutions have only limited usefulness.

Great emphasis should be directed toward water, soap and water, and the principles of decontamination. Mass decontamination should be considered and practiced for. Limited time should be spent on the use of technical solutions due to the problems that are created and considerations thereof. Options on the use of technical solutions should be planned, advising use once the preceding criteria have been met, detection strategies have been put in place, and the limitations of such use are acknowledged. Technical solutions have their place in decontamination, however limited their use may be.

The predominant action is to decontaminate with water as soon as possible, disrobing the individuals to the level with which they are comfortable. Once this has been accomplished and a systematic decontamination corridor is set in place, then the more sophisticated soap and water process is initiated. Technical

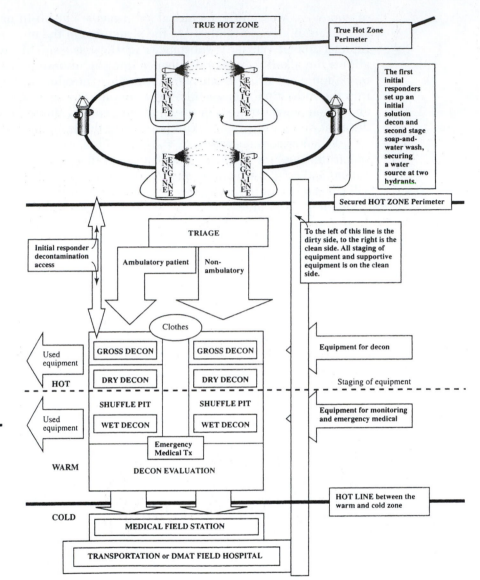

Figure 6–12 *A complete decontamination corridor. (From A. Bevelacqua and R. H. Stilp,* Terrorism Handbook for Operational Responders *[Albany, NY: Delmar, 1998, p. 256].)*

solution may be needed, but let the chemical warfare agent direct the type of decontamination solution required, associated with the concentration of the material. Detection will direct you in this decision.

The scene of a hazardous materials emergency can have either dynamic or static qualities. It is up to the emergency responder to decide at which point the incident is proceeding. Are you in a predominately rescue mode or at mitigation? Is the scene classified as static or dynamic?

If populations are affected, then you are in the dynamic rescue mode and critical decisions will have to be made. Base these decisions on the science rather than the emotion of the situation. Rescue may be as simple as shelter in place or as complicated as an evacuation. If a static scene presents, take the time to handle the incident appropriately. Identify the environmental factors involved and how the incident can be managed without it changing into a dynamic situation. You can make the best decisions by understanding the science involved and the information the references are disclosing.

BIOLOGICAL WARFARE AGENTS

Biologicals have been part of the DOT classification for many years; however, hazmat members traditionally had not educated themselves on this area of response until October 2001, when a community in south Florida had a confirmation of weaponized anthrax only a few weeks after 9/11. Within days, the Northeast was hit and panic occurred in all communities. History is the guide towards the future, and thus it is important to study the biological component of hazardous materials response.

Biologicals are the group of organisms visible only with a microscope; hence the name microorganisms. Responders have a difficult time with this issue in that these organisms are alive and pose a threat to one's well being. There are bacteria, rickettsiae, fungi, algae, protozoa, and viruses, to name a few. Most contribute to the balance of life by maintaining natural environments and controlling and recycling elements that are critical to life. However, some cause disease, reproduce very fast, and overwhelm the body's defense mechanisms.

Microorganisms have a system of classification. This system of nomenclature is known as taxonomy and is the classification of order for living things. The classifications descend from the overall kingdom, which is the broadest classification, to phylum, class, order, and then family. The name includes the genus and the species. For example, tularemia is called *Francisella tularensis,* where *Francisella* is the genus and *tularensis* is the species. There are several criteria used to establish this type of naming, such as the form of the organism (morphology), cellular features, staining characteristics, nutritional requirements, chemical requirements, and genetics, to name a few.

The two major groups of microorganisms are the prokaryotes and eukaryotes. Prokaryotes include simple organisms such as bacteria, fungi, and protozoa. The more sophisticated microorganisms such as algae have a nucleus and organelles (structures that function in a specific manner). Viruses do not fall into either of these two groups because they are basically bits of genetic material (DNA or RNA) housed in a protein coat or envelope. Viruses must become part of a living cell, or host (living organism), because they do not have their own metabolism. During reproduction they destroy the cells in which they are housed. Bacteria, on the other hand, can exist on their own within a growth medium and will reproduce given the right environment. Bacteria look like rods (bacilli), spheres (cocci), or spirals (spirochetes).

Bacteria are very small and viruses are even smaller. Because microorganisms are invisible to the naked eye, we need a microscope to look at these organisms. One type of microscope that is gaining popularity in the response community is the phase contrast microscope. It has filters and light condensers that cast light at angles, passing through the organism and highlighting its internal features such as mitochondria and lysosomes. Because the internal fluid (cytoplasm) of bacteria is transparent, along with some of the internal structures, if we did not use the phase contrast microscope descriptive features of the organism would be lost and identification would be difficult. The bacteria would have to be stained in order to see the structures. In some cases the organism is stained as a tool of classification. Some organisms stain differently, and so there are different techniques for two separate groups: Gram positive and Gram negative.

In the Gram positive technique, the cell on a slide has crystal violet applied, followed by mordant iodine. The slide with the bacterial sample is washed with alcohol. The Gram positive bacteria retain the crystal-violet stain (blue in color) and the result is called a Gram positive; when the stain is washed, a safranin dye is applied and the Gram positive bacteria has a red color under the microscope. Bacillus anthracis (anthrax) is a Gram positive rod, hence the bacilli; when we Gram stain the culture we see red rods (or spores).

The chemical compounds that make up the structures of microorganisms are called organic compounds. These compounds are associated with all life, and in general can be placed into five categories: carbohydrates, lipids, proteins, enzymes, and nucleic acids.

In order to field test these compounds we use a variety of methods, some of which exploit these five basic building blocks of microorganisms. First we look at the pH range of viable life for the microorganisms that we are concerned with; 4–8 is the pH range that they live in. Some of these organisms (sporilating bacteria such as a bacillus and clostridium) require some medium for life. Testing the environment to see the pH range starts the process of presumptive testing. Next we rule out chemicals utilizing a variety of chemical monitors and technologies. Once we have identified the unknown as a possible biological, we exploit the five basic components of organic life. Positive protein tests, for example, look for the proteins that both viruses and bacteria have. Next are the handheld assessments that look at very specific antibody/antigen reactions utilizing the enzyme reactions that occur, and phase contrast microscopy. Nucleic acids can be recognized through the use of polymerase chain reaction (PCR) technology, which looks at enhancing any nucleic acids in the sample to read.

Nomenclature in its entirety is beyond the scope of this book; however, the Center for Disease Control and Prevention (CDC) has placed microorganisms into categories in accordance with their ability to cause harm and their potential threat to society:

Category A Diseases and Agents

High-priority agents include organisms that pose a risk to national security because they:

- May be easily disseminated or transmitted from person to person
- May cause high mortality
- May have the potential for major public health impact
- Might cause public panic and medical infrastructure collapse
- Require special action for public health preparedness

Examples are:

- Anthrax (*Bacillus anthracis*)
- Botulism (*Clostridium botulinum toxin*)
- Plague (*Yersinia pestis*)
- Smallpox (*Variola major*)
- Tularemia (*Francisella tularensis*)
- Viral hemorrhagic fevers (filoviruses; Ebola, Marburg, and Arenaviruses; Lassa, Machupo)

Category B Diseases and Agents

Second-highest priority agents that pose a risk to national security include those that:

- Are moderately easy to disseminate
- Cause moderate morbidity and low mortality
- May require specific enhancements of the CDC's diagnostic capacity and enhanced disease surveillance

Examples are:

- Brucellosis (*Brucella species*)
- Epsilon toxin of Clostridium perfringens Food safety threats (e.g., *Salmonella species, Escherichia coli O157:H7, Shigella*)
- Glanders (*Burkholderia mallei*)
- Tickborne encephalis (nee encephalitis)
- Melioidosis (*Burkholderia pseudomallei*)
- Psittacosis (*Chlamydia psittaci*)
- Q fever (*Coxiella burnetii*)
- Ricin toxin from *Ricinus communis* (castor beans)
- Staphylococcal enterotoxin B
- Typhus fever (*Rickettsia prowazekii*)
- Viral encephalitis (Alphaviruses; Venezuelan equine encephalitis, eastern equine encephalitis, western equine encephalitis)
- Water safety threats (*Vibrio cholerae, Cryptosporidium parvum*)

Category C Diseases and Agents

Third-highest priority agents that pose a risk to national security include emerging pathogens that could be engineered for mass dissemination in the future because of:

- Availability
- Ease of production and dissemination potential for high morbidity and mortality and major health impact

Examples are:

- Hantaviruses
- Multidrug-resistant tuberculosis
- Nipah virus
- Tickborne encephalitis viruses
- Tickborne hemorrhagic fever viruses
- Yellow fever

Biological Summary

ANTHRAX (*Bacillus anthracis*), gram-positive rod

Transmission Inhalation of spores, cutaneous contact with spores, gastrointestinal: ingesting contaminated food.

Clinical Presentation Inhalation caused by bioterrorism attack. Fever, malaise, fatigue, dry cough, and chest pain. Abrupt onset of dyspnea and cyanosis, resulting in respiratory failure. Bacteremia develops, followed by septic shock and disseminated disease.

Diagnosis Chest x-ray, wide mediastinum, effusions; blood culture, gram-positive rods (days 2–3), anthrax toxin (ELISA). Nasal swabs are not useful for diagnostics.

Treatment Decontamination—if exposure has just occurred, remove clothing and place in a plastic bag. Wash with soap and water.

Adults:	Ciprofloxacin 500 mg p.o. every 12 hours orally *or* Doxycycline 100 mg p.o. every 12 hours
Children:	Ciprofloxacin 15–20 mg/kg p.o. every 12 hours (not to exceed 1 gm/day) *or* Doxycycline <8 years old and >45 kg: 200 mg loading dose, *then* 100 mg every 12 hours; <8 years old and <45 kg: 4.4 mg/kg/ loading dose, *then* 2.2–4.4 mg/kg/day in two divided doses. If susceptible, amoxicillin >20 kg, 500 mg p.o. every 8 hours; <20 kg: 40 mg/kg divided into three doses every 8 hours

Prophylaxis Ciprofloxacin 500 mg p.o. two times a day × 8 weeks *or* doxycycline 100 mg p.o. two times a day × 8 weeks.

Vaccine One dose at 0, 2, and 4 weeks, then at 6, 12, and 18 months with annual booster vaccination. Has not been given to children under the age of 18. Only available through the CDC.

Isolation Standard precautions for respiratory anthrax. No droplet or respiratory isolation is needed. Cutaneous—avoid contact with wounds. No room or transport restrictions. Routine disinfection of surfaces with bleach and water. Routine linen and waste handling.

CHOLERA (*Vibrio cholerae*), short-curved, motile, gram-negative, non-sporulating rod

Transmission Ingestion is the only form of transmission. To be used as a biological weapon, large quantities would be needed to contaminate water or food sources.

Clinical Presentation Severe onset of headache, intestinal cramps, and voluminous diarrhea occur 12–72 hours after exposure. Little or no fever will be detected. Loss of 5–10 liters of fluid per day.

Diagnosis Microscopic examination of stool samples noting no red or white cells and almost no protein. The organism can be identified by dark-field or phase-contrast microscopy.

Treatment Fluid and electrolyte replacement. The following antibiotics will shorten the duration of diarrhea:

Tetracycline 500 mg every 6 hours for 3 days *or*
Doxycycline 300 mg once or 100 mg every 12 hours for 3 days

Prophylaxis Vaccination.

Vaccine Licensed killed vaccine for those at risk. Provides only a 50% protection that lasts for no more than 6 months. Schedule for vaccine is an initial dose followed by another dose 4 weeks later with a booster every 6 months.

Isolation Transmission via person-to-person contact is rare, but standard and contact precautions should be used.

PLAGUE (*Yersinia pestis*), gram-negative bacillus

Transmission Pneumonic: Bioterrorism dispersion of an aerosol. Bubonic: Contact with infected fleas. Person-to-person via large aerosol droplets.

Clinical Presentation Pneumonic: Incubation period is 2–3 days. The infection presents as an acute fulminant pneumonia, a fever, rare cervical buboes, chills, sepsis, a bloody sputum, or respiratory failure, progressing to shock and organ failure.

Diagnosis Radiologic evidence of bronchopneumonia. Blood cultures, sputum, and aspiration of buboes all show gram-negative bacillus (safety pin appearance) with bipolar staining immunofluorescent stain. Serum antigen detection and serum antibody titer rise.

Treatment Pneumonic plague requires immediate therapy. Mortality is high if treatment is delayed more than 24 hours. The following antibiotics are used for treatment:

> Gentamicin 5 mg/kg IM or IV every day *or*
> Chloramphenicol 2 gm IV, every day × 2 weeks *or*
> Doxycycline 200 mg (initial dose) *then* 100 mg IV every 12 hours × 2 weeks

Bubonic plague is treated with the same antibiotics.

Prophylaxis Doxycycline 100 mg p.o., two times a day × 7 days.

Vaccine Available through the CDC for persons who have been exposed and who have a high risk of infection.

Isolation Pneumonic plague: standard and droplet precautions. Private room in hospital or cohort. Surface disinfection with bleach and water. Routine linen and waste handling.

TULAREMIA (*Francisella tularensis*), gram-negative rod

Transmission Inhalation of aerosolized bacteria from a bioterrorism attack. Cutaneous inoculation by contact of damaged skin with infected animals. No person-to-person transmission.

Clinical Presentation Inhalation causes respiratory distress, chest pain, cough, septicemia, and typhoidal fever with rapid wasting.

Diagnosis Cultures of blood, sputum, urine, skin lesion. Serologic testing.

Treatment   Gentamicin 3–5 mg/kg IV per day × 14 days *or*
        Doxycycline 100 mg IV two times a day *or*
        Ciprofloxacin 400 mg IV two times a day

Prophylaxis   Doxycycline 100 mg p.o. two times a day × 14 days
        Ciprofloxacin 500 mg IV two times a day

Vaccine Live attenuated vaccine provided through the state public health office.

Isolation Standard precautions.

Q FEVER (*Coxiella burnietii rickettsia*)

Onset of symptoms Symptoms appear in 10–20 days.

Types of symptoms Q fever lasts from 2 days to 2 weeks and consists of fever, headache, fatigue, pneumonia, and pleuritic pain. It rarely causes endocarditis and death.

Types of dispersion Aerosolized material is inhaled.

Personal protection and decontamination Patients do not pose a risk to medical care-givers, but a contaminated environment of equipment can be decontaminated with soap and water.

SALMONELLAE (*Salmonella typhimurium*)

Onset of symptoms Symptoms occur in 8–48 hours.

Types of symptoms Symptoms include fever, headache, abdominal pain, and watery diarrhea. Localized infection may result in endocarditis, meningitis, pericardi-tis, pneumonia, or abscesses.

Types of dispersion The bacteria may be dispersed by contaminating raw food or water sources.

Personal protection and decontamination Infected patients may contaminate others. Exercise body fluid precautions and clean contaminated objects with hypochlo-rite solution.

SMALLPOX (variola virus)

Transmission Inhalation of aerosol. Person-to-person transmission from airborne and droplet exposure and contact with skin lesions.

Clinical Presentation Inhalation: viremia, rash malaise, fever, chills, headache, delir-ium. Rash is centripetal. All lesions are in the same stage. Scabs contain virus, patient is infectious until all scabs fall off. Variola pneumonia is associated with 50% mortality. Strict quarantine with airborne isolation of all cases and contacts of index case for 17 days.

Diagnosis Electron microscope exam of vesicular or pustular fluid—CDC.

Treatment Ribovirin IV, investigational for smallpox. Consult infectious disease specialists.

Prophylaxis Vaccine available through CDC. Dryvax currently in use. Synthetic vaccine under investigation.

Isolation Airborne precautions and contact precautions should be used with stan-dard precautions. Special handling of soiled linen is required. Needs dedicated equipment. In hospital private, negative pressure room. Door must be closed. Limit transport of patient (must wear N95 mask) and have all skin covered. Ter-minal disinfection of the room is required.

VIRAL ENCEPHALITIS (Venezuelan equine encephalitis (VEE), eastern equine encephalitis (EEE), western equine encephalitis (WEE))

Transmission Mosquito-borne infection in nature. Aerosol transmission through a bioterrorism attack.

Clinical Presentation Viral syndrome: fever, headache, myalgia, general malaise, then reurological disease progressing to coma.

Diagnosis Serologic testing—ELISA.

Treatment Supportive, including vector control, fever management, hydration.

Vaccine Vaccines are available and new live attenuated viral vaccines are actively being investigated.

VIRAL HEMORRHAGIC FEVERS (VHFs)

Includes: Arena viruses re: Lassa fever, Argentine (AHF) and Bolivian hemorrhagic fever (BHF), Congo-Crimean hemorrhagic fever (CCHF), Rift Valley fever (RVF), filoviridae viruses re: Ebola, Marburg, yellow fever, Hanta, new world hemorrhagic fever

Transmission Transmitted by contact in nature. Aerosol through bioterrorism attack.

Clinical Presentation Viral syndrome: malaise, fever, myalgia, conjunctivitis, hypotension, petechial hemorrhages, mucosal hemorrhages, and shock. Brain hemorrhages, liver abnormalities, and renal failure.

Diagnosis ELISA and PCR for antigen detection and IgM antibodies.

Treatment Ribavirin for Lassa fever, BHF & AHF, CCHF, RVF, and New World. Convalescent plasma may be used.

Vaccine Available for yellow fever, RVF, BHF, and AHF. Other vaccines are being investigated.

Isolation Contact precautions are used along with standard precautions.

BOTULISM (*Clostridium botulinum* toxin)

Transmission Inhalation of aerosolized toxin. Ingestion of contaminated food source. No person-to-person transmission.

Clinical Presentation Food-borne: gastrointestinal symptoms. Inhalational and food-borne share symptoms (dose dependent, progressive symptoms), ocular palsies, midriasis, diplopia ptosis, skeletal muscle paralysis, respiratory paralysis, bulbar paralysis.

Diagnosis Ag detection—ELISA.

Treatment Equine antitoxin provided by the CDC. Early administration. Antitoxin can only neutralize circulating toxin in patients with symptoms that continue to progress. Available only through the health departments or the CDC.

Vaccine DOD toxoid.

Isolation Standard precautions.

STAPHYLOCOCCAL ENTEROTOXIN B (SEB) (produced by *Staphylococcus aureus*)

Transmission Ingestion of preformed toxin in contaminated food. Inhalation through aerosolized toxin in a bioterrorism attack.

Clinical Presentation Symptoms appear 3–12 hours after ingestion or inhalation of toxin. Symptoms include high fever, headache, myalgia, chills, cough, dyspnea, severe diarrhea, and hypotension.

Diagnosis Antigen detection in urine—toxin assay in nasal swabs.

Treatment Hydration fluid and electrolyte replacement, respiratory support.

Vaccine Not available but currently being tested.

RICIN (*Ricinus communis*) (biological toxin obtained from castor bean seeds)

Transmission Inhalation of aerosolized product from a bioterrorism attack. Can also be ingested or injected. No person-to-person transmission.

Clinical Presentation Rapid onset respiratory distress, followed by pulmonary edema and respiratory failure.

Diagnosis Leukocytosis, bilateral interstitial infiltrates. Serum and convalescent sera provide retrospective diagnosis.

Treatment If contamination is present, decontaminate with soap and water. Supportive care, including treatment of pulmonary edema. Gastric lavage and cathartics for ingestion. Charcoal is not effective.

Isolation Standard precautions.

CDC Emergency Response Office
(770) 488-7100

	BACTERIALS								VIRUSES				BIO. TOXINS			
	Anthrax	Brucellosis	Cholera	Glanders	Bubonic plague	Pneumonic plague	Tularemia	Q Fever	Smallpox	Ven. Equine Enceph.	Viral Encephalitis	Viral Hemorr. Fever	Botulism	Ricin	T-2 Mycotoxins	Staph. Enterotoxin B
ISOLATION PRECAUTIONS																
Standard precautions for all aspects of patient care	X	X	X	X	X	X	X	X	X	X	X	X	X	X	X	X
Contact precautions		X							X		X	X				
Airborne precautions			X	X					X							
Use N95 mask when entering room									X							
Droplet precautions						X			X	X						
Wash hands with antimicrobial soap		X	X						X							
PATIENT PLACEMENT																
No restrictions	X												X	X	X	X
Cohort "like" patients when private room is unavailable					X	X	X				X	X				
Private room			X	X	X	X			X	X	X	X				
Negative pressure room			X	X					X							
Door must be closed at all times			X	X					X							
PATIENT TRANSPORT																
No restrictions	X						X	X					X	X	X	X
Limit movement to essential medical purposes only					X	X			X	X	X	X				
Place mask on patient to minimize dispersal of droplets					X	X			X	X						
CLEANING, DISINFECTION OF EQUIPMENT																
Routine terminal cleaning of room with hospital-approved disinfectant on discharge	X		X	X	X	X	X	X	X	X	X	X	X	X	X	X
Disinfect surfaces with bleach-and-water solution 1:9 (10% solution)	X	X														
Dedicated equipment that is disinfected prior to leaving room									X		X	X				
Linen management as with all other patients	X	X	X	X	X	X	X	X	X	X	X	X	X	X	X	X
Biohazardous waste—no special handling	X	X	X	X	X	X	X	X	X	X	X	X	X	X	X	X
DISCHARGE MANAGEMENT																
No special discharge instruction necessary	X	X	X	X	X	X	X	X	X	X	X		X	X	X	X
Home care providers need to be taught principles of standard precautions	X	X														
Not discharged from hospital until it is determined that patient is no longer infectious					X	X			X			X				
Patient usually is not discharged until completing 72 hours of antibiotics treatment				X	X	X										
POSTMORTEM CARE																
Follow principles of standard precautions	X	X	X	X	X	X	X	X	X	X	X	X	X	X	X	X
Droplet precautions						X			X							
Airborne precautions									X							
Use N95 mask when entering room									X							
Negative pressure room									X							
Contact precautions									X			X				
Routine terminal cleaning of room with hospital-approved disinfectant on autopsy	X		X	X	X	X	X	X	X	X	X	X	X	X	X	X
Disinfect surfaces with bleach-and-water solution 1:9 (10% solution)	X															

STANDARD PRECAUTIONS: Standard precautions prevent direct contact with all body fluids (including blood), secretions, excretions, nonintact skin (including rashes), and mucous membranes. Standard precautions routinely practiced by healthcare providers include: **handwashing, gloves,** and when contact with above, **mask, eye protection,** or **face shields** while performing procedures that cause splash or spray, and **gowns** to protect skin and clothing during procedures.

Linens used by patients with smallpox must be handled wearing N-95 mask, gown, and gloves.

Figure 6–13 *Bioterrorism: Infection control practices for patient management.*

TRICHOLTHECENE MYCOTOXIN (T2)

Onset of symptoms Symptoms appear within minutes to hours.

Types of symptoms Symptoms include burning, itching, reddened skin, burning in the nose and throat, sneezing, burning of the eyes, and conjunctivitis. Gastrointestinal symptoms include nausea, vomiting, diarrhea, and abdominal pain.

Types of dispersion T2 is dispensed in aerosol form (reportedly as yellow rain).

Personal protection and decontamination Complete skin and respiratory protection is necessary. Patient isolation is not required. Wash contaminated equipment with hypochlorite solution.

REFERENCE SOURCES

Books

The following is a short list of valuable chemical reference documents. Remember to always obtain *three separate sources* of chemical information.

> *A Guide to Fire Hazard Properties of Flammable Liquids, Gases and Volatile Solids* (NFPA).
>
> *Bioterrorism: Biological and Chemical Agents Emergency Response Guide* (JJ Kellert Associates, Inc., 2002).
>
> *Chemical Hazards Response Information System (CHRIS)* (Washington, DC: U.S. Government Printing Office).
>
> *Coopers Toxic Exposures* (Boca Raton, FL: CRC Press).
>
> *CRC Handbook of Chemistry and Physics* (Boca Raton, FL: CRC Press).
>
> *Dangerous Properties of Industrial Materials,* N. Irving Sax. (New York: Van Nostrand Reinhold).
>
> *Documentation of the Threshold Limit Values,* American Conference of Governmental Industrial Hygienists (Cincinnati, OH: ACGIH).
>
> *Emergency Action Guide* (American Association of Railroads).
>
> *Emergency Care for Hazardous Materials Exposure,* A. C. Bronstein and P. L. Currance (St. Louis, MO: Mosby).
>
> *Emergency Handling of Hazardous Materials in Surface Transportation* (Association of American Railroads).
>
> *Emergency Medical Response to Hazardous Materials Incidents,* R. H. Stilp and A. S. Bevelacqua (Albany, NY: Delmar, 1997).
>
> *Farm Chemicals Handbook* (Willoughby, OH: Meister).
>
> *Genium's Handbook of Safety, Health, and Environmental Data for Common Hazardous Substances* (Genium).
>
> *Handbook of Emergency Chemical Management,* D. R. Quigley (Boca Raton, FL: CRC Press).
>
> *Handbook of Reactive Chemical Hazards,* L. Bretherick (Boston: Butterworth-Heinemann).
>
> *Handbook of Toxic and Hazardous Chemicals and Carcinogens,* M. Sittig (Noyes Publications).
>
> *Hawley's Condensed Chemical Dictionary,* N. Irving Sax and R. J. Lewis (New York: Van Nostrand Reinhold).

Hazardous Chemicals Data (NFPA).

Hazardous Chemicals Desk Reference, R. J. Lewis (New York: Van Nostrand Reinhold).

Hazardous Materials Field Guide, A. S. Bevelacqua and R. H. Stilp (Albany, NY: Delmar, 1998).

Merk Index and Encyclopedia of Chemicals and Drugs (Rahway, NJ: Merck).

NIOSH Pocket Guide to Chemical Hazards, DHHS (Washington, DC: U.S. Government Printing Office).

North American Emergency Response Guide (U.S. Department of Transportation, 2000).

Pesticide Fact Handbook, EPA (Noyes Publications).

Terrorism Handbook for Operational Responders, A. S. Bevelacqua and R. H. Stilp (Albany, NY: Delmar, 1998).

Threshold Limit Values for Chemicals and Physical Agents and Biological Exposure Indices, American Conference of Governmental Industrial Hygienists (Cincinnati, OH: ACGIH).

Databases

Toxicity Databases

RTECS—Registry of Toxic Effects of Chemical Substances (NIOSH)

TRI—Toxic Release Index (Inventory) (EPA)

HSDB—Hazardous Substances Database (National Library of Medicine)

TOMES—Toxicological Database (Micromedex Inc.)

POISONDEX—Toxicological Database (Micromedex Inc.)

CAMEO—Computer-Aided Management for Emergency Operations (NOAA)

Toxicity Databases on CD-ROM and Disc

Hawley's Condensed Chemical Dictionary (Van Nostrand Reinhold)

Dangerous Properties of Industrial Materials (Van Nostrand Reinhold)

ATSDR's Toxicological Profiles (Lewis Publishers)

Pocket Guide to Chemical Hazards (NIOSH)

Merck Index (on CD-ROM)

Websites

http://www.chemfinder.com

Chemfinder search engine: Enter either name (full, contains, starts with, ends with), molecular formula, molecular weight, or CAS registry number.

http://webbook.nist.gov/chemistry/name-ser.htm

National Institute of Standards and Technology (NIST) search engine: Enter either formula, name, partial formula, CAS registry number, structure, ion energetics, vibrational and electronic spectra, or molecular weight.

http://www.epa.gov/epahome/search.html

Search engine for related documents as they pertain to chemicals in question.

http://siri.org/msds/index/grep.cgi

Search engine: Enter either manufacturer's name, product name, or CAS registry number.

http://ntp-server.niehs.nih.gov/cgi/iH_Indexes/Chem_H&S/ iH_Chem_H&S_Frames.html

Search engine for MSDS-type information.

http://pubs.acs.org/chemcy/

Search engine for manufacturers' MSDS.

http://response.restoration.noaa.gov/chemaids/react.html

This is a downloadable database that references the reactivity of the chemical that you pick from the mix. Updates are available on the Web.

http://www.ilpi.com/msds/index.html

This is a hotlink page with a huge variety of research options. Not for use at the scene of an emergency.

Emergency Management and Planning Sites

http://www.epa.gov/swercepp/

U.S. Environmental Protection Agency: Chemical Emergency Preparedness

http://hazmat.dot.gov

U.S. Department of Transportation Office of Hazardous Materials

http://www.osha.gov/

OSHA

http://www.fema.gov

FEMA

http://www.coastside.net/USERS/bkinsman/ema

Emergency Management Associates

http://epix.hazard.net

EPIX Emergency Preparedness Information

http://www.haznet.org

HazNet by IDNDR

Weather Sites

http://www.weather.com

Weather Channel

http://ww2010.atmos.uiuc.edu/(Gh)/wx/surface.rxml

Weather Visualizer

http://www.nhc.noaa.gov/

National Hurricane Center

http://www.nssl.noaa.gov

National Severe Storms Laboratory

http://www.noaa.gov

NOAA

News Sites

http://www.cnn.com

CNN News

http://abcnews.go.com

ABC News

http://ap.org

AP News

http://www.nbc.com

NBC News

http://nytimes.com

The New York Times

http://usatoday.com

USA Today

http://wiser.nlm.nih.gov/

A wide range of information on hazardous substances Operational version of WISER for Palm™ OS

http://www.freewebs.com/hazmatspecops/informationresearch.htm

A wide range of Palm™ OS programs

Worksheet 1

NOMENCLATURE FLOW CHART

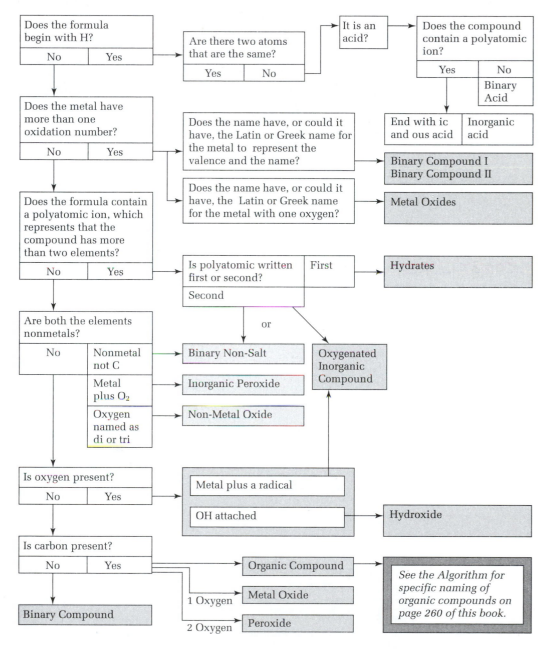

Worksheet 2

Algorithm Identification

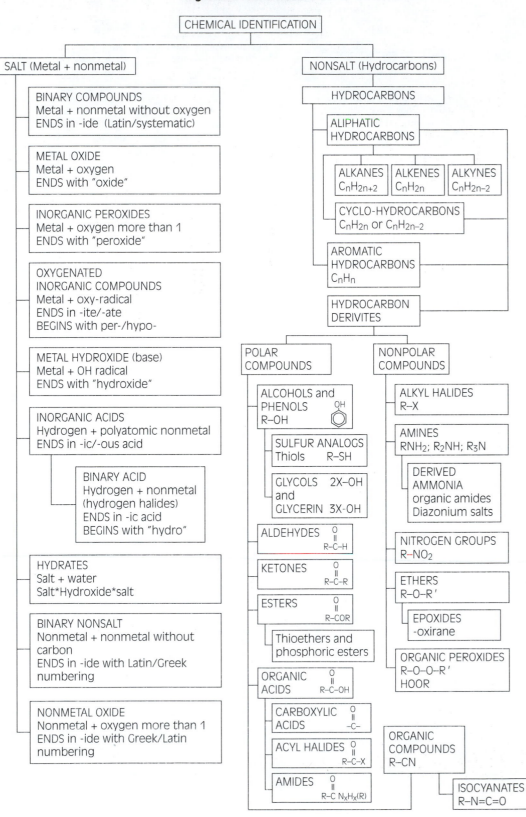

CHEMICAL IDENTIFICATION

SALT (Metal + nonmetal)

BINARY COMPOUNDS
Metal + nonmetal without oxygen
ENDS in -ide (Latin/systematic)

METAL OXIDE
Metal + oxygen
ENDS with "oxide"

INORGANIC PEROXIDES
Metal + oxygen more than 1
ENDS with "peroxide"

OXYGENATED
INORGANIC COMPOUNDS
Metal + oxy-radical
ENDS in -ite/-ate
BEGINS with per-/hypo-

METAL HYDROXIDE (base)
Metal + OH radical
ENDS with "hydroxide"

INORGANIC ACIDS
Hydrogen + polyatomic nonmetal
ENDS in -ic/-ous acid

BINARY ACID
Hydrogen + nonmetal
(hydrogen halides)
ENDS in -ic acid
BEGINS with "hydro"

HYDRATES
Salt + water
Salt*Hydroxide*salt

BINARY NONSALT
Nonmetal + nonmetal without
carbon
ENDS in -ide with Latin/Greek
numbering

NONMETAL OXIDE
Nonmetal + oxygen more than 1
ENDS in -ide with Greek/Latin
numbering

NONSALT (Hydrocarbons)

HYDROCARBONS

ALIPHATIC
HYDROCARBONS

ALKANES C_nH_{2n+2}

ALKENES C_nH_{2n}

ALKYNES C_nH_{2n-2}

CYCLO-HYDROCARBONS
C_nH_{2n} or C_nH_{2n-2}

AROMATIC
HYDROCARBONS
C_nH_n

HYDROCARBON
DERIVITES

POLAR
COMPOUNDS

NONPOLAR
COMPOUNDS

ALCOHOLS and
PHENOLS
R–OH

SULFUR ANALOGS
Thiols   R–SH

GLYCOLS   2X–OH
and
GLYCERIN 3X–OH

ALDEHYDES
$R-\overset{\overset{\text{O}}{\|}}{C}-H$

KETONES
$R-\overset{\overset{\text{O}}{\|}}{C}-R$

ESTERS
$R-\overset{\overset{\text{O}}{\|}}{C}OR$

Thioethers and
phosphoric esters

ORGANIC
ACIDS
$R-\overset{\overset{\text{O}}{\|}}{C}-OH$

CARBOXYLIC
ACIDS
$-\overset{\overset{\text{O}}{\|}}{C}-$

ACYL HALIDES
$R-\overset{\overset{\text{O}}{\|}}{C}-X$

AMIDES
$R-\overset{\overset{\text{O}}{\|}}{C}\ N_xH_x(R)$

ALKYL HALIDES
R–X

AMINES
$RNH_2; R_2NH; R_3N$

DERIVED
AMMONIA
organic amides
Diazonium salts

NITROGEN GROUPS
$R–NO_2$

ETHERS
R–O–R'

EPOXIDES
-oxirane

ORGANIC PEROXIDES
R–O–O–R'
HOOR

ORGANIC
COMPOUNDS
R–CN

ISOCYANATES
R–N=C=O

Worksheet 3

TOXICOLOGICAL AND MEDICAL REFERENCE TACTICAL WORKSHEET

#1 Exclusion Zone:_____ Level Found:___ Instruments:_____ Entry Team 1,2,3 in A, B, C
Chemical Name:_____ Synonyms:_____ Chemical Formula:_____ DOT:_____ CAS:_____
State of Matter:_____ State Found:_____ Water Reactive: Y N Oxidizer: Y N Other Hazards:_____
Wind Direction:_____ Wind Speed:_____ Temp.:_____ Humidity:_____ Adjusted Temp.:_____ Dew Point:_____
Manufacturer:_____ Contact Person:_____ Phone:_____ Fax:_____

#2 Exclusion Zone:_____ Level Found:___ Instruments:_____ Entry Team 1,2,3 in A, B, C
Chemical Name:_____ Synonyms:_____ Chemical Formula:_____ DOT:_____ CAS:_____
State of Matter:_____ State Found:_____ Water Reactive: Y N Oxidizer: Y N Other Hazards:_____
Wind Direction:_____ Wind Speed:_____ Temp.:_____ Humidity:_____ Adjusted Temp.:_____ Dew Point:_____
Manufacturer:_____ Contact Person:_____ Phone:_____ Fax:_____

INORGANIC	COMPOUND STRUCTURE and HAZARDS	ORGANIC

INORGANIC — SALTS

-ide
General hazards:
Toxic gas with water
contact with following
Carbides = Acetylene
Hydrides = Hydrogen
Nitrides = Ammonia
Phosphides = Phosphine

METAL OXIDES

-oxide
Hydroxides when in contact with
water; all water reactive; heat
generation and caustic solutions

HYDROXIDES

Hydroxide solids and solutions
are corrosive

INORGANIC

Peroxide
Water reactive; strong oxidizers;
heat production with high
ignition potential

OXYGENATED INORGANICS

per-; -ate; -ate; -ite; hypo-; ite
Normal oxygenated states

−1	−2	−3
ClO_3	CO_3	PO_4
BrO_3	SO_4	BO_3
IO_3		
NO_3		

More than 1: per -ate
Normal: -ate
Less than 1:   -ite
Less than 2: hypo -ite

COMPOUND STRUCTURE and HAZARDS

#1 COMPOUND STRUCTURE:

BP: _____   IT: _____
FP: _____   LEL: _____
VP: _____   UEL: _____

TLV: _____
PEL: _____   STEL: _____

#2 COMPOUND STRUCTURE:

BP: _____   IT: _____
FP: _____   LEL: _____
VP: _____   UEL: _____

TLV: _____
PEL: _____   STEL: _____

PPM = %   1,000,000 ppm = 100%
             75,000 ppm = 7.5%
             10,000 ppm = 1%
             100 ppm = 0.01%

ENTRY SUIT COMPATIBILITY:

DECON SUIT COMPATIBILITY:

DECON SOLUTIONS:

ANTIDOTES: #1   #2

ORGANIC — NONPOLAR

ALKYL HALIDES

R–X
R radical + halogen
Toxic and flammable

NITROGEN GROUPS

R—N with O groups
R radical + nitrogen group
Explosive

AMINE GROUP

R_XNH_X
R radical + H or other ion
Toxic and flammable

ETHERS

R–O–R′
R and R′ may be same
or different radicals
anesthetic and flammable

ORGANIC PEROXIDES

R–O–O–R′
R and R′ may be same
or different radicals
Explosive and oxidizer potentials

Hospital
Contacted:_____

		POLAR COMPOUNDS—Organic				UV Lamp Potentials	
ENTRY	DECON	ALCOHOL	ESTER	ALDEHYDE	KETONE	ORGANIC ACIDS	
LEVEL		R–OH	R–C–OO–R	R–C–O–H	R–COR′	R–COOH	9.5 eV Mgf
A	A	-ol	-ate	Aldehyde	Ketone	Acid	10.0 eV CaF
B	B	Flammable	Flammable and	Flammable,	Flammable	Toxic	10.2 eV H_2
C	C	and toxic	polymerization	toxic	and CNS	and	10.6 eV N_2
D	D	qualities		polymerization	effects	corrosive	11.7 eV Ar_2
						pH___	11.8 eV LiF

AMB: PREPARED

Team Members : Entry : On Air Time:_____ Off Air Time:_____ Back-up : On Air Time:_____ Off Air Time:_____

Chapter 7

The Art of Infrared Spectroscopy

Learning Objectives

Upon completion of this chapter, you should be able to:

- Explain the advantages and limitations of infrared spectroscopy.
- Identify how orbital theory influences the spectrum.
- Discuss how energy interacts with a dipole moment change.
- Explain how frequency and wave number are used.
- Identify the approach toward reading a spectrum.
- Identify the regions and their specifics towards chemical identification.
- Explain the spectra of: alkanes, alkenes, alkynes, aromatics, and functional groups.

INTRODUCTION

Detection and air monitoring of an environment has been an art form for hazardous materials teams for many years. Although most do not see this segment of the operation as an art form, it truly has been and will continue to be into the future. It is the technician's job to view, analyze, and interpret the information that the instrumentation displays. The reality of current-day operations is that responders must have the ability to logically identify the answers to questions dealing with reoccupation of a facility or office space or relocation of a population. In order to provide the answers to a variety of difficult questions, the first responder must understand the advantages and limitations of each instrument. All technology has its advantages and limitations; it will be this understanding that the technician must apply towards the event, device information, and field observations. Issues such as sick building syndrome, accidental chemical release, and the intentional use of hazardous substances to thwart society have all led response agencies to look at equipment that was once considered part of a laboratory practice as being viable in the emergency response arena.

With the move of high-tech equipment into hazmat response, the tactical application of specific equipment should not lie in the technology but rather in the responder's ability to translate the information into useful action. As hazmat responders, we can, when given the name of a chemical, DOT number, CAS number, or physical property, identify the chemical family or group from which the chemical in question is derived, and thus determine the hazards. Instrumentation gives the hazmat community the ability to rule out one family or group over another. How and why one would use a specific piece of instrumentation is based on the incident, the chemicals involved, and the family or group the chemicals belong to. As with any tactical deployment, the understanding of the chemistry and science behind the technology should guide the technician.

This chapter will give the hazardous materials responder a sense of understanding that will encourage further studies into this technology, and more importantly, the interpretation of the data—not just IR interpretation. We will present the basic science along with a decision tree as the crux of our discussion. It is primarily the understanding and logical application of a tactical approach that is presented within the confines of hazardous materials "street" chemistry.

INFRARED TECHNOLOGY

Hazmat technicians are neither chemists nor industrial hygienists. We are a group of individuals who can look at a scene and analyze divergent conditions of an incident, and then make some logical deductions (see Figure 7–1). This reasoning is gained through experience, learned facts, and intense training. However, some common sense should always be part of this process.

> ■ **Note**
> The spectra found in this book are produced by Smiths Detection, and are actual spectra from the HazMat ID™.

Infrared technology, and specifically the HazMat ID, is one technology that we can use at the scene of a hazmat incident. However, just as with all technologies, there are positive attributes and some limitations. The true strength of infrared technology is its ability to classify chemicals into families. More importantly, it is most useful for identifying covalently bonded chemicals. Energies

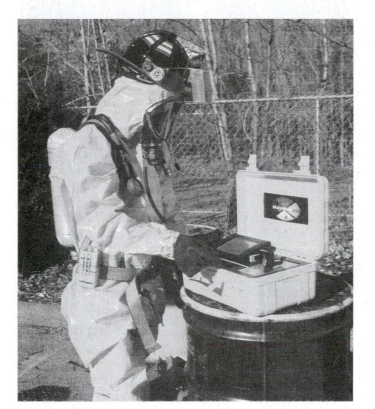

Figure 7–1 *Photo of HazMat ID™.*

that cause specific markers in a spectrum are at very specific wave numbers for certain chemical families. Even though the instrument has a vast library of chemical spectra, it is the technician's job to use the results, correlated to the incident, in order to make the final determination of possibilities. The instrument will function as it was designed; it is up to the technician to interpret the results. If a spectrum doesn't make sense, the sample should be run two or three times, each time making note of the wave numbers at which peaks and valleys occur. By correlating each run with the previous sample run, the general chemical group or family should eventually be evident. Each facet of the waveform has a specific meaning: the height of the wave is an indicator of the bonds' strength or the concentration of the chemical; a large number of peaks and valleys indicates potentially complex bonds and complex mixtures. If the wave form is consistent with ketones, for example, the combustible gas indicator identifies an LEL and flammable liquid containers are observed at the scene, then the possibility of a ketone is obviously high.

Advantages and Limitations

As with any instrument that is used in the emergency response arena, understanding the technology's capabilities and limitations is one of the key issues affecting application. This approach is as simple as understanding how and when a specific tool is used in construction. You wouldn't use a circular saw when you need a scroll saw, or pick up a hammer if you need a screwdriver.

Monitoring for and identification of an unknown in hazardous materials response is an art, with scientific analysis and some street logic thrown in. It involves a logical process of thinking through the problem and looking for possibilities, while at the same time synthesizing the facts that are currently presented. All equipment has procedures developed for it and techniques used to gain these facts must be scrutinized. The logical approach is to look at possible end states along with the facts presented.

The overall advantage is that instrumentation in general can lead the responder towards defined tactical objectives. The HazMat ID™ can quickly analyze covalently bonded molecules. Because covalently bonded (primarily organic) chemicals represent the majority of transported materials, the HazMat ID™ gives a quick method for group or family identification. These instruments come with a wide variety of chemical spectra in the computer library, which can be rapidly analyzed through search engines recognizing digitized reference spectra. However, with advantages come limitations. One limitation is the ability of the first responder to interpret the data the instrument is portraying. Currently, responders rely on the computer interface to give the complete interpretation and the logical tactical assumptions. This provides, as with so many issues in hazardous materials response, a false sense of security. Sometimes what is *not* found is more important than what *is* identified. Being able to look at infrared spectra and logically deduce from comparing this and other information that has been gathered provides the clues towards family identification, and in most cases scene mitigation.

As with all tools, **infrared spectroscopy (IR)** has its advantages and its limitations. It is the basic science of IR spectroscopy that gives the responder the ability to develop facts. With the advent of field-expedient equipment, this device gives the first responder a fast and effective analytical technique. The primary advantage of IR spectroscopy is that it can analyze a vast majority of the compounds that are found in emergency response situations. The reference spectra that come with the IR spectroscopy unit are related to the physical properties of the chemicals that they represent. The rapid assessment of the potential chemical occurs in the library of digitized spectra. This allows the first responder to analyze potential chemical families and compounds. The area of chemistry in which this technology is predominately useful is the identification of organic functional groups.

Within organic (covalent) chemical bonds a beam of electromagnetic energy that is passed through such a substance; the energy can be absorbed or transmitted. The energy that is absorbed is at a specific frequency and is unique for that compound and functional group. In other words, if a carbon chain is placed on an amine, or a halogen is attached, or is part of the ketone, ester, and so on, the area of carbon chain absorption will occur at the same frequency. This analysis can be thought of as looking at a small photograph of a larger object. Each frequency band (at a specific wave number that is correlated to a specific frequency) represents a photograph of a larger object. Consider looking at separate photographs of a truck, the truck's rearview mirror, and its front bumper. Separately, these photos do not tell the whole story; however, if the tire is a large tire, the rearview mirror is a truck-style mirror, and the bumper looks similar to a fire engine's bumper, and we are told that these photos are pieces of a larger object, we can make the assumption that the photos represent a fire engine. With IR spectroscopy, each wave and valley represents part of a chemical object. Each segment of a graph identifies parts of a larger picture. This is what makes IR spectroscopy such a powerful tool.

The limitations of IR spectroscopy are based in the science that certain bonds react to **electromagnetic radiation** in such a fashion that either the movement is notable or does not occur. It is the movement of the elements in the compound that causes **absorption**. Single atoms, the noble gases, and diatomic molecules (identical atoms in combination) do not produce an infrared spectrum.

infrared spectroscopy (IR)

the specific method by which an infrared light beam is directed into a molecule, enhancing the vibration of the elements, their bonds, and thus the electrons, to show very specific peaks at specific wave numbers; used for identification of functional groups

electromagnetic radiation

whenever atoms are exposed to energy such as light, they become excited as they take in the energy, and the electrons start to move into different orbital states

absorption

the process by which energy is transferred to a molecule

This is due in part to the symmetry of the molecule, which does not allow the molecule to create a positive or negative charge. Complex mixtures, solutions that contain water (aqueous solutions), and ionic salts do not produce comprehensible spectra. In these cases, laboratory techniques such as separation (chromography) are the key to achieving an understandable spectrum. Procedures using other technologies in combination with IR spectroscopy can give the responder an understanding of what is being dealt with. IR spectroscopy is useful in the field when the following two conditions are present:

- Pure substances (nonaqueous solutions)
- Organic compounds (functional groups present or covalent bonds)

IR spectroscopy is useful in the field because:

- Spectra can give important covalent compound information if the above two conditions are met.
- It is considered a universal laboratory technique—a spectrum in the lab will have the same characteristics as a spectrum taken in the field. This fact assists the first responder when the program is accessed.

It is the basic principles of chemistry and physics that make IR spectroscopy useful to the emergency responder when dealing with unknowns. Sometimes it is more beneficial to find what the substance is not than what it might be.

absorbance
measured, observable parameter, defined as the logarithm of transmittance and directly proportional to concentration

transmittance
observable measurement, defined as the ratio of transmitted radiation to the incident radiation

Beer-Lambert Law
an optical law of physics that states that absorbance and concentration are linearly proportional

dipoles
the ability of electron movement in a molecule to create a predominately positive and negative end in the molecule

frequency
the number of wavelengths or cycles that occur over a given period of time

---

■ Note

Spectra can be represented as **absorbance** or **transmittance**. The difference in these two but similar measurements has to do with the instruments' configuration, but more important is what we are trying to achieve. In transmittance, the frequency peak points down from the top of the graph. Transmittance is a percentage of frequency response and is not directly proportional to concentration. Absorbance, on the other hand, is directly proportional to concentration, and thus can be measured quantitatively. The frequency response is at the bottom of the graph and moves towards the top. This has to do with the **Beer-Lambert Law**, which states that absorbance of radiation and the concentration of molecular species are linearly proportional. So we can determine the concentration to some degree by measuring the amount of light that a sample absorbs.

Concentrations are described by chemists in moles per liter. When we shine light through the material, a percentage of light (photons) is absorbed by the material and a percentage makes its way to the detector. At low concentrations, more photons reach the detector. When the concentration is increased, the number of photons reaching the detector decreases. What this is telling us is that low-concentration materials absorb less light allowing a higher percentage of the light moving through the material to strike the detector. The higher the concentration, the less photons are detected once they pass through the material, and thus the more that are absorbed. Thus, the Beer-Lambert law gives us a relationship between the absorbance and concentration of a material.

ORBITAL THEORY

As mentioned in Chapter 1, all atoms strive towards their most stable state or configuration. In part this is true; however, atoms are constantly moving, and during this motion the molecules vibrate, sometimes creating a small charge on the molecule (a positive end and a negative end, or what is referred to as a **dipole** moment). Although chemically stable, a small vibration exists. When we shoot a beam of infrared light through a molecule, the atoms in the molecule tend to vibrate at a specific **frequency**, thus producing a level of absorption or transmission. Covalent bonding is the sharing of two or more pairs of electrons between atoms in such a way that it seems to each atom that the shared electrons belong to it. This is evident in resonance bonding, in which the electron or electron pair races between several atoms, satisfying all the atoms in the structure.

We previously described how electrons rotate around an atom in very simple terms. In actuality, electrons have **orbital**s and suborbitals. In general, electron configurations can and do overlap to form molecular orbitals. These orbitals have different shapes and sizes depending on the energy level they represent. Orbitals, which are actually spinning electrons in a cloud around the nucleus, occur as levels of energy. The shape of these orbitals and their placement in space respective to each other define the arrangement and the chemical behavior of the compound. More specifically, the reaction to infrared radiation can change the movement of these electrons in sections of the molecule and within each orbital.

The orbital names are based on the energy spectra of light. The first orbital level is called the *s* (sharp) level; the second is the *p* (principal) level; third *d* (diffuse) level; and the fourth *f* (fundamental) level. The *s* orbital is spherical in shape and is at the lowest energy level, which makes sense because it is closest to the nucleus. This level contains two electrons and is represented as the 1*s* orbital.

The second level is the second orbital, in which we again see an *s*-shaped orbital that contains two electrons, and a *p* orbital. The *p* orbital is dumbbell-shaped with sections in the *x*, *y*, and *z*-axis. Each suborbital, or axis, possesses two electrons, for a total of six electrons in the *p* orbital. Adding the two in the *s* orbital, there are a total of eight electrons in the second orbital.

The *d* orbital includes the characteristics of the first two orbitals, with the addition of a node that looks similar to a donut, making the physical representation of this orbital level very complex. As we move to the *f* orbital, another level of complexity arises.

When atoms are brought together to form a molecule, it is their atomic orbitals that are interacting with each other. In the formation of a covalent bond, the electrons in the orbitals of one atom interact with electrons in the orbitals of the approaching atom to accomplish the sharing of electrons between the atoms (see Figure 7–2). The basic idea is that a covalent bond is formed by the overlap of atomic orbitals. The two electrons that are spinning within this shared orbital are shared by the two bonded atoms. The greater the degree of overlap of these orbitals, the greater the degree of sharing, and the stronger the bond (see Figure 7–3).

orbital

the areas of high statistical probability in which an electron will be found; the energy level at which that electron spins

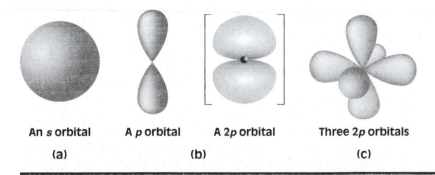

An *s* orbital   A *p* orbital   A 2*p* orbital   Three 2*p* orbitals

(a)   (b)   (c)

Figure 7–2   *Representations of s and p orbitals. (a) The s orbitals are spherical, and (b) the p orbitals are dumbbell-shaped. The lobes of p orbitals are often drawn for convenience as "teardrops," but their true shape is more like that of a doorknob, as indicated by the computer-generated representation. (c) The three p orbitals in a given shell are oriented along mutually perpendicular directions.*

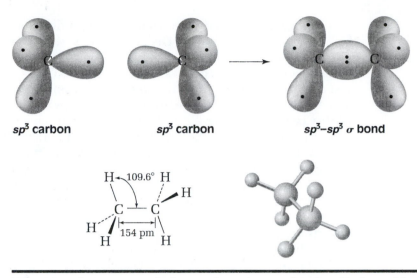

Figure 7–3   *The structure of ethane. The carbon–carbon bond is formed by the overlap of two carbon sp^3 hybrid orbitals. (For clarity, the smaller lobes of the hybrid orbitals are not shown.)*

These orbitals are in configurations that give rise to the geometry of the molecule. For example, there might be an overlap that is part *s* orbital and part *p* orbital. When this occurs we call it hybridization. It is the vibration that occurs within this hybridization that infrared spectroscopy takes advantage of.

Geometry and energy are hidden in these combination orbitals. For example, when we have one *s* orbital and one *p* orbital on a central atom, this gives rise to two *sp* orbitals. Hybridization as an *sp* orbital gives two areas of attachment, and the molecule is linear; that is, the orbitals are 180° apart. If the central atom is vibrated by IR light entering this molecule, then the central atom can move either right or left, and an overall charge difference (dipole moment) may or may not be produced. The hybridization of one *s* orbital and two *p* orbitals on a central atom gives rise to three sp^2 orbitals, and the molecule is planar, with 120° between each attachment. The hybridization of one *s* and all three *p* orbitals on a central atom gives rise to four identical sp^3 orbitals, which are aligned in a tetrahedral configuration and are 105° apart. This is the typical configuration we find in the methane molecule, for example.

The energy frequencies enhanced by IR beams are of diagnostic value. In a compound with sp^3 orbitals (alkanes, for example), the C–H absorption will always occur at frequencies less than 3000 cm^{-1} (3000–2840 cm^{-1}). C–H absorption for compounds with sp^2 orbitals as observed in the alkenes will occur at frequencies greater than 3000 cm^{-1}. The *sp* hybridization of C–H in terminal alkynes will show a predominant peak at about 3300 cm^{-1}. As the beam of IR energy is directed into the molecule, and more specifically into the orbital, vibration of the molecule is related to the absorption of light at that specific frequency. Each orbital configuration, and thus each family of covalently bonded chemicals, will vibrate at a very specific frequency or wave number, and this will occur each time a specific bond configuration has the IR beam directed at it. This is the greatest advantage of this technology. Orbital theory also gives us an opportunity to understand how some compounds are created; however, it is more important to understand the subtle differences that an instrument reveals.

There are many chemical compounds that do not adhere to the traditional shell theory (see Chapter 1). The shell theory supposes that the electrons fill each shell consecutively as described by the Bohr model: two electrons in the first shell and eight in each succeeding shell (see Chapter 1). This explains how most elements combine. However, there are several elemental combinations that cannot be explained by the traditional shell theory, including resonance and stability of the benzene ring, and multivalance of the transitional metals. In addition, some atoms have far more shells or bonds than the simple Bohr model suggests. Diatomic oxygen, as an example, has two magnetic states: one is dangerous, and the other is a major component of the air we breathe (paramagnetic versus diamagnetic). However, the configurations of these two states look very similar according to the Bohr model. Suborbital theory also has a bearing on the limitations of photoionization and flame ionization detectors. Suborbital theory explains the degrees of oxidation that occur in the oxidizer classification. It also explains bond angles, which describes how infrared spectra are identified.

Suborbital theory is mostly concerned with the magnetism that a molecule creates in space. Water molecules, for example, act similar to a magnet in that they have a positive end and a negative end. Oxygen has a larger electronegativity than hydrogen. The electron pairs from the hydrogen atoms are pulled closer towards the oxygen atom, leaving a partial positive charge on the hydrogen atoms and a partial negative charge on the oxygen atoms: a polar covalent bond. Each of these ends can attract other water molecules, thus giving rise to water's unique properties. These are called dipole moments, or more commonly, dipoles.

When infrared light energy is directed into a molecule, the atoms in the molecule start to vibrate. The movement of the electrons is still occurring in the molecule's orbitals, displacing the overall electrical charge, and thus causing this vibration. When this occurs, there is a more predominate negative end and positive end of the molecule. This has a direct bearing on the spectra of the molecule, and is the reason why some molecules produce an identifiable spectra and others do not.

Electron configurations in an atom lead to the atomic structure of unpaired electrons. This presence of unpaired electrons gives rise to the chemical reactivity process, bond angle, and bond strength. We can determine through magnetic properties the unpaired electrons that are available for bonding, reactions, and so on.

■ Note

When we started our discussion of atomic theory we used a very elementary way of thinking about how electrons travel around the nucleus of the atom. Now that we are looking at the electron configurations to study bonds and how they vibrate we must correct our ideas of electron movement.

If we could dissect each electron in an atom away from its shell, what we would see is that each electron exists within a cloud of probability. In other words, it is moving around the center of the atom, but the opportunity to catch the electron occurs only when we get close enough to where it may be found. This is further complicated by the fact that electrons like certain levels of energy and are placed within this system (the atom) dependent on the energy level. We call these levels suborbitals, and hence the designations 1s, 2s, 2p, 3s, and so on, which are defined as the quantum number for each electron.

The geometry of a molecule depends on how these electrons move into the energy levels, i.e., the cloud of electron probability, and the suborbital they occupy. It is this geometry, the energy within the molecule or the energy that can be placed into the molecule, that causes a reaction and the creation of different molecules and structures. Where the electron is found, or what suborbital the electron occupies, is specific for a certain atom. For example, in a lithium atom, three electrons are found in two energy levels: two electrons in 1s and one electron in 2s. When in combination with the chlorine molecule, which has 17 electrons, the outer suborbital has one space for the 2s electron from lithium to occupy. The overlapping of each atom's electron cloud is what gives rise to the type of orbital or bond. It is these molecular bonds that produce the oscillation or vibration in a molecule. Each molecular bond has energy; this energy is the vibration of the molecule. These bonds move, rotate, vibrate, bend, and shorten depending on the energy placed into the system. As with any dynamic system, as energy moves into a molecule, the electrons move, creating increased vibration, rotation, and transitional movement. It is the vibrational movement that we are interested in. We want to introduce exactly the right type of energy with the right magnitude into the molecule to cause vibrational movement. The infrared light that moves through the molecule, and specifically the electrical carrier in the IR beam, is exactly the right magnitude to move the bonds in such a way that we see frequency signatures in the spectra. Because of this, we can use IR as an analytical tool for family identification.

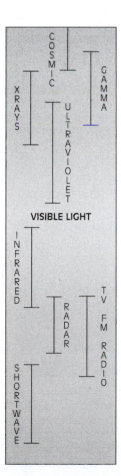

Infrared energy has an electrical carrier component. It is this carrier of energy that interacts with the molecule and creates the absorption of IR energy. This electrical carrier has both a negative and a positive nature—a polarity of a sort. Alternating cycles of positive and negative energy occur over time. When we have a molecule that has an atom with high electronegativity, electrons are pulled closer to this specific atom, leaving the entire structure polar; in other words, a partial negative and partial positive charge on the molecule, or dipole moment.

When the electrical carrier found in infrared energy passes through a molecule, the positive portion of the energy repels the partial positive charge on the atom that contains the positive charge, and the bond length shortens. As the energy influx continues, the carrier creates a negative field, and the portion of the atom that had the partial positive charge now becomes attached to the negative field, and the bond lengthens. This cycle of expansion and contraction continues over time, and enhanced vibration of the molecule occurs. This infrared absorption will occur at a very specific frequency, and is related to the energy of the light that is transferred to the molecule. This detection of absorption is the decrease in light intensity at a specific frequency of light, resulting in infrared spectra of the specific molecule.

Chemical bonds that are Infrared-active have a naturally occurring dipole moment, the light energy that is passed through the molecule enhances this naturally occurring moment. These dipoles are enhanced when exposed to Infrared: the positive energy carrier will pull the electrons of an atom that are found within the molecule "producing" a moment. When the negative portion of energy move through the electrons it is repelled and the positive moment is enhanced. In essence, the IR has augmented the dipole moment and vibration occurs. The bond length has changed from natural length to shorter, longer and back to its original position, or in essence vibrates. This vibration causes spectra that can be produced and identified at a specific frequency.

ELECTRON SPECTRA AND LIGHT

What we as humans see is but the very small subsection of electromagnetic radiation that we call visible light. Visible light is in the middle of the electromagnetic radiation spectrum; on one side of it is ultraviolet light, and on the other side is infrared light, which lies below the visible red spectrum and has lower energy.

The energies that are found in the infrared region of the spectrum are very low compared to a photoionization detector, which emits a light source that will ionize the chemical in question. The energies are so low that photon delivery

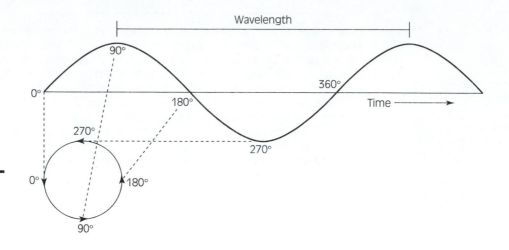

Figure 7–4 *Visual description of wavelength and frequency.*

into the molecule will stretch or shorten a bond but will not release electrons. Bonds are not a rigid structure, but a fluid dynamic interaction between the concentrations of electrons in the electron clouds of one or more atoms. These bonds are sometimes influenced by magnetic charges.

The reason we are able to use the electrical field of a compound's electrons to view that compound can be explained by some basic physics. Let's start by rotating an object around in a circle. We can translate this uniform motion of the object as a linear component of time on a graph (see Figure 7–4). The circle translates to a sine wave with time plotted on the horizontal axis. As we start at zero and move through the circle 90°, we have moved from zero to the crest of the first wave. As we translate through the circle another 90° (for a total of 180°), we return to zero. Another 90° (270°), and we are at the trough of the second wave. At 360°, we return to zero.

All waves, whatever the medium, have some very distinct interdependent properties. First is the **wavelength**, which is usually measured in meters, and is the distance between two identical points on two adjunct waves. In the previous example it would be the distance from the top of the first wave (90°) to the top of the next wave (450° from origin). By counting these wavelengths, or cycles, that occur over a given period of time, the frequency is determined. The frequency is defined as the number of wave cycles that occur per unit of time (usually measured in cycles or waves per second, or Hertz). This relationship can be written as follows:

$$\text{Speed} = \text{Wavelength} \times \text{Frequency}$$

The speed of light is a fundamental constant of approx 3×10^8 m/s. Therefore frequency of 1 Hz has a wavelength of 3×10^8 meters. If the frequency is one cycle per minute, or 1/60 cycles per second, then the wavelength is 1.8×10^{10} meters. The wave number, which is used in the context of IR spectroscopy, is the reciprocal of the wavelength (1/wavelength), and is given in centimeters^{-1} (cm^{-1}). Hence the numbers found at the bottom of a spectroscopy graph. Usually the graph ranges from 4000 cm^{-1} on the left to 400 cm^{-1} on the right, which represents only a sliver of the mid-infrared energy placed into a molecule.

Infrared light energy actually consists of oscillating electric and magnetic fields perpendicular to each other (see Figure 7–5). It is the electric field, or carrier, that is used in IR spectroscopy. As the light wave is propagated through a

wavelength

typically the distance between two identical points on two adjunct waves, measured in meters

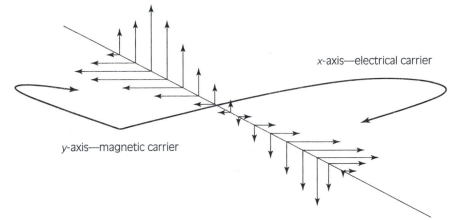

Figure 7–5 *As light moves forward, two carriers are produced: one electrical and the other magnetic. It is the electrical carrier that influences the molecular vibration.*

molecule, the field that is generated creates positive and negative fields. As the wave moves through time, this positive-to-negative oscillating field pushes and pulls the electrons in the molecule. The result is an enhanced vibration of the components of the molecule. Each component in the molecule vibrates at a very specific frequency, producing a set of photographs that are specific to a certain chemical family type. These frequency relationships, when considered in total, can provide a snapshot of the components of the molecule.

Condition of Absorption

Molecules absorb light at very specific frequencies. These frequencies represent a defined pattern, the molecular structure of the molecule. These frequencies match the normally occurring frequencies at which the molecule vibrates. The introduction of light or an electromagnetic field enhances this vibration such that the instrument can evaluate it. In a complex molecule, many vibrations occur simultaneously, but only the most intense are represented, due to polarity, strength of polarity, masking from other more intense vibrations, and so on. The primary condition that must be present for the absorption to take place is a change in the naturally occurring dipole moment when exposed to an electric field.

When an electromagnetic field passes through a molecule, the elements in the molecule move along its orbital axis. An already-existing dipole is enhanced. The element is excited and has a more dramatic positive or negative side, with enhanced vibration occurring. Figure 7–6 shows various types of molecular movement.

The masses of the elements in the molecule have an influence on this vibration. Elements that have more mass tend to enhance this vibration, and thus there is a stronger force for the instrument to sense. This mass relationship is observed when we have a carbon–hydrogen bond. The carbon has more mass than the hydrogen, and thus a strong vibration occurs between the two atoms because of the difference in mass.

Vibrations sometimes will complement each other and change the observed frequencies. This is observed in the aldehydes, which give rise to a splitting of the frequency band either above or below the area that is expected. Here, fundamental and overtone or combination vibrations occur with similar molecular symmetry and frequency, and each vibration interacts with the other. This is referred to as Fermi resonance.

Stretching

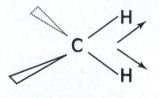

Symmetric stretch—Each
hydrogen atom is moving away
from the carbon atom. Occurs at
about 2850 cm^{-1}

Asymmetric stretch—One hydrogen atom
is moving away from the carbon atom, while
the other hydrogen atom is moving toward
the carbon atom. Occurs at about 2920 cm^{-1}

Bending

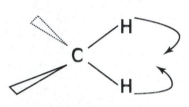

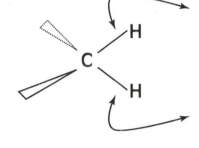

Scissoring—Both hydrogen atoms
move up and down as if they are the
blades on a scissors. Occurs at about
1450 cm^{-1}

Wagging—The hydrogen atom
moves back and forth an equal
distance out of the plane. Occurs
at about 1250 cm^{-1}

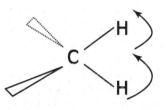

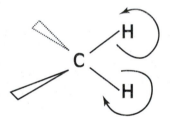

Figure 7–6   *Types
of molecular
movement.*

Rocking—Both hydrogen atoms move
up or down, swaying within the
plane. Occurs at about 720 cm^{-1}

Twisting—Two hydrogen atoms move
away from each other out of the plane
in opposite directions. Occurs at about
1250 cm^{-1}

INTRODUCTION TO SPECTRA: SYSTEMATIC APPROACH

A detailed approach to interpretation must take place with all responses. Infrared
spectroscopy application is no different. The steps should include:

1. Ensure that baseline corrections have been made with smoothing if necessary.
2. Look at the information in total.
   A. Other chemical analysis (PID, FID, acid/base, electrochemical, etc.)
   B. Placards, labels, container shape and size
   C. DOT and CAS numbers
3. Read the spectrum from left to right, looking at each region to get a general
   feel for the spectra.
   A. Read Region I.
      1. Alkanes—the main absorption for the C–H stretch at 3000 cm^{-1} and
         another band at 1450 cm^{-1}

Table 7–1   *Spectra regions.*

	Region I		Region II	Region III	Region IV	Region V
Absorbance	Hydrogen stretching C–H, OH, and N–H		Triple bonds C≡C C≡N	Double bonds C=C Carbonyl C=O C=N	Fingerprint region	Aromatic rings Alkyl halides

3700   2500   2300   2000   1600   1000   400

Wave Number (cm^{-1})

Converting wavelength (λ) to wave number (w)

$W = 1/\lambda$

If $\lambda = 2.5\ \mu m$ $W = ?\ cm^{-1}$
$\lambda = 2.5\ \mu m \times (1\ m/10^6\ \mu m) = 2.5 \times 10^{-6}\ m$
$2.5 \times 10^{-6}\ m\ (10^2\ cm/1\ m) = 2.5 \times 10^{-4}\ cm$
or $w = 1/\lambda = 1/2.5 \times 10^{-4}\ cm = 4000\ cm^{-1}$
or $\nu\ (cm^{-1}) = 10^4/1\ (mm)$

Where ν (pronounced "new") is the common spectroscopic symbol for wavenumber. And λ (lambda) is the micrometers. It's always important when discussion spectroscopy terms to clarify the units (cm^{-1} or Hz).

Table 7–2   *Step 3A identifies.*

Bond Type	Characteristic	IR Range (cm^{-1})	
C–H	C^3_{sp}–H	2800–3000	← Alkenes
	C^2_{sp}–H	3000–3100	← Alkenes (see Region III)
	C_{sp}–H	3300	← Alkynes (see Region II)
N–H	RNH_2, R_2NH	3400–3500 (two)	} Amines (see Region IV)
	RNH_3^+, $R_2NH_2^+$, R_3NH^+	2250–3000	
O–H	ROH	3610–3640	} Alcohols and phenols (see Region IV)
		3200–3400 (H bonds)	
S–H	S–H (thiols)	2550–2600	

Table 7–3   *Step 3B identifies.*

Bond Type	Characteristic	IR Range (cm^{-1})	
C–C	C≡C	2100–2260	← Alkynes (see Region I)
C–N	C≡N	2210–2260	
	RCN	2280–2200	← Nitriles
Phosphine	P–H	2280–2440	← Phosphine (see Region V)

2. Alkenes—weak absorption for the double bond near 1650 cm^{-1} and a CH stretch near 3000 cm^{-1}

3. Alkynes—absorption at 3300 cm^{-1} with a band in Region II

4. Amines—absorption between 2250 cm^{-1} and 3500 cm^{-1}, and a C–H stretch at 3000 cm^{-1} due to R groups

5. Alcohols and phenols: OH band—if alcohols are present, look at Region III for a carbonyl band (organic acid, in particular)

B. Read Region II.

    1. Triple bonds—alkynes (see Regions I and V)

    2. Nitriles—isocyanates, isothiocyanates, and azides

C. Read Region III.

    1. Alkenes—weak absorption for the double bond near 1650 cm^{-1}, and a C-H stretch near 3000 cm^{-1} (see Region I)

Table 7–4 *Step 3C identifies.*

Bond Type	Characteristic	IR Range (cm^{-1})	
C–C	C=C	1600–1670	← Alkenes
C–N	C=N	1640–1690	
C–O	C=O	1650–1800	← Carbonyl group: aldehydes,
Aldehydes	CHO	1720–1740	ketones, carboxylic acids, esters
	ArCHO	1690–1700	
Ketones	R$_2$C=O	1710–1720	
	ArC=O	1690	
Carboxylic acids	RCOOH	1760	
Esters	RCOOR	1735–1750	
Nitro	N=O	1530–1550	

Table 7–5 *Step 3D identifies.*

Bond Type	Characteristic	IR Range (cm^{-1})
Alkyl halides	C–F	1000–1350
	C–Cl	800–850
	C–Br	500–680
	C–I	400–500
Alkynes	C≡C	600–700
Alkenes	RCH=CH$_2$	910, 990
	R$_2$C=CH$_2$	890
Trans	RCH=CHR	970
Cis	RCH=CHR	725, 675
	R$_2$C=CHR	790–840
Aromatic		730–770, 690–710
	Ortho	735–770
	Meta	750–810, 690–710
	Para	810–840
	1,2,3	760–780, 705–745
	1,3,5	810–865, 675–730
	1,2,4	805–825, 870, 885
	1,2,3,4	800–810
	1,2,4,5	855–870
	1,2,3,5	840–850
Ester	ArCOOR	720
Phosphoric esters	P–OR	900–1050
Phosphine	P–H	950–1250
Sulfur ester	S–OR	700–900

      2. Carbonyl group—depending on the R group, several other regions may be affected

      3. Nitro group

    D. Read Region V.

      1. Alkyl halides

      2. Alkynes

      3. Alkenes

      4. Aromatic rings

      5. Esters

      6. Phosphine

    E. Read the fingerprint region. In this region there will be very specific wave forms that are indicative of certain groups.

  4. Write down the possible functional groups observed within the spectrum.

  5. Correlate the spectrum with Step 2.

  6. List all possibilities.

    A. Do the spectra correlate to the observations of the material?

    B. Does this information correlate to information from other instruments?

HYDROCARBONS

So far we have identified significant areas of the IR spectra that are specific to certain types of molecules. These areas are only "snapshots" of the molecule; each snapshot needs to be interpreted. It would not be feasible to have a molecular spectrum assigned to every molecule; rather, it is the combination of bands that gives us a telltale look at specific groups or families of chemicals.

The first set of chemical families we will consider are the straight-chained hydrocarbons, which are the backbone for the functional groups. Hydrocarbons make up a vast quantity of chemicals used in society (see Chapter 3). Everything from automobile fuel to lubricants is derived from crude oil or is synthetically manufactured oil.

Alkanes

methylene group
a CH_2 group found between two methyl groups

methyl group
a CH_3 radical, or the terminal ends of a carbon chain

The alkanes, the first group in the aliphatic hydrocarbon family, consist of basic units of saturated carbons, with carbon–carbon and carbon–hydrogen single bonds. Methyl and **methylene groups** of atoms are found in these chains, or $-CH_3$ and $-CH_2$, respectively (methylene should not be confused with the second group of saturated hydrocarbons—alkenes).

Each constituent of the molecule of these components of the alkane family can vibrate. These vibrations occur both symmetrically and asymmetrically. If we look at the **methyl group**, either the hydrogen atoms can stretch away from the carbon atom (symmetric), or one or two hydrogen atoms can move towards the carbon atom while the other one or two hydrogen atoms moves away from the carbon atom (asymmetric). With the methylene component ($-CH_2$), we see movement of the hydrogen atoms away from the carbon atom (symmetric), and one hydrogen atom moving towards the carbon atom while the other hydrogen atom moves away (asymmetric). Each of these vibrations occurs at a specific area on the spectra, very close to each other, and spaced between 3000 cm^{-1} and 2800 cm^{-1}.

Not only can there be stretching and shortening of each bond between the carbon atom and the hydrogen atoms, but there can also be bending vibrations—

predominately with the methyl group (at each end of the chain). When this occurs, the hydrogen atoms push or move slightly in either a counterclockwise or clockwise direction (symmetric bend), or one hydrogen atom moves clockwise while two hydrogen atoms move in a counterclockwise direction (asymmetric bend). Here we have a symmetric bend (all the hydrogens moving in one rotational direction of bond) and an asymmetric bend (one hydrogen moving one way while the other hydrogens are rotating in a bend the opposite direction). The methylene groups also have a bend; however, in this case it is either a scissors vibration, with both hydrogen atoms moving towards one another, or a **rocking** vibration, with each hydrogen atom moving in either a left or right direction. These bends or rotations, in relation to each methylene group, can rotate in phase with each other (where each methylene group rotates around the axis of the molecule in one direction), or out of phase (where each methylene group moves in a direction opposite of the other methylene group). The symmetric and asymmetric bends and the scissors vibration occur roughly around $1350–1470$ cm^{-1}, and the rocking vibration occurs at 720 cm^{-1}. The phase rotations occur around $720–750$ cm^{-1}, and it is very hard to distinguish between the rocking and phase motions. Table 7–6 summarizes the types of vibration that occur in the alkanes.

rocking

when both hydrogen atoms are moving up or down, swaying within the plane; occurs at about 720 cm^{-1}

Table 7–6   *Alkane vibrations.*

Vibration	Occurrence in Spectra
Symmetric and asymmetric stretch	3000 cm^{-1} and 2800 cm^{-1}
Rotational symmetric and asymmetric bend and scissors	$1350–1470$ cm^{-1}
Rocking and phase rotational bend	720 cm^{-1}

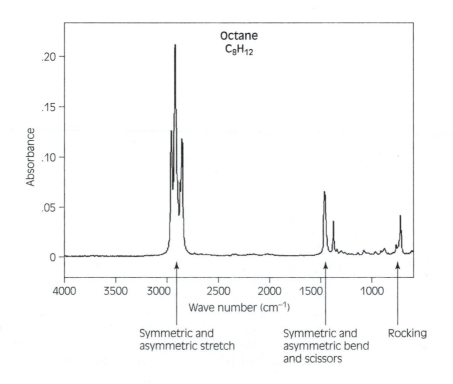

Figure 7–7   *Octane* C_8H_{12}

Alkenes

The next group in the hydrocarbon family is the alkenes. These are unsaturated, which means that every possible bond with hydrogen is not accommodated. The sites that hydrogen would normally occupy are taken by the double bond, making this group an unsaturated hydrocarbon with a carbon–carbon double bond.

As with the alkanes, the straight chain portion of the molecule has stretching bands, but due to the double bond, these stretches are translated towards the left portion of the spectra, all falling above 3000 cm^{-1}. The saturated portions of the molecule fall below 3000 cm^{-1}.

However, with the alkenes we can have stationary vibrations surrounding the double bond. Additionally, portions of the double bond can influence the carbon–carbon and carbon–hydrogen bonds on the saturated portion of the molecule. Surrounding the double bond are attachments that can occur either on the same side of the double bond (cis isomer) or opposite from each other (trans isomer) (see page 102). It is this area of the alkene molecule that makes the wave numbers significant. Whereas vibrations generally occur in the range of 1630–1680 cm^{-1}, this is a highly variable range, with very weak band above and below this range due to the substituted chains on the molecule. With the movement of the

Table 7–7 *Alkene vibrations.*

Vibration	Occurrence in Spectra
Saturated stretches	Below 3000 cm^{-1}
Unsaturated stretches	Above 3000 cm^{-1}
C=C double bond	1630–1680 cm^{-1}
Carbon–hydrogen bending	720–1000 cm^{-1}

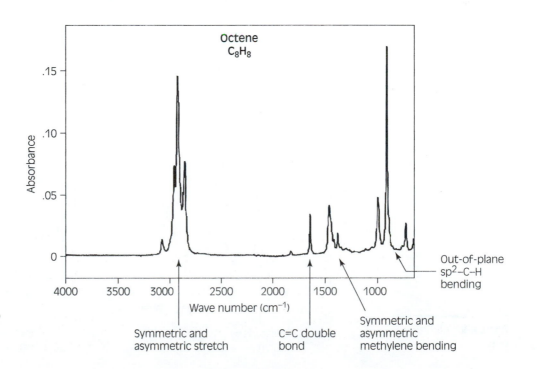

Figure 7–8 *Octene* C$_8$H$_8$

C=C bond being very small or nonexistent (with the dipole moment at zero), we are looking at the area surrounding the double bond—in other words, the R groups that are attached to the carbon atom at the double bond, in relation to the double bond itself.

The out-of-plane carbon–hydrogen bends, along with correlation to the carbon double bond and stretching, provide clues for identifying the alkene group of hydrocarbons. More specifically we are looking for the bends surrounding the double bonds, or the next carbon atom to either side of the double bond and their movement. These occur in the range of 720–1000 cm^{-1}.

Alkynes

In the alkynes group, the hydrocarbon chain has a triple bond, and the molecules are considered unsaturated. Alkynes are rarely seen in hazmat response, but are frequently used in industrial process and laboratories. They are sometimes called substituted acetylenes. There are three general areas to look for with alkynes: the carbon–hydrogen stretch, the carbon triple bond, and the carbon–hydrogen wag. The wag is actually a bending motion, but is significant for the triple-bonded carbon–hydrogen bond, occurring in the range of 600–700 cm^{-1}.

The carbon–hydrogen stretch is normally found in the range of 3350–3250 cm^{-1}. The carbon–carbon triple bond also has a dipole moment of zero, so again we are looking at the atoms surrounding the carbon on either side of the

Table 7–8   *Alkynes vibrations.*

Vibration	Occurrence in Spectra
Carbon–hydrogen wag	600–700 cm^{-1}
Unsaturated region	3350–3250 cm^{-1}
Carbon triple bond	2200–2100 cm^{-1}

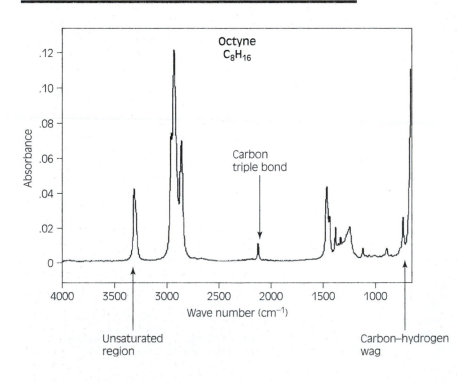

Figure 7–9 *Octyne* C_8H_{16}

triple-bonded carbon. The stretches for these atomic bonds are in the 2200–2100 cm^{-1} range.

Aromatic Hydrocarbons

The aromatic hydrocarbons are a ring of carbon atoms with the electron cloud appearing above and below the carbon plane. This provides a strong resonance bond associated with a flat plane of carbons. We call this plane of electron activity delocalized, because the clouds of electrons are satisfying all the carbon atoms in the structure and are not associated with any one or two carbons in the structure. It is this movement of electrons that gives the benzene ring its chemical and physical properties and unique IR spectra.

The benzene ring creates different chemical compounds with certain functional groups. The arrangement of these additional groups leads to a large variety of compounds. It is not our intent to provide an explanation for all variables, due to the complexity of this area of interpretation, but rather to identify the benzene ring and the functional group.

Benzene molecules consist of twelve atoms: six carbon and six hydrogen. This arrangement creates a liberal quantity of possible spectra, especially when attachments to the ring are considered. In this case, the symmetry of the molecule changes, and we can identify the presence of the ring due to some predominate bands in the spectra. In specific cases, the symmetry leads towards more symmetry, and bands that would normally be present are missing. It is this change in the dipole moment that causes the identification of benzene and its analogs to be very confusing. When we look for the benzene ring, we are looking for its presence in the unknown sample, and associate these IR spectra with what we have identified at the scene as known samples.

In general, the benzene ring produces a spectrum that has three bands above or around 3000 cm^{-1}, with an area of weaker bands in the 2000 cm^{-1}–1700 cm^{-1} range. These two areas are due to the level of hydrocarbon saturation (the number of hydrogen atoms attached to the carbon ring), and the number of possible electron configurations due to the electron clouds. From 1620–1000 cm^{-1}, we see the stretching and contracting of the carbon–carbon bonds, or the "ring breathing" band (1478 cm^{-1}). These spectra bands should be used only to distinguish between a straight-chained hydrocarbon and the presence of the benzene ring. From 1000–700 cm^{-1}, the hydrogen–carbon bonds vibrate outside the plane of the ring. These depend on the attached groups and the dipole moments that are created.

Table 7–9   *Aromatic hydrocarbon vibrations.*

Vibration	Occurrence in Spectra
Carbon–hydrogen stretches	3100–3000 cm^{-1}
Breathing ring	1604 and 1495 cm^{-1}
Hydrogen–carbon out-of-plane bend and overall ring bend	728 and 694 cm^{-1}

Alkyl Halides

The simplest hydrocarbon derivative involving the attachment of the halogens creates a very polar bond, which is created because the halogens are the most electronegative elements found on the periodic table. It is this electronegative quality

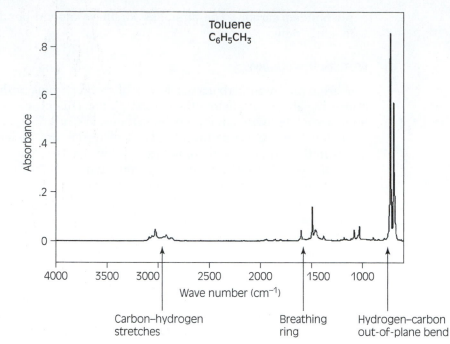

Figure 7–10   *Toluene*
$C_6H_5CH_3$

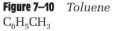

Table 7–10   *Alkyl halide vibrations.*

Vibration	Occurrence in Spectra
Carbon–fluorine stretch	1300–1000 cm^{-1}
Carbon–chlorine stretch	800–600 cm^{-1}
Carbon–bromine stretch	650–600 cm^{-1}
Carbon–iodine stretch	570–500 cm^{-1}

that creates a strong dipole moment, and the stretches are very intense. These moments have an intense band between 1300 and 500 cm^{-1}, and are dependent on the halogen and the quantity of halogens on the carbon–carbon chain.

A problem arises when we try to determine signature patterns in the spectra that will lead us toward a family or group of families. The patterns that are most predominate for the alkyl halides fall in the same range as some of the bands found in the benzene ring. Intensities of some carbon–hydrogen stretches occur in this range, leaving a poor set of wave numbers to distinguish between the halides and other potential compounds.

Amines

The functional group known as the amines contains the base derivatives of ammonia, and contain nitrogen and one or more hydrocarbon chains. We have named the different varieties of amines, but have not yet identified their actual differences.

Primary amines have one carbon chain attached to the nitrogen atom. Secondary amines have two carbon chains, and tertiary amines have three chains attached. The reason we have to identify this simple fact is due to the polarity

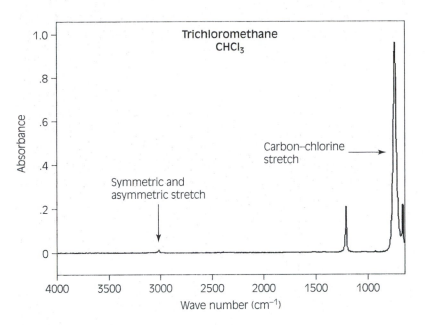

Figure 7–11

Trichloromethane
CHCl₃

these compounds have. Hydrogen bonding occurs between the amine molecules just as with hydrogen bonding in water (see page 152).

There is a similarity between the amine molecule and the water molecule, and thus the same type of stretching that has previously been mentioned. In the primary amine, a carbon chain and two hydrogen atoms are attached to the central nitrogen. Looking just at the nitrogen and the two hydrogen atoms attached, there is the possibility of stretching and scissors types of movement. The two hydrogen atoms can move to the left or right (**asymmetric stretch**), each hydrogen atom can move away from the nitrogen atom (**symmetric stretch**), or both nitrogen atoms can move towards one another (scissors movement).

The movement of the amine in this case can be in the plane or out of the plane. The in-plane movement is the **scissoring** effect, with the hydrogen atoms' bend towards one another occurring within $1650–1580$ cm^{-1}. The out-of-plane movement is a **wagging** effect above and below the carbon–nitrogen plane, and occurs between 850 and 750 cm^{-1}. The primary amine has only one carbon–nitrogen bond, and thus has a stretching vibration in the range of $1250–1020$ cm^{-1}. As with the halides, the carbon–nitrogen bond falls in the same range as the carbon–oxygen bonds, masking out the much weaker carbon–nitrogen bond.

Secondary amines have two carbon chains attached to the nitrogen atom. As a consequence there is only one nitrogen–hydrogen bond, and there is only one stretching band that occurs around 3400 cm^{-1}, with an out-of-plane wag in

asymmetric stretch

one hydrogen atom moving away from the carbon atom, while the other hydrogen is moving toward the carbon atom; occurs at about 2920 cm^{-1}

symmetric stretch

each hydrogen atom moving away from the carbon atom; occurs at about 2850 cm^{-1}

scissoring

when both hydrogen atoms move up and down as if they are the blades on a scissors; occurs at about 1450 cm^{-1}

wagging

when the hydrogen atom moves back and forth an equal distance out of plane; occurs at about 1250 cm^{-1}

Table 7–11   *Amines vibrations.*

Vibration	Occurrence in Spectra
Asymmetric stretch	$3380–3350$ cm^{-1}
Symmetric stretch	$3310–3280$ cm^{-1}
Scissors motion	$1650–1580$ cm^{-1}
Carbon–nitrogen stretch	$1250–1020$ cm^{-1}
Out-of-plane nitrogen/hydrogen	$850–750$ cm^{-1}

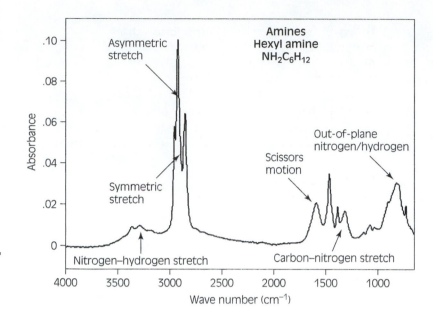

Figure 7–12 *Amines: Hexyl amine* $NH_2C_6H_{12}$

Table 7–12 *Secondary amine vibrations.*

Vibration	Occurrence in Spectra
Nitrogen–hydrogen stretch	3320–3280 cm^{-1}
Asymmetric carbon and nitrogen carbon stretch	1180–1130 cm^{-1}
Nitrogen–hydrogen wag	750–700 cm^{-1}

the range of 750–700 cm^{-1}. (The out-of-plane wag occurs between 850–750 cm^{-1} with the primary amine, providing a useful means of identification.) An asymmetric stretching occurs due to the two carbon–nitrogen bonds, and is identified by a band in the range of 1180–1130 cm^{-1}.

Tertiary amines have three carbon chains attached to the central nitrogen. This means that there are no nitrogen–hydrogen bonds, and thus no bands that would identify bending or stretching. Carbon–nitrogen stretches occur in the range of 1250–1020 cm^{-1}, and also in a range in which other functional groups vibrate. Tertiary amines are very difficult to distinguish using IR spectra alone.

NITROGEN GROUP

The nitrogen group is a family of organic compounds that are energetic materials. These compounds have a very distinct structure, and thus a specific electron configuration. Molecules in the nitrogen group consist of a nitrogen atom bonded to two oxygen atoms. If we add up the valences of each and try to accommodate the electron configuration, we see that it does not work. However if we look at this structure from the standpoint of the oxygen that is attached to the nitrogen, we see a resonance structure in which the two oxygens and the nitrogen have electrons that are spread out across both the nitrogen and oxygens, satisfying all the electron orbital configurations. However, this bond is weak and is subject to initiation (see page 147), which is why we call these compounds energetic materials. Given the appropriate energy (primary detonation, for example a

Table 7–13 *Nitrogen group's vibrations.*

Vibration	Occurrence in Spectra
Asymmetric stretch	1390–1330 cm^{-1}
Symmetric stretch	1500 cm^{-1}
Nitrogen/oxygen scissors	890–835 cm^{-1}

Table 7–14 *Ethers vibrations.*

Vibration	Occurrence in Spectra
Asymmetric stretch	1300–1000 cm^{-1}
Symmetric stretch	900–800 cm^{-1}

blasting cap or fuse), the cascade of moving bonds (electrons) creates a sudden release of heat, energy, and light—an explosion.

Because of these two bonds that are in resonance between the oxygen atoms and the nitrogen atom, there is both an asymmetric stretch and a symmetric stretch in the molecule. The asymmetric stretch occurs in the range of 1390–1330 cm^{-1}, and the symmetric stretch occurs at 1500 cm^{-1}. Both of these bands are fairly intense. There is also an associated bending scissors action that occurs in the range of 890–835 cm^{-1} (which can be masked by carbon–hydrogen vibrations in benzene). If the compound is an alkane or alkene, the intense bands are readily identifiable; however, if the compound has an aromatic ring (trinitro-toluene), the bands are masked and thus almost impossible to identify.

Ethers

The ether family is identified by a central oxygen atom that is attached to two carbon chains, which gives rise to an intense stretching band that can be either

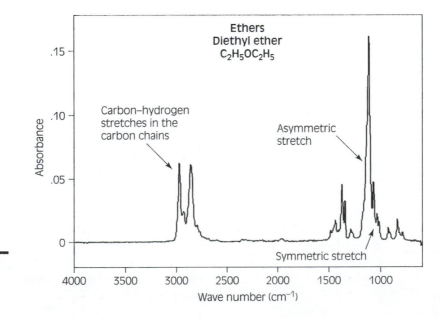

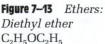

Figure 7–13 *Ethers: Diethyl ether* $C_2H_5OC_2H_5$

asymmetric or symmetric. The asymmetric stretching occurs intensely in the range of 1300–1000 cm^{-1}; a less-intense symmetric band is found between 900 and 800 cm^{-1}.

Alcohols

There is a stretching that occurs in alcohols between 1300 and 1100 cm^{-1}. This is due to the chemical structure of an alcohol, which like an ether, also contains a carbon–oxygen bond. The hydroxyl group attached to the carbon chain identifies a compound as an alcohol. As with the amines and water, we also see hydrogen bonding. This is due to the highly electronegative oxygen atom attached between the hydrogen atom and the carbon chain. This produces a strong dipole moment on the oxygen and the hydrogen (negative and positive, respectively).

Alcohols can produce these hydrogen bonds whenever the environment is conducive (for example, in water). Because of this and the ability of alcohols to form these hydrogen bonds in pure substances, the bands that are associated with this family are quite broad and stand out as the largest bands on the spectra (3342–667 cm^{-1}). The reason for this very large spectral range is due to the concentration of the alcohol. In solutions that are dilute and in which hydrogen bonding is less frequent, the band, width, intensity, and position of the spectra are affected.

Alcohols have characteristic absorption associated with the oxygen–hydrogen and carbon–oxygen stretches. The oxygen–hydrogen stretch that is found around 3350 cm^{-1} is the most useful diagnostically. When bending of the

Table 7–15   *Alcohol vibrations.*

Vibration	Occurrence in Spectra
Carbon–oxygen stretch	1260–1000 cm^{-1}
Oxygen–hydrogen stretch	3350 cm^{-1}
Oxygen–hydrogen bend	1350 and 650 cm^{-1}

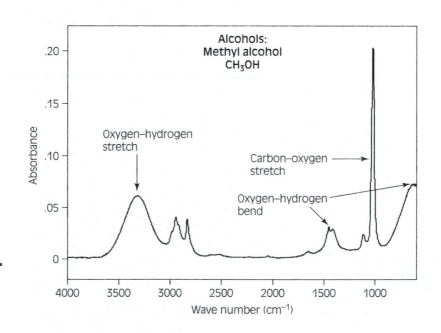

Figure 7–14
Alcohols: Methyl alcohol CH$_3$OH

Table 7–16 *Phenols vibrations.*

Vibration	Occurrence in Spectra
Carbon–oxygen stretch	1260–1200 cm^{-1}
Oxygen–hydrogen stretch	3345 cm^{-1}
Oxygen–hydrogen bend	1350 and 650 cm^{-1}

oxygen–hydrogen bond occurs in plane (1359 cm^{-1}) and out of plane (650 cm^{-1}) in association with the oxygen–hydrogen stretch, the substance is clearly identified as an alcohol.

Phenols are a subgroup of the alcohols in which the hydroxyl group is attached to a benzene ring. This gives rise to a multitude of variables, as with the aromatics. Although the benzene ring can give rise to masking elements, with the phenols OH attached to benzene, the vibrations that occur around the O–H group are not influenced by the surrounding carbon–hydrogen bonds. The distinct bands for phenols is an OH stretch at 3345 cm^{-1}, in-plane bends at 1367 cm^{-1}, and a carbon–oxygen stretch between 1260 and 1200 cm^{-1}.

CARBONYL GROUP

The carbonyl group includes aldehydes, ketones, esters, and carboxylic acids. All have a carbon atom double bonded to oxygen. This specific organization of atoms gives a partial positive and a partial negative charge. This is due to the electronegativity that oxygen has with respect to the attached carbon atom. Normally this group has a large dipole moment, which is strongly influenced by infrared light. The dipole moment is very intense due to the vibration of the carbon and oxygen atoms moving at the same time. This occurs in the range of 1800 and 1600 cm^{-1}.

Aldehydes

Aldehydes are a carbonyl group with one carbon chain attached, thus leaving a hydrogen atom attached to the carbonyl group. This hydrogen atom attached to the carbonyl carbon atom poduces a stretching vibration that occurs between 2850 and 2700 cm^{-1}, which is the most important band for the aldehydes. The predominate factor here is the appearance of the carbon–hydrogen stretch at a lower point in the spectra compared to the carbon–hydrogen stretch that occurs with straight-chained hydrocarbons. A phenomenon known as Fermi resonance also occurs with aldehydes. This is due to vibrations that interact with each other, thus producing peaks of wave numbers above and below the point they would normally be expected.

Table 7–17 *Aldehyde vibrations.*

Vibration	Occurrence in Spectra
Carbon–hydrogen stretch	2860–2800 and 2760–2700 cm^{-1} (one or two bands)
C=O stretch	1740–1725 cm^{-1}
Carbon–hydrogen bend	1390 cm^{-1}

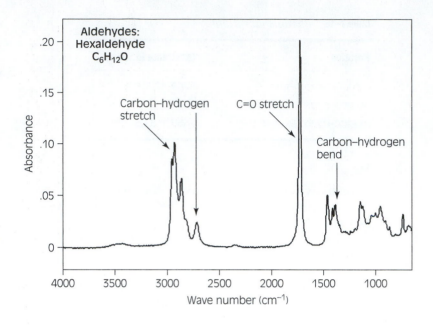

Figure 7–15
Aldehydes:
Hexaldehyde
$C_6H_{12}O$

Table 7–18   *Ketone vibrations.*

Vibration	Occurrence in Spectra
Carbon–oxygen stretch (C=O)	1715 cm^{-1}
Carbon–carbon stretch	1230–1100 cm^{-1}
Dissimilar–carbon chain stretch	1300–1230 cm^{-1}

Ketones

In the ketone group, the carbonyl carbon atom (double-bonded to an oxygen atom) is attached to two carbon chains. The predominate feature of ketones is that the two carbon chains attached to the carbonyl group can be either the same or different. Because of this variety we see carbon–carbon stretches when the two R groups are different, as well as the carbon–oxygen stretch.

Esters

A carbon–oxygen double bond characterizes the esters. These are compounds that have some degree of polarity, and are basically derivatives of the carboxylic acids. All of the esters have three bands that occur at 1700, 1200, and 1100 cm^{-1}. The 1700 cm^{-1} peak is due to the carbon–oxygen double bond, which occurs at this frequency whenever this component occurs in a molecule. The 1200 cm^{-1} peak reflects the stretching of the carbon–carbon bonds and the oxygen atom that is attached to the double-bonded carbon. The 1100 cm^{-1} peak occurs because of the carbons that are attached to the oxygen atom off of the carbonyl carbon atom. This latter peak is predominate when two or more carbon atoms exist on this side of the molecule. One key observation to make when looking at the ester spectrum is to rule out the presence of an alcohol.

The carbon–oxygen stretch in the saturated ester will occur slightly higher than in the ketones or aldehydes. Frequency peaks around 1735 cm^{-1} should be investigated further for the presence of an ester. The carbon–carbon–oxygen bond is very strong, and sometimes can look more intense than the C=O band.

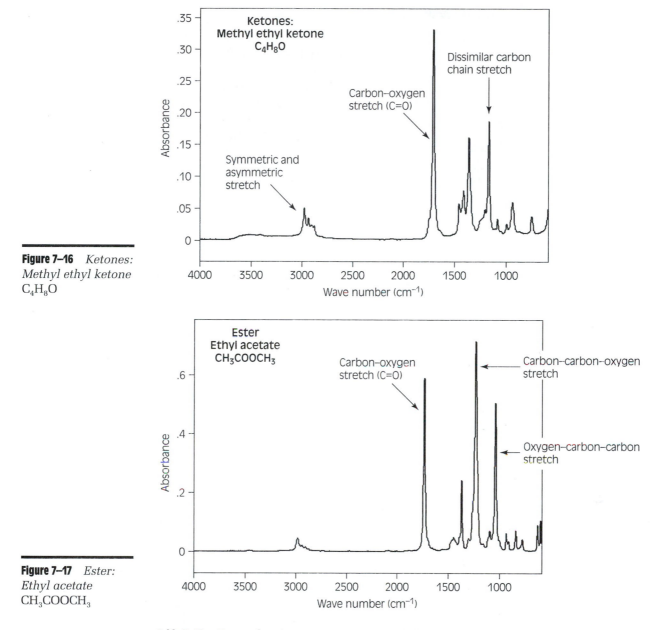

Figure 7–16   *Ketones: Methyl ethyl ketone* C_4H_8O

Figure 7–17   *Ester: Ethyl acetate* CH_3COOCH_3

Table 7–19   *Ester vibrations.*

Vibration	Occurrence in Spectra
Carbon–oxygen stretch (C=O)	1750–1735 cm^{-1}
Carbon–carbon–oxygen stretch	1210–1160 cm^{-1}
Oxygen–carbon–carbon stretch	1100–1030 cm^{-1}
Phosphoric–carbon stretch	1300–1240 cm^{-1}

With aromatic esters, the C=O stretch is lowered to the range of 1730–1715 cm^{-1}. The oxygen–carbon–carbon bond is observed in the range of 1130–1100 cm^{-1}. With aromatic esters, the overall bands are seen approximately 30 cm^{-1} lower than with saturated esters. With phosphoric esters, a band occurs in the range of 1300–1240 cm^{-1}.

Carboxylic Acids

The carboxylic acids are soluble in lower-weight compounds, in part due to the polarity they possess. Strong hydrogen bonding occurs and will affect boiling points and solubility, as well as infrared spectroscopy. With carboxylic acids, the carbonyl group has a carbon atom with a hydroxyl group attached to it, which makes these compounds acid. As with inorganic acids, these compounds will lose hydrogen, thus creating acidic-type qualities.

Because of the attached OH (hydroxyl) group, a very intense response is observed in the range of 3500–2500 cm^{-1}. Because of the placement of this band, the carbon–carbon chain is often masked. Once water has been ruled out, this band becomes very characteristic of the carboxylic acid family.

As with alcohol OH bending, both in-plane and out-of-plane bending occurs, along with stretching. Carboxylic acids contain the carbon–oxygen bond, which appears in the range of 1320–1210 cm^{-1}, a range very similar to that for esters.

Table 7–20 *Carboxylic acid vibrations.*

Vibration	Occurrence in Spectra
Hydroxyl stretch (very strong and wide)	3500–2500 cm^{-1}
Carbon–oxygen stretch (C=O)	1730–1700 cm^{-1}
Carbon–oxygen stretch	1320–1210 cm^{-1}
Hydroxyl in plane	1440–1395 cm^{-1}
Hydroxyl out of plane	960 – 900 cm^{-1}

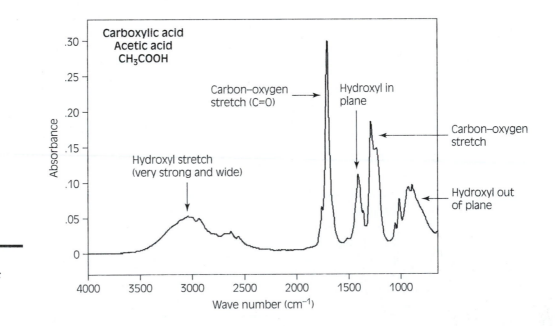

Figure 7–18

Carboxylic acid: Acetic acid CH_3COOH

SUMMARY

Infrared spectroscopy is an analytical technique used in chemical laboratories all over the world. The simplicity of this popular analytical platform is such that even the most inexperienced scientist can run samples and produce spectra after only a minimum level of instruction. This ease of operation is also true in the mobile application of infrared spectroscopy. First-response agencies across the country are seeing the value in having this instrument available for use by their hazmat technicians.

Most responders rely on the onboard computer to perform the analysis, however some responders have taken it upon themselves to learn the science behind IR. These are the pioneers in the hazmat response community. The analogy is the EKG machines in the early days of EMS. In those days, you were taught how to connect to your patient, look for aberrations (loose electrodes), and recognize a handful of critical arrhythmias. The "strip" would be sent to the emergency department, where a physician would read the strip and provide the orders for medication. The current state of IR is similar what field EKGs were in the late 1960s and early 1970s. Over time the field responder learned the hows and whys of EKGs. They learned 12 lead and over time became proficient in the interpretation of EKGs. Similarly, IR technology has been available to first responders for only few years, and responders are beginning to learn the hows and whys of infrared spectroscopy. As with the EKGs in the early days of EMS, IR will eventually become an everyday tool for the hazmat responder.

The learning curve for first responders must begin with the following: Learn the basic spectra of the various family types; each have their signature profile in spectra. Use the quick-check spectra numbers that are found in this book and FOG as reminders of where you would expect to find certain parameters on the spectra. Use the systematic approach presented as your base of knowledge, and in a short time you will start to recognize the spectra of families occurring in the regions identified. Within a few years, first responders will view this chapter as a basic primer and will want to understand more.

Review Questions

1. The advantage of infrared technology is to look at the bond when the chemical has:
   a. a coordinate covalent bond
   b. an ionic bond
   c. a covalent bond
   d. an ionic dipole

2. By using the infrared spectrum we are looking at the entire molecule.
   a. true
   b. false

3. The HazMat ID™ uses visible light to penetrate the molecule.
   a. true
   b. false

4. When a light wave enters a molecule:
   a. each bond vibrates at a specific frequency
   b. each bond aligns itself with the beam of light
   c. the molecular orbital moves toward the electrical carrier
   d. both A and C

5. In infrared spectroscopy, the following are correct:
   a. Only diatomic molecules vibrate, thus giving off a spectrum.
   b. Ionic salts are visualized by the movement of the transferred electrons in the orbital.

Table 7-21 *IR spectra analysis approach.*

Step	Family Functional Group	Molecule Component	Vibration Wave Numbers (cm⁻¹)	Intensity	Description
1	Alcohols, acids	O–H	3000–2500	Broad	First step is to look at the spectra in a general sense. Do not try to analyze each waveform, but rather look for predominate peaks of known wave numbers. By determining the presence or absence of these general functional groups, immediate information will be gained.
	Alkenes	C=C	1660–1600	Stretch	
	Alkynes	C≡C	2150	Stretch	
	Nitrogen products, amines	N–H	1640–1550	Bend stretch	
	Esters	C–O	1300–1000	Medium	This is not a rule-out process, but rather a puzzle of known spectra wave numbers.
	Cyanide	C≡N	2250	Sharp	
	Nitrogen oxides	NO_2			
2	Carbonyl group	C=O	1820–1660	Strong	This band is the most predominate in the spectra; very intense with medium width. If present, work though Step 2; if absent, move to Step 3.
	Acids	O–H	3400–2400	Broad strong	O–H is a very broad band that can overlap the C–H stretch in aliphatic hydrocarbons.
		O–H	3650–3600	Sharp peak	Band occurs here when an alcohol is dissolved in a solvent (neat solvent) if diluted band is shifted to a lower frequency.
	Carboxylic acids	C–O	1320–1210	Medium stretch	Check for the O–H band and the C=O stretch that occurs in the range of 1730–1700 cm⁻¹.
	Amides	N–H	3400	Medium doublet	Look for C=O stretch in the range of 1680–1630 cm⁻¹, N–H₂ at 3350 cm⁻¹ and 3180 cm⁻¹, a secondary one at 3300 cm⁻¹, and a bend in the range of 1640–1550 cm⁻¹.
		N–H	3500–3300	Stretch	Look at the amines for further direction.
	Esters	C–O	1300–1000	Strong	No O–H band present.
		RCOOR	1750–1735	Stretch	Occurs for straight-chained molecules.
		C=C–COOR	1740–1715	Combination (conjugation)	C=O with C=C.
			1640–1625	Combination (conjugation)	C=O and C=C; two bands are possible for cis and trans isomers.
		Ar–COOR	1740–1715	Combination (conjugation)	Ar.
			1600–1450	Combination (conjugation)	C=O and Ar.
		P=O	1300–1240	Stretch	One very strong band.
		R–O	1088–920	Stretch	One or two strong bands.
		P–O	845–725	Stretch	Medium.

			Wavenumber	Band	Comments
	Anhydrides	C=O	1830–1800 1775–1740	Two bands	The intensity is dependent on absorption to a lower frequency. Straight chains move frequency to lower wave numbers; rings increase frequency to higher wave numbers.
		C–O	1300–900	Multiple stretch	Two strong bands in the regions 1830–1800 cm^{-1} and 1775–1740 cm^{-1}. These bands are a result of symmetric and asymmetric stretch, along with combination (conjugation). Combination (conjugation) moves waves to a lower frequency; ring strain moves the frequency higher. Predominate and broad stretch of the C–O vibration.
	Aldehydes	C–H	2860–2800 and 2760–2700	Weak	These are at a lower frequency of the straight chain C–H absorption (3000 cm^{-1}). This is a weak pair of bands, and can be blurred by the hydrocarbon chain.
		RCHO	1740–1725	Stretch	Normal straight chain.
		C=C–CHO	1700–1680 and 1640	Combination (conjugation)	C=O and C=C.
					C=C.
		Ar–CHO	1700–1660 and	Combination (conjugation)	C=O and C=C.
			1600–1450		Benzene ring (Ar).
	Ketones	C=O	1725–1705	Strong	1720–1708 for straight-chained ketones.
		C=C–COR	1700–1675 and	Combination (conjugation)	C=O with C=C.
			1644–1617		C=C.
		Ar–COR	1700–1680 and	Combination (conjugation)	C=O with Ar.
			1600–1450		Benzene ring (Ar).
3	Alcohols, phenols	R–OH Ar–OH	3650–3600 3400–3300	Broad and sharp	In association with the broad band identifying the OH, a secondary band from the C–O absorption is located at 1300–1000. This is dependent on the level of dilution with a solvent or water.
		C–OH	1440–1220	Bend	Broad and weak, blurred by the –CH$_3$ bend.
		C–O	1260–1000	Stretch	Placement of the OH on the carbon chain.
	Amines	N–H	3400	Medium	Amides should have been ruled out.
		N–H	3500–3300	Stretch	Primary amines have two bands; secondary amines have one band; tertiary amines have no N–H stretches.
		N–H	1640–1560 1500	Broad bend	Primary amines. Secondary amines.
		N–H	800	Out-of-plane bend	Very noticeable and wide.
		C–N	1350–1000	Stretch	Medium to intense.

(continues)

Table 7–21 (Continued)

Step	Family Functional Group	Molecule Component	Vibration Wave Numbers (cm^{-1})	Intensity	Description
	Ethers	C–O	1300–1000	Stretch	Check for the absence of O–H around 3400 cm^{-1}. Also look for the C=O and O–H stretch to ensure that the C–O stretch is not due to an ester.
		Ar–C–O	1250 and 1040	Strong	Straight-chained ethers have a strong band at around 1120 cm^{-1}.
4	Alkenes	=C–H	Greater than 3000	Stretch	C–H stretch occurs at greater than 3000 cm^{-1}, along with peaks below 3000 cm^{-1}.
		=C–H	1000–650	Out-of-plane	Bending.
		C=C	1660–1600	Stretch	When combination (conjugation) occurs, the stretch is of a higher intensity at lower frequencies. Cis symmetry is a stronger band; Trans is weaker. Any symmetric molecule with attachments of equal character does not absorb energy due to the loss of a dipole moment.
5	Aromatic rings	=C–H	Greater than 3000	Stretch	C–H stretch occurs at greater than 3000 cm^{-1} along with out-of-plane bending.
		=C–H	900–690	Out-of-plane	Bending is dependent on the attached (substituted) molecule(s).
		C=C	1600 and 1475	Ring stretch	Combination bands are sometimes observed at 2000 cm^{-1} and 1667 cm^{-1}, and are dependent on the attached (substituted) molecule(s).
6	Alkynes	≡C–H	3300	Stretch	Predominating stretches with intense patterns when the triple bond is at the end of the molecule.
		C≡C	2150	Stretch	Combination (conjugation) can move the stretch frequency lower. When the triple bonds are symmetrical, the band is very weak (and can be almost no band at all.
7	Nitrogen groups	R–N^{+}–(O)$_2^{-}$	1600–1530 and 1390–1300	Strong stretch / Medium stretch	Hydrocarbon nitrogen compounds.
		Ar–N^{+}–(O)$_2^{-}$	1550–1490 and 1355–1315	Strong stretch	Aromatic nitrogen compounds.

#					
8	Alkanes	C–H	3000	Stretch	In the alkanes, the stretch occurs in the range of 3000–2840 cm^{-1}; if there is a vinyl, acet, cyclo, or aromatic molecule, the stretch occurs at greater than 3000 cm^{-1}.
		CH$_2$	1465	Bend	Characteristic rocking motion is observed when there are four or more CH$_2$ molecules in a chain. This motion is observed at 720 cm^{-1}.
		CH$_3$	1375	Bend	Characteristic of the CH$_3$ group.
9	Halides	R–F	1400–1000	Strong stretch	Multiple F atoms will produce multiple bands 1350–1100 cm^{-1}.
		R–Cl	1300–1230 785–540	Bend stretch	The number of substitution halides will give multiple bands.
		R–Br	1250–1190 650–510	Bend stretch	
		R–I	1200–1150 600–485	Bend stretch	
10	Mercaptans	R–S–H	2550	Weak	One weak stretch and few other absorptions occur at this frequency.

   c. Pure substances are captured by the technology.

   d. Infrared spectroscopy cannot see the organic functional groups.

6. Infrared spectroscopy is useful when:

   a. covalent bonds are present

   b. considered a universal laboratory technique

   c. the spectra in the field look the same as the laboratory spectra

   d. all the above

7. When we take a spectrum with the HazMat ID™, we visualize the molecule:

   a. as an entire compound

   b. as if it were a group of small photos

   c. as if it were one large photo

   d. as one segment of the compound

8. When using infrared spectroscopy, the electron orbital has an influence on the wave number.

   a. true

   b. false

9. Hybridization of the sp^3 orbital between the C–H bond in alkanes results in a wave number of:

   a. 3000 cm^{-1}

   b. greater than 3000 cm^{-1}

   c. less than 3000 cm^{-1}

   d. $3000\text{--}2640 \text{ cm}^{-1}$

10. Hybridization of the sp^2 orbital in the C–H bond in alkenes results in a wave number of:

   a. 3000 cm^{-1}

   b. greater than 3000 cm^{-1}

   c. less than 3000 cm^{-1}

   d. $3000\text{--}2640 \text{ cm}^{-1}$

11. Hybridization of the sp orbital in the terminal C–H bond in alkynes results in a wave number of:

   a. 3000 cm^{-1}

   b. greater than 3000 cm^{-1}

   c. less than 3000 cm^{-1}

   d. 3300 cm^{-1}

12. The infrared light that passes through a molecule causes the entire vibration of the molecular bonds.

   a. true

   b. false

13. Each vibration is enhanced by the IR beam.

   a. true

   b. false

14. How many regions are there on a spectrum?

   a. 3

   b. 4

   c. 5

   d. 6

15. Which region identifies the carbonyl group?

   a. I

   b. II

   c. III

   d. IV

16. The alkanes show a vibration at frequencies greater than 3000 cm^{-1}.

   a. true

   b. false

17. The functional groups all have vibration in the third region from the right.

   a. true

   b. false

18. Place the types of bonds one could observe in the appropriate region in the following table.

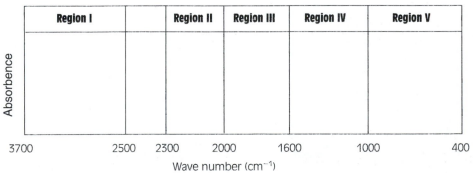

Glossary

A the mass number of an element

absolute pressure the true pressure for gases, it is the sum of the gauge pressure and the atmospheric pressure

absorbance measured, observable parameter, defined as the logarithm of transmittance and directly proportional to concentration

absorption the incorporation of one material into another, or the passage of a material into and through the tissues; the ability of a material to draw within it a substance that becomes a part of the original material; the process by which energy is transferred to a molecule

acetylcholine a chemical neurotransmitter that stimulates the heart, skeletal muscles, and glands

acetylcholinesterase the enzyme that hydrolyzes acetycholine, thereby stopping its activity

acid a compound that forms hydronium ions in water

actinides *see* inner transition metals

activation energy the amount of energy required for reactants to combine, progress over the hill, and generate the final product.

acute exposure a short-term exposure or the rapid onset of symptoms after an exposure or several exposures occurring within 24 hr

acyl halides members of the carbonyl group, in which the hydrogen has been replaced with a halogen

aerosol a liquid or solid composed of finely divided particles suspended in air

alcohol an organic compound composed of an alkyl or aryl group and a hydroxyl group

alcohol foam a film-forming fire extinguishing agent that is less dense than water that persists in the presence of polar compounds

aldehydes organic compounds composed of a carbonyl group, one hydrogen atom and an R group

aliphatic hydrocarbons hydrocarbons whose molecules are *not* composed of benzene or benzenelike structures

alkali metals (group IA, 1, or 1A) the elements that are in the first column or group to the left on the periodic table: lithium, sodium, potassium, rubidium, cesium, and francium

alkaline earth metals (group IIA or 2) the elements that are in the second column or group from the left on the periodic table: beryllium, magnesium, calcium, strontium, barium, and radium

alkane a hydrocarbon composed of carbon-to-carbon single bonds

alkene a hydrocarbon composed of at least one double bond in the carbon–carbon chain

alkyl group the group of atoms remaining after a hydrogen atom such as an alkane is removed from a molecular structure

alkyne a hydrocarbon composed of at least one triple bond in the carbon–carbon chain

allotropes the forms of an element each possessing different chemical and physical properties

alpha particle the radiation particle consisting of two neutrons and two protons that is emitted from certain radioisotope decay

aluminum family (group IIIB, 13, or IIIA) a periodic-table group of metals and one nonmetal: boron, aluminum, gallium, indium, and thallium

amides derivatives of ammonia in which one, two, or all three hydrogens can be replaced with a hydrocarbon

amine an analog of ammonia that has been alkylated

anions negatively charged ions

analogue chemical compounds that are similar in structure and chemistry

appearance the form, size, and color of a material

aromatic hydrocarbons   benzene or benzenelike compounds primarily composed of carbon and hydrogen atoms in one or more ring structures

aryl group   the aromatic group that remains after a hydrogen atom is removed from benzene or benzenelike structure

asymmetric stretch   one hydrogen atom moving away from the carbon atom, while the other hydrogen is moving toward the carbon atom; occurs at about 2920 cm^{-1}

atmospheric pressure   the force exerted on matter by the mass of the overlying air

atomic number   the number of protons in the nucleus of an atom

atomic radius   the distance halfway across an individual atom

atomic weight   the mass of an atom, which equals the sum of protons and neutrons

base   a compound that forms hydroxide ions in water

becquerel   the unit of measurement for radioactivity equivalent to one disintegration of a nucleus per second

Beer-Lambert law   an optical law of physics that states that absorbance and concentration are linearly proportional

beta particle   an electron (negatively charged) or positron (positively charged) emitted from a radioactive isotope during beta decay

binary compound   a compound composed of only two elements

boiling point   the temperature at which a liquid's vapor pressure equals the atmospheric pressure

Boyle's law   at constant temperature, the volume of a gas is inversely proportional to its absolute pressure

brisance   the shattering power of an explosive

carbamates   organic compounds that can temporarily inhibit the enzymatic activity of acetylcholinesterase by binding to acetylcholine

carbonyl group   a ketone or aldehyde functional group

carboxyhemoglobin   the compound formed when carbon monoxide reacts with hemoglobin, depriving the hemoglobin of its ability to carry oxygen and thus causing chemical asphyxiation

carboxylic acids   organic acids that contain a carboxyl group

catalyst   a substance that increases the rate of reaction, without being consumed in reaction

cations   positively charged ions

Charles's law   at a constant pressure, the volume of a gas is directly proportional to its absolute temperature

chemical properties   intrinsic characteristics of a substance described by its tendency to undergo chemical change, such as heat of combustion, reactivity, and so on

chronic exposure   a long-term exposure, usually recurring during 80% of the total life span of the exposed organism

Class A fire   ordinary combustibles

Class B fire   flammable liquids

Class C fire   electrical

Class D fire   combustible metals

colorimetric tubes   glass tubes containing a chemically treated substance that reacts with specific chemicals, chemical families to produce a stain within the substrate; the tubes are calibrated to indicate approximate concentrations in air

combustible   an ignitable and free-burning material

combustible liquid   a liquid with a flash point at or above 100°F; *see also* flammable liquid

combustible metal   the alkali metals, alkaline earth metals, and transitional metals such as aluminium, titanium, and zirconium that burn in air and are denoted as Class D fires

combustion   the rapid oxidation of a material in the presence of an oxidizer

compound   a substance composed of two or more elements in a fixed proportion that are bonded chemically

compressed gas   a gas within a container having an absolute pressure exceeding 40 psi at 70°F

concentration   the relative amount of the lesser component of a mixture or solution

conduction   the mechanism by which heat is transferred between materials that are in contact with one another

convection   the mechanism of heat transference by the movement of a gas or liquid

covalent bond   the force holding together atoms that share electrons

cracking   a chemical process in which organic molecules are broken down into smaller molecules by heating

critical pressure   the minimum pressure required to cause a gas to liquefy at its critical temperature

critical temperature   the temperature above which a gas cannot be liquefied

cryogenic liquids   substances that have been cooled to very low temperatures

curie   the outdated unit of measurement of radioactivity as compared to 1 g of radium whereby 1 Ci = 3.700×10^{10} decays per second

cyclo compound   a hydrocarbon chain in a ring formation

Dalton's law   the partial pressure of several gases within a container is the sum of all the pressures of each individual gas

"Dangerous When Wet"   describes a material that becomes spontaneously flammable or evolves a flammable or toxic vapor when water interacts with it

decomposition   a chemical reaction involving the breakup of a compound into a more basic compound or substance

deflagration   combustion that occurs at less than 1086 ft/s, which creates a shock wave moving at less then 3300 ft/s

density   the mass of a substance per a given volume

detonation   combustion that occurs at greater than 1086 ft/s, which creates a shock wave moving at greater then 3300 ft/s

diene   an organic compound with two double bonds

dipoles   the ability of electron movement in a molecule to create a predominately positive and negative end in the molecule

distillation   the physical process of converting a liquid to a vapor and condensing it back into a liquid in order to separate compounds of the original liquid

dose-response   the range of effects that can be observed within a given population that has been exposed to a chemical

dry chemical   an extinguishing agent that works by inhibiting the production of free radicals and heat during a fire and blocks oxygen from the fire

dusts   fine particles of matter, such as coal dust, sawdust, and grain dust

effective dose (ED)   dose that results in minimal observable effects in 50% of an unprotected population

electromagnetic radiation   whenever atoms are heated or exposed to energy such as light they become excited as they take in the energy, and the electrons start to move into different layers orbitals

electrons   negatively charged fundamental particles orbiting the nucleus of an atom

electronegativity   a measure of how strongly an atom holds on to its electrons or attracts electrons from other atoms

enantiomer   a pair of chemical compounds having molecular structures that are mirror images of each other; each can have different boiling points, melting points, densities, and overall chemical and physical properties; they are identical in structure but they cannot be superimposed due to the mirror-image quality the compounds possess

energy   the capacity to do work

endothermic   a reaction that absorbs heat

epoxide   an ether in a cyclic configuration

esters   organic compounds in which an alkyl or aryl group is attached to one side of the carbonyl group, and oxygen with an alkyl or aryl group to the other side

ether   an organic compound bonded in the middle with oxygen and two alkyl or aryl groups

evaporation   the escape into the vapor state of molecules on the surface of a liquid

exothermic   a reaction that evolves heat

fire point   the temperature of a flammable liquid at which the self-sustained combustion of its vapors occurs in air

fissile   the ability of an atomic nucleus to undergo simultaneous emission of one or more neutrons and energy

flammable   capable of being easily ignited and susceptible to burning intensely

flammable gas   any product that is a gas at 68°F or less and a pressure of 14.7 psi and is ignitable at 14.7 psi when the mixture of 13% or less, or the vapors of this material possessing a flammable range of at least 12 percent regardless of the LEL

flammable ranges or limits   represents the minimum and maximum concentration of a mixture in air favorable for ignition

flammable liquid   a liquid with a flash point below 100°F; *see also* combustible liquid

flammable solid   there are three types of a flammable solids: wetted explosives, self-reactive, and readily combustible; according to the DOT regulation, there are three divisions of flammable solids: Flammable Solid, Spontaneously Combustible, and Dangerous When Wet

flash point   the minimum temperature at which the vapor of a liquid or solid ignites when in contact with an ignition source, but there are not enough vapors to sustain combustion

foam   a frothy firefighting–extinguishing agent that forms a barrier to limit the vapors being produced

fractional distillation   a process of evaporation and recondensation used for separating a mixture into its various constituents based on the individual boiling points of those constituents

free radical   a reactive molecular species having an odd or even number of electrons

freezing point   the temperature at which a liquid becomes a solid

frequency   the number of wavelengths or cycles that occur over a given period of time

functional group   any of the groups of atoms that identify a particular organic compound or represent a characteristic chemical entity

gamma radiation   electromagnetic radiation of energy emitted from radioisotopes

gas   a chemical that exists as a molecular identity in air

germanium family (group IVB, 14, or IVA)   a group of elements on the periodic table represented by carbon, silicon, germanium, tin, and lead

Gray   the unit of absorbed dose of radiation

groups   vertical columns of the periodic table containing families of elements of similar properties

Group I A: alkali metals

Group II A: alkaline earth metals

Group IIIA: aluminum family

Group IVA: germanium family

Group V A: nitrogen family

Group VI A: sulfur family

Group VIIA: the halogens

Group VIII A: the noble gases

half-life   the length of time for half the atoms in a radioactive substance to decay

halogenated agent   halogen used in combination with methane or ethane to produce a nonflammable colorless gas that can be used as an extinguishing agent or refrigerant

halogenated hydrocarbon   a derivative of a hydrocarbon in which one or more of the hydrogens has been replaced by a halogen

halogens (group VIIB, 17, or VIIA)   a group of elements on the periodic table consisting of fluorine, chlorine, bromine, iodine, and astatine

Henry's law   the amount of gas absorbed by a given volume of a liquid at a given temperature is directly proportional to the pressure of the gas

heterogeneous   a material or solution in which there are visual differentiating parts

homogeneous   a material or solution having no differentiating parts

hydrocarbons   compounds whose molecules are composed predominately of carbon and hydrogen

hydroxyl group   the –OH functional group

hypergolic   chemicals that when mixed will self-ignite

ignition temperature   the minimum temperature at which a material will ignite and sustain combustion without a continuing outside source of ignition

immiscible   incapable of mixing

infrared spectroscopy   the specific method by which an infrared light beam is directed into a molecule, enhancing the vibration of the elements, their bonds, and thus the electrons, to show very specific peaks at specific wave numbers; used for identification of functional groups

ingestion   the incorporation of a material into the gastrointestinal tract

inhalation   intake and absorption into the body via the respiratory system

inhibitor   a substance that controls or delays a reaction

injection   the forced introduction of a substance into the underlying body tissue

inner transition metals   a region of the periodic table underneath the transition metals that includes two main groups: lanthanides and actinides

ion   an atom or group of atoms that have lost or gained one or more electrons, giving the entire structure an overall valence

ionic bond   the electrostatic attraction of oppositely charged particles; atoms or groups of atoms can form ions or complex ions

ionic radius   the distance halfway across an individual ion

ionization   the process by which an electron is removed from the outermost shell of an atom

ionization energy   the amount of energy required to remove an electron from the outermost shell of an atom

ionization potential (IP)   the minimum energy required to release an electron or photon from a molecule; measured in electron volts (eV)

isomer   *see* stereoisomer; structural isomer

isotope   one form of an element having the same number of protons but a different number of neutrons

ketone   an organic compound composed of a carbonyl group bonded with two alkyl or aryl groups

lanthanides   *see* inner transition metals

Lewis dot structure   a visual representation of elements, atoms, and ions

liquefied petroleum gas (LPG)   a compressed and liquefied mixture of alkanes and alkenes: propane, propylene, butane, isobutane, and butylenes

lower explosive limit (LEL)   the lowest ratio of a gas and air that will permit ignition

mass number   the total number of neutrons and protons in the nucleus of an atom

maximum safe storage temperature (MSST)   the highest storage temperature above which a reaction and explosion may ensue

melting point   the temperature at which a material changes from a solid to a liquid

metalloids   a region of the periodic table that includes the elements between the nonmetals and the right-side representative metals

methyl group   the radical designated as CH_3- or the terminal ends of a carbon chain

methylene group   a CH_2 group found between two methyl groups

miscibility   the ability of materials to dissolve into a uniform mixture

mists   liquids that have been atomized, for example, spray paint, high-pressure oil leaks, and aerosolization of nerve agents; a mist and aerosol can be thought of as the same

mole   the sum of the atomic masses of the molecules components expressed in grams pounds, etc; the quantity of a substance that contains the same number of particles (atoms, ions, molecules)

molecular weight (MW)   the sum of the atomic weights of all the atoms in a molecular formula

molecule   composed of two or more atoms, the smallest unit of a substance that retains all the properties of the substance

monomers   molecules considered a single unit, the combination of which results in a polymer

neutralization   the process of bringing an acid or alkali back to a pH of 7

nitrogen family (group VB, 15, or VA)   a group of elements on the periodic table consisting of nitrogen, phosphorus, arsenic, antimony, and bismuth

noble gases (group VIII or 18)   a group of elements on the periodic table made up of helium, neon, argon, krypton, xenon, and radon; sometimes referred to as *inert gases*

nonflammable gas   any material or mixture of gas that exerts in its container a pressure of 41 psi or greater at 68°F, or a material that does not meet the definition of DOT 2.1 or 2.3

nonmetals   a region of the periodic table that includes hydrogen and the elements on the right side of the chart

nucleus   the center of an atom, containing the protons and neutrons

neutrons   fundamental particles without electrical charge that are part of the nucleus of an atom

octet rule   the observation that simple atoms achieve an electronic stability by attaining eight electrons in the outer shell through ionic or covalent bonding, each trying to attain the noble gas state of eight electrons in the outer shell

orbital the areas of high statistical probability in which an electron will be found; the energy level at which that electron spins; *see* shell

organic peroxide a functional group in which the hydrogens in hydrogen peroxide have been replaced with alkyl or aryl groups

outage an amount expressed in percentage by total volume of the container, by which a container of a liquid falls short of being completely filled

oxidation the chemical process or reaction in which there is a loss of electrons, or an increase in the oxidation number

oxidation number a number assigned to an element's atom in a compound or the valence number in an ionic compound

pascal a unit of measurement for pressure

periodic table a general framework of the elements that shows relationships between the chemical groups

periods the seven horizontal rows of the periodic table; the first six periods end with a noble gas; the rare earth metals (or inner transition metals) are placed within rows 6 and 7; for easy reading these elements are pulled to the lower portion of the table

phase diagram a graph that shows the relationship between temperature, pressure and the state of matter

pH the scale used to measure how acidic or basic a substance is based on hydrogen ion concentration

phenols a group of aromatic compounds with the hydroxyl group attached directly to the benzene nucleus

physical properties intrinsic characteristics of a substance that can be observed and measured, such as appearance, melting point, density, and so on

polarity the quality of possessing two opposing tendencies, as in the existence of a negative and a positive end on a molecule

polyatomic ion an ion having more than one constituent atom

polymerization the chemical reaction (extremely violent if uncontrolled) that occurs when smaller compounds are linked together to form larger compounds

polymers compounds of high molecular weight made from repetitive linking together of monomers

product the result of a chemical reaction

protons a positively charged fundamental particles that are part of the nucleus of an atom and carry a positive charge

pyrolysis from the Greek words meaning "fire breakdown," it is the dividing of covalent bonds found in hydrocarbons; it occurs near the surface of the material (not directly on it) at a distance in which the flammable range is reached through the heating of the material and the mixing with air to produce a combustion reaction

pyrophoric a liquid or solid that will ignite without an external ignition source within 5 min without an external ignition source after coming in contact with air

racemic mixture an optically inactive mixture of exactly equal amounts of two enantiomers

RAD (radiation absorbed dose) a unit, slightly greater than 1 roentgen, that is used to measure dosage

radiation the mechanism by which heat is transferred between two materials that are not in contact through the outward movement of heat from the source via electromagnetic waves

radioactive decay the process by which an unstable nucleus releases particles or energy in order to become stable; sometimes referred to as radioactivity or spontaneous transmutation

rare earth metals *see* inner transition metals

reactant one of the starting materials in a chemical reaction

REM (roentgen equivalent man) a measurement of biological effect, it represents the amount of absorbed radiation of any type that produces the same effect on the human body as 1 roentgen of gamma radiation

resonance a description of a molecule consisting of a combination of more than one structural representation

representative metals a region of the periodic table that includes the first two columns or groups and seven elements from groups IIIB–IVB (or 13–14, or IIIA–IVA)

rocking when both hydrogen atoms are moving up or down, swaying within the plane; occurs at about 720 cm^{-1}

roentgen the measurement of radioactivity for gamma radiation; it is the amount of ionization per cubic centimeter of air; it is a measurement of exposure and represents the amount of gamma radiation that produces two billion ion parts in dry air

saturated hydrocarbon   a hydrocarbon with only single bonds between the carbons

scissoring   when both hydrogen atoms move up and down as if they are the blades on a scissors; occurs at about 1450 cm^{-1}

self-accelerating decomposition temperature (SADT)   the temperature at which an organic peroxide or synthetic compound will react to heat, light, or other chemicals and release oxygen, energy, and fuel in the form of an explosion or rapid decomposition

shell   the region in space where electrons are located around the nucleus of an atom; the principal energy level of an electron

Sievert   a dose equivalent equal to 100 REM

solubility   the ability of a material to blend uniformly with another to form a solution

solute   the substance added to a solvent

solvent   a substance that has the capability of dissolving another substance to form a solution

specific gravity   the weight of a solid or liquid as compared to an equal volume of water

spontaneously combustible material   a self-heating or pyrophoric material due to an exothermic reaction with air

standard temperature and pressure (STP)   a temperature of 32°F (0°C or 273 K) and a pressure of 1 atm (760 mm Hg or 760 torr) for measured volumes of gas

stereoisomers   compounds with the same molecular formula and connectivity, but different arrangement of atoms in space

structural isomers   compounds with the same molecular formula but different structure, chemical properties, and physical properties

subchronic exposure   an exposure that recurs during approximately 10% of the organism's life span

sublimation   a physical change in which a substance passes from a solid into the gaseous state without having a liquid state

sulfur family (group VIB, 16, or VIA)   also known as the *chalcogen family;* a group of five moderately reactive elements on the periodic table: oxygen, sulfur, selenium, tellurium, and polonium

symmetric stretch   each hydrogen atom moving away from the carbon atom; occurs at about 2850 cm^{-1}

temperature   a relative measure of heat, utilizing the Fahrenheit, Celsius, and Kelvin scales

thiols   compounds similar to alcohol with the oxygen replaced by a sulfur atom; generally denoted by R–SH and commonly referred to as mercaptans

transition metals   a region of the periodic table that includes the groups in the lower portion of the main chart, identified by the groups IIIA–IIB (or 3–12, or IIIB–IIB)

transmittance   observable measurement, defined as the ratio of transmitted radiation to the incident radiation

unsaturated hydrocarbons   molecules that have a carbon chain with at least one double or triple bond

upper explosive limit (UEL)   the highest ratio of a gas and air that will permit ignition

valence electrons   the electrons in the outermost shell of an atom

vapor   the diffused state of matter that is released from a liquid substance, which when combined with air forms a potential ignitable mixture

vapor density   the weight of a vapor or gas as compared to an equal volume of air

vapor pressure (VP)   the pressure exerted by a vapor; in particular, the pressure a gas exerts against the sides of an enclosed container

vesicant   an agent that causes blistering at a particular temperature

viscosity   a measure of flow

volatility   description of a chemical's capacity to become vaporized

wagging   when the hydrogen atom moves back and forth an equal distance out of plane; occurs at about 1250 cm^{-1}

wavelength   typically the distance between two identical points on two adjunct waves, measured in meters

wetting agents   liquid "soap" concentrates that, when mixed with water, reduce the surface tension of the water, thus allowing for deeper penetration into the material

Z   the atomic number, or number of protons, of an element

Acronyms

The hazardous materials field has blossomed with acronyms, words formed by the initial letters of each word in a phrase or within terminology, and abbreviations. Some seem meaningless without the historical significance. The following is a list of some of the terminology shortcuts used in connection with emergency response.

AC	hydrogen cyanide
ACGIH	American Conference of Governmental Industrial Hygienists
Ach	Acetylcholine
AchE	Acetylcholinesterase
AFFF	aqueous film-forming foam
AIHA	American Industrial Hygiene Association
ALARA	as low as is reasonably achievable
ALD	average lethal dose
AMU	atomic mass unit
BLEVE	boiling liquid expanding vapor explosion
CAM	chemical agent monitor
CAMEO	computer-aided management of emergency operations
CEA	clean extinguishing agent
CG	phosgene
CGI	combustible gas indicator
CK	cyanogen chloride
CL	ceiling level; chlorine
CN	mace
CNG	compressed natural gas
CR	tear gas
CS	tear gas
CWA	chemical warfare agent
CX	phosgene oxime
DM	adamsite
ED	effective dose

EEL	emergency exposure limit
EL	excursion limit
EPA	Environmental Protection Agency
ERPG	Emergency Response Planning Guideline
eV	electron volt
FFFP	film-forming fluoroprotein
FID	flame-ionization detector
FP	flash point
GA	tabun
GB	sarin
GD	soman
GRS	German Research Society
H	mustard agent
HD	distilled mustard
HN-1	nitrogen mustard
HN-2	nitrogen mustard
HN-3	nitrogen mustard
IDLH	immediately dangerous to life and health
IH	inhalation hazard
IP	ionization potential
IT	ignition temperature
IUPAC	International Union of Pure and Applied Chemistry
K	equilibrium constant
L	Lewisite
LC	lethal concentration
LC_{50}	lethal concentration fifty
LCT_{50}	lethal concentration fifty (statistically derived)
LClo	lethal concentration low
LD	lethal dose
LD_{50}	lethal dose fifty
LEL	lower explosive limit

LNG	liquefied natural gas
LPG	liquefied petroleum gas
MAC	maximum allowable concentration (also MAK)
MEC	minimum exposure concentration
MSDS	material safety data sheet
MSST	maximum safe storage temperature
NIOSH	National Institute for Occupational Safety and Health
OC	pepper spray
OSHA	Occupational Safety and Health Administration
PEL	permissible (or personal or published) exposure limit
PID	photoionization detector
pH	positive hydronium ion (used as a scale for acids and bases)
ppb	parts per billion
PPE	personal protective equipment
ppm	parts per million
RAD	radiation absorbed dose
REM	roentgen equivalent man
REL	recommend exposure limit
RgasD	relative gas density
RD	respiratory depression

RD_{50}	respiratory depression fifty percent
SADT	self-accelerating decomposition temperature
SD	specific density (equal to vapor density)
sg	specific gravity
SOG	standard operating guidelines
STEL	short-term exposure limit
STP	standard temperature and pressure
TCL	toxic concentration low
TDL	toxic dose low
TEEL	temporary emergency exposure limit
TGD	thickened soman
TLV	threshold limit value
TLV-c	threshold limit value–ceiling
TLV-s	threshold limit value–skin
TLV-TWA	threshold limit value–time-weighted average
TWA	time-weighted average
UEL	upper explosive limit
VD	vapor density
VP	vapor pressure
VX	persistent nerve agent
WEEL	workplace environment exposure limit

Index

Note: Page numbers in *italic* type refer to figures or tables.